Delphi Cookbook
Third Edition

Recipes to master Delphi for IoT integrations, cross-platform, mobile and server-side development

Daniele Spinetti
Daniele Teti

BIRMINGHAM - MUMBAI

Delphi Cookbook
Third Edition

Commissioning Editor: Merint Mathew
Acquisition Editor: Alok Dhuri
Content Development Editor: Priyanka Sawant
Technical Editor: Ruvika Rao
Copy Editor: Safis Editing
Project Coordinator: Vaidehi Sawant
Proofreader: Safis Editing
Indexer: Tejal Daruwale Soni
Graphics: Jason Monteiro
Production Coordinator: Shraddha Falebhai

First published: September 2014
Second edition: June 2016
Third edition: July 2018

Production reference: 1310718

Published by Packt Publishing Ltd.
Livery Place
35 Livery Street
Birmingham
B3 2PB, UK.

ISBN 978-1-78862-130-4

www.packtpub.com

`mapt.io`

Mapt is an online digital library that gives you full access to over 5,000 books and videos, as well as industry leading tools to help you plan your personal development and advance your career. For more information, please visit our website.

Why subscribe?

- Spend less time learning and more time coding with practical eBooks and Videos from over 4,000 industry professionals

- Improve your learning with Skill Plans built especially for you

- Get a free eBook or video every month

- Mapt is fully searchable

- Copy and paste, print, and bookmark content

PacktPub.com

Did you know that Packt offers eBook versions of every book published, with PDF and ePub files available? You can upgrade to the eBook version at `www.PacktPub.com` and as a print book customer, you are entitled to a discount on the eBook copy. Get in touch with us at `service@packtpub.com` for more details.

At `www.PacktPub.com`, you can also read a collection of free technical articles, sign up for a range of free newsletters, and receive exclusive discounts and offers on Packt books and eBooks.

Contributors

About the authors

Daniele Spinetti is a software architect living in Rome. He is an Embarcadero MVP. Delphi/Object Pascal is his favorite tool/programming language, and he is a lead and active member of several projects in the open source community. In his tutoring activities (conferences and training), he likes to talk about innovative topics in software architectures. He's a huge fan of design patterns and TDD.

Daniele Teti is a software architect, trainer, and consultant with over 20 years of experience. He drives the development of the most popular Delphi open source project on GitHub—DelphiMVCFramework. He's also a huge fan of design patterns, machine learning, and AI. Daniele is the CEO of BIT Time Professionals, an Italian company specializing in high-level consultancy, training, development, and machine learning systems.

About the reviewer

Primož Gabrijelčič started coding in Pascal on 8-bit micros in the 1980s, and he never looked back. In the last 20 years, he was mostly programming high-availability server applications used in the broadcasting industry. He's also an avid writer and has written several hundred articles, and he is a frequent speaker at Delphi conferences, where he likes to talk about complicated topics, ranging from memory management to creating custom compilers.

Packt is searching for authors like you

If you're interested in becoming an author for Packt, please visit `authors.packtpub.com` and apply today. We have worked with thousands of developers and tech professionals, just like you, to help them share their insight with the global tech community. You can make a general application, apply for a specific hot topic that we are recruiting an author for, or submit your own idea.

To my family, to the memory of my grandfather, and to my friends. I want to thank each of them for their love, support, and inspiration.

I would like to extend my gratitude to Daniele Teti. I work closely with Daniele for several years and I learned many things from him, but the most important it's like thinking, asking the right questions and correctly approaching a problem. I am proud to work with him.

I want to leave you a funny anecdote from when I was a boy:

Mom: Daniele, dinner's ready!! C'mon! Daddy's hungry!
Me: I'm coming, just a moment!
Mom: Daniele hurry up, all the day that you're at computer, it does not feed you!

Table of Contents

Preface

If you've been a software developer for a long time, you certainly know how useful a conversation can be with a colleague who has already done something similar to what you are doing and can explain it, as they will have faced the same problem. It is not possible to put all the possible situations that a developer can face in a book, but many problems are similar, at least in principle. This is the reason this book is organized as a cookbook: just like a combination of foods can be adapted and modified to be appropriate for different types of dinner, a programming recipe can provide the idea to solve many different problems. This book is an advanced guide that will help Delphi developers get more skilled in their everyday job. The everyday job, and the quality of your deliverables, is what is contributing to the quality of your professional life. If it does not make sense, reinventing the wheel repeatedly, especially when working with a well-established tool, such as Delphi. The focus of the book is to provide readers with comprehensive and detailed examples on how effectively the Delphi software can be designed and written. All the recipes in the book are the result of years of development, training, and consultancy activities in many different fields of IT, from small systems with thousands of installations to large systems commissioned by big companies or by governments. It is not a magic book that will solve all your development problems (if you find it, tell me, please!), but it could be helpful to get a different point of view on a specific problem, or a hint on how to solve problems. Armed with the knowledge of advanced concepts, such as high-order functions and anonymous methods, generics and enumerable, extended RTTI and duck typing, LiveBindings, multi-threading, FireMonkey, mobile development, server-side development, and IoT, you will be pleasantly surprised how quickly and easily you can use Delphi to write high-quality, clean, readable, fast, maintainable, and extensible code. I read too many boring programming books, so I tried to maintain a relaxed and light exposition. A small applicability scenario that describes a situation where a particular technology, approach, or design pattern can be used successfully introduces all the recipes. The recipes are not very complex, because otherwise the book may become thousands of pages long, but also not trivial, because the IT books landscape is already full of simple examples with few direct applicabilities. I tried to do a good tradeoff, and I hope to be able to do it. Every time I start reading a new book, I ask myself, *Will the author have something interesting to say?*, *How much will this book change my point of view about the topics mentioned?*, *Is it worth the time spent to read it?* Now, in spite of being from the other side of the river, I worked harder to put as much good quality contents in my books as possible; I hope that will match your expectations. One last note: writing hundreds of pages about advanced programming is not an easy task. However, I am very pleased to have done it, and I hope you will enjoy reading it at least as much as I enjoyed writing it.

In this third edition, the book still respects the structure and the style previously applied, which Daniele Teti explains perfectly in his preface. I just want to add that this approach is the best method for learning: a minimum of theory to understand the topic and a pragmatic approach with the resolution of a real problem so that the reader can have practical feedback and see it implemented.

Who this book is for

This book aims to help professional Delphi developers in their day-to-day job. This book will teach you about the newest Delphi technologies and its hidden gems. It is not a book for a newbie, but the practical approach will help you reach a new level with your Delphi skills. The experienced developer can benefit from this book, because non-trivial problems are solved using best practices. Where more than one way is available or the topic is too broad to be explained in the available pages, references are provided to allow you to go deeper into that field. It is a book to have on your desk for the next few years.

What this book covers

Chapter 1, *Delphi Basics*, talks about a set of general approaches that should not be ignored by any Delphi programmer. Some topics are simple and immediate and some are not, but all of them should be well understood. By the end of this chapter, the reader is able to use some of the fundamental Delphi techniques related to RTL, VCL, and OS integration.

Chapter 2, *Becoming a Delphi Language Ninja*, focuses on the Object Pascal language. The programming language is the way you talk to the machine, so you must be fluent and know all the possibilities offered. This chapter talks about higher-order functions, practical utilization of the extended RTTI, regular expressions, and other things useful to augment the power of your code and to lower the amount of time spent on debugging.

Chapter 3, *Knowing Your Friends – The Delphi RTL*, focuses on the Delphi's RTL. There isn't a detailed description of all the Delphi's RTLs (you would need 10 books like this one, which would be particularly boring, I guess), but you can find some recipes that explain some of the most important RTL features and some less known but really useful classes. You'll learn how to use regular expressions, the most popular encoding format used by HTTP base applications, and how to use the built-in data de/compression-related classes.

Chapter 4, *Going Cross-Platform with FireMonkey*, is dedicated to the FireMonkey framework in general. What you will learn from this chapter can be used in many of the platforms that FireMonkey supports. Moreover, you will learn about non-trivial LiveBindings utilizations.

Chapter 5, *The Thousand Faces of Multithreading*, talks about thread synchronization and the mechanisms used to obtain this synchronization, such as TMonitor, thread-safe queues, and TEvent. It is also one of the most complex chapters. By the end of this chapter, the reader will be able to create and communicate with background threads, leaving your main thread free to update your GUI (or to communicate with the OS).

Chapter 6, *Putting Delphi on the Server*, focuses on how well Delphi can behave when running on a server. Some people think that Delphi is a client-only tool, but that's not true. In this chapter, we'll show you how to create powerful servers that offer services over a network. We'll also implement a JavaScript client that brings the database data into the user browser. The techniques explained in this chapter open a range of possibilities, especially in the mobile and web area.

Chapter 7, *Linux Development*, is dedicated to development with Delphi on the most commonly-used operating system for server environments—Linux. This chapter covers several key points about Linux development: from the foundation, such as handling signals, forking, and daemonizing processes to the construction of RESTFul Server with database access and JavaScript client.

Chapter 8, *Riding the Mobile Revolution with FireMonkey*, explores the mobile development with Delphi and FireMonkey. If you are interested in mobile development, I think that will be your favorite chapter! Mobile is everywhere, and this chapter will explain how to write software for your Android or iOS device, what are the best practices to use, how to save your data on the mobile, how to retrieve and update remote data, and how to integrate with the mobile operating system.

Chapter 9, *Using Specific Platform Features*, shows you how to integrate your app with the underlying mobile operating systems beyond what FireMonkey offers. You will learn how to import Java and Objective C libraries in your app and how to use the SDK classes from your Object Pascal code.

Chapter 10, *Delphi and IoT*, talks about how the two most popular boards on the market—Arduino and Raspberry Pi—can interact with Delphi. You will learn how a Delphi application can control them, how can you control the components associated with them and recover data that they hold. The approach and techniques explained in this chapter open Delphi to IoT.

To get the most out of this book

This book talks about Delphi, so you need it. Not all the recipes are available in all the Delphi editions. Most of the projects, including the mobile, can be compiled with Delphi Community Edition or Professional (depending on license term) and higher, while for some server development parts, especially Linux, the Architect or Enterprise version is required. All the projects are compiled and tested with the latest Delphi version at the time of writing, but many recipes can also be compiled on older versions.

If you want to run the mobile app on a phone or a tablet, you can use the Android emulator or the iOS simulator, but we strongly suggest an actual device to see how the app really behaves. To deploy an iOS app on your device, you also need an Apple computer with MacOSX. For the development with Linux, a machine (also a Virtual Machine is okay) with the Linux OS is necessary to be installed, and for the IoT section, the reference boards are Arduino and Raspberry Pi.

Download the example code files

You can download the example code files for this book from your account at `www.packtpub.com`. If you purchased this book elsewhere, you can visit `www.packtpub.com/support` and register to have the files emailed directly to you.

You can download the code files by following these steps:

1. Log in or register at `www.packtpub.com`.
2. Select the **SUPPORT** tab.
3. Click on **Code Downloads & Errata**.
4. Enter the name of the book in the **Search** box and follow the onscreen instructions.

Once the file is downloaded, please make sure that you unzip or extract the folder using the latestversion of:

- WinRAR/7-Zip for Windows
- Zipeg/iZip/UnRarX for Mac
- 7-Zip/PeaZip for Linux

The code bundle for the book is also hosted on GitHub at `https://github.com/PacktPublishing/Delphi-Cookbook-Third-Edition`. We also have other code bundles from our rich catalog of books and videos available at `https://github.com/PacktPublishing/`. Check them out!

Download the color images

We also provide a PDF file that has color images of the screenshots/diagrams used in this book. You can download it here: `https://www.packtpub.com/sites/default/files/downloads/DelphiCookbookThirdEdition_ColorImages.pdf`.

Conventions used

There are a number of text conventions used throughout this book.

`CodeInText`: Indicates code words in text, database table names, folder names, filenames, file extensions, pathnames, dummy URLs, user input, and Twitter handles. Here is an example: "When we selected the `Iceberg Classico` style as the default style, the Delphi IDE added a line just before the creation of the main form, setting the default style for all the applications using `TStyleManager.TrySetStyle` static methods."

A block of code is set as follows:

```
begin
  Application.Initialize;
  Application.MainFormOnTaskbar := True;
  TStyleManager.TrySetStyle('Iceberg Classico');
  Application.CreateForm(TMainForm, MainForm);
  Application.Run;
end
```

Any command-line input or output is written as follows:

```
$ systemctl disable unit
```

Bold: Indicates a new term, an important word, or words that you see onscreen. For example, words in menus or dialog boxes appear in the text like this. Here is an example: "If you want to open a separate Terminal window for running and debugging your application, you have to enable the **Use Launcher Application** checkbox (you can find it in **Project** | **Options** | **Debugger**)."

 Warnings or important notes appear like this.

Tips and tricks appear like this.

Sections

In this book, you will find several headings that appear frequently (*Getting ready*, *How to do it...*, *How it works...*, *There's more...*, and *See also*).

To give clear instructions on how to complete a recipe, use these sections as follows:

Getting ready

This section tells you what to expect in the recipe and describes how to set up any software or any preliminary settings required for the recipe.

How to do it...

This section contains the steps required to follow the recipe.

How it works...

This section usually consists of a detailed explanation of what happened in the previous section.

There's more...

This section consists of additional information about the recipe in order to make you more knowledgeable about the recipe.

See also

This section provides helpful links to other useful information for the recipe.

Get in touch

Feedback from our readers is always welcome.

General feedback: Email `feedback@packtpub.com` and mention the book title in the subject of your message. If you have questions about any aspect of this book, please email us at `questions@packtpub.com`.

Errata: Although we have taken every care to ensure the accuracy of our content, mistakes do happen. If you have found a mistake in this book, we would be grateful if you would report this to us. Please visit `www.packtpub.com/submit-errata`, selecting your book, clicking on the Errata Submission Form link, and entering the details.

Piracy: If you come across any illegal copies of our works in any form on the internet, we would be grateful if you would provide us with the location address or website name. Please contact us at `copyright@packtpub.com` with a link to the material.

If you are interested in becoming an author: If there is a topic that you have expertise in and you are interested in either writing or contributing to a book, please visit `authors.packtpub.com`.

Reviews

Please leave a review. Once you have read and used this book, why not leave a review on the site that you purchased it from? Potential readers can then see and use your unbiased opinion to make purchase decisions, we at Packt can understand what you think about our products, and our authors can see your feedback on their book. Thank you!

For more information about Packt, please visit `packtpub.com`.

1
Delphi Basics

In this chapter, we will cover the following recipes:

- Changing your application's look and feel with VCL styles and without any code
- Changing the style of your application at runtime
- Customizing TDBGrid
- Using owner-draw combos and listboxes
- Making an owner-draw control aware of the VCL styles
- Creating a stack of embedded forms
- Manipulating JSON
- Manipulating and transforming XML documents
- I/O in the 21st century—knowing the streams
- Creating a Windows Service
- Associating a file extension with your application on Windows
- Being coherent with the Windows look and feel using TTaskDialog
- The amazing TFDTable—indices, aggregations, views, and SQL
- ETL made easy—TFDBatchMode
- Data integration made easy—TFDLocalSQL

Introduction

This chapter will explain some of the day-to-day needs of a Delphi programmer. These are ready-to-use recipes that will be useful every day and have been selected ahead of others because, although they may be obvious for some experienced users, they are still very useful. Even if there isn't any specific database-related code, many of the recipes can be used when you are dealing with data.

Changing your application's look and feel with VCL styles

Visual Component Library (**VCL**) styles are a major new entry in the latest versions of Delphi. They were introduced in Delphi XE2 and are still one of the lesser-known features for good old Delphi developers. However, as business people say, *looks matter*, so the look and feel of your application could be one of the reasons to choose your product over another from a competitor. Consider that, with a few mouse clicks, you can apply many different styles to your application to change the look and feel of it. So, why not to give it a try?

 A **style** is a set of graphical details that define the look and feel of a VCL application. A style allows you to change the appearance of every part and state of VCL controls.

Getting ready

VCL styles can be used to revamp an old application or to create a new one with a non-standard GUI. VCL styles are a completely different beast to FireMonkey styles. They are both styles, but with completely different approaches and behaviors.

To get started with VCL styles, we'll use a new application. So, let's create a new VCL application and drag and drop some components onto the main form (for example, two **TButton** controls, one **TListBox**, one **TComboBox**, and a couple of **TCheckBox**).

You can now see the resulting form that is running on my Windows 10 machine:

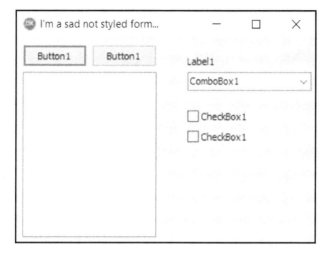

Figure 1.1: A form without style

How to do it...

Now, we've got to apply a set of nice styles by following these steps:

1. Go to **Project** | **Options** from the menu. Then, in the resultant dialog, go to **Application** | **Appearance** and select all the styles that we want to include in our application.

2. If you use the **Preview** button, the IDE shows a simple demo form with some controls, and we can get an idea about the final result of our styled form. Feel free to experiment and choose the style or set of styles that you like. Only one style at a time can be used, but we can link the necessary resources inside the executable and select the proper one at runtime.

3. After selecting all the required styles from the list, we've got to select one in the combobox at the bottom. This style will be the default style for our form and it will be loaded as soon as the application starts. You can delay this choice and make it at runtime using code, if you prefer.

4. Click on **OK**, hit *F9* (or go to **Run** | **Run**), and your application will be styled:

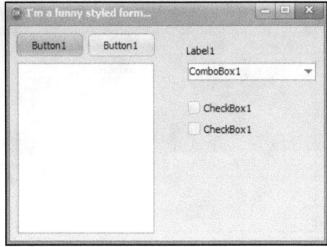

Figure 1.2: The same form as Figure 1.1 but with the Iceberg Classico style applied

How it works...

Selecting one or more styles from **Project** | **Options** | **Application** | **Appearance** will cause the Delphi linker to link the style resource into your executable. It is possible to link many styles into your executable, but you can use only one style at a time. So, how does Delphi know which style you want to use when there are more than one? If you check the Project file (the file with the .dpr extension) by going to **Project** | **View Source Menu** (for shortcut lovers, *Ctrl + V* with the project selected in **Project Manager**), you can see where and how this little bit of magic happens.

The following lines are the interesting section:

```
begin
  Application.Initialize;
  Application.MainFormOnTaskbar := True;
  TStyleManager.TrySetStyle('Iceberg Classico');
  Application.CreateForm(TMainForm, MainForm);
  Application.Run;
end
```

When we selected the `Iceberg Classico` style as the default style, the Delphi IDE added a line just before the creation of the main form, setting the default style for all the applications using `TStyleManager.TrySetStyle` static methods.

`TStyleManager` is a very important class when you deal with VCL styles. We'll see more about it in an upcoming recipe, where you'll learn how to change styles at runtime.

There's more...

Delphi and C++Builder 10.2 Tokyo come with 39 VCL Styles available in the folder (with a standard installation) at `C:\Program Files (x86)\Embarcadero\Studio\19.0\Redist\styles\vcl\`.

Embarcadero provides an additional eight premium styles that are available in the VCL premium style pack: `https://cc.embarcadero.com/item/30492`.

Moreover, it is possible to create your own styles or modify existing ones using the Bitmap Style Designer. You can access it by going to **Tools | Bitmap Style Designer Menu**.

For more details on how to create or customize a VCL style, visit `http://docwiki.embarcadero.com/RADStudio/en/Creating_a_Style_using_the_Bitmap_Style_Designer`.

The Bitmap Style Designer also provides test applications to test VCL styles.

Changing the style of your VCL application at runtime

VCL Styles are a powerful way to change the appearance of your application. One of the main features of VCL styles is the ability to change the style while the application is running.

Getting ready

Because a VCL style is simply a particular kind of binary file, we can allow our users to load their preferred styles at runtime. We could even provide new styles by publishing them on a website or sending them by email to our customers.

In this recipe, we'll change the style while the application is running using a style already linked at design time, or let the user choose between a set of styles deployed inside a folder.

How to do it...

Style manipulation at runtime is done using the class methods of the TStyleManager class. Follow these steps to change the style of your VCL application at runtime:

1. Create a brand new VCL application and add the Vcl.Themes and Vcl.Styles units to the main form's implementation uses section. These units are required to use VCL styles at runtime.

2. Drop on the form a **TListBox**, two **TButton**, and a **TOpenDialog**. Leave the default component names.

3. Go to **Project | Appearance** and select eight styles of your choice from the list. Leave the **Default style** as **Windows**.

4. The TStyleManager.StyleNames property contains the names of all the available styles. In the FormCreate event handler, we have to load the already linked styles present in the executable to the listbox to let the user choose one of them. So, create a new procedure called StylesListRefresh with the following code and call it from the FormCreate event handler:

```
procedure TMainForm.StylesListRefresh;
var
  stylename: string;
begin
  ListBox1.Clear;
  // retrieve all the styles linked in the executable
  for stylename in TStyleManager.StyleNames do
  begin
    ListBox1.Items.Add(stylename);
  end;
end;
```

5. In the `Button2Click` event handler, we set the current style according to the one selected from the `ListBox1` using the code that follows:

```
TStyleManager.SetStyle(ListBox1.Items[ListBox1.ItemIndex]);
```

6. The `Button1Click` event handler should allow the user to select a style from the disk. So, we have to create a folder named `styles` at the level of our executable and copy a `.vsf` file from the default `styles` directory, which in RAD Studio 10.2 Tokyo is `C:\Program Files (x86)\Embarcadero\Studio\19.0\Redist\styles\vcl`.

7. After copying, write the following code under the `Button1Click` event handler. This code allows the user to choose a `style` file directly from the disk. Then, you can select one of the loaded styles from the listbox and click on `Button1` to apply it to `application`:

```
if OpenDialog1.Execute then
begin
  if TStyleManager.IsValidStyle(OpenDialog1.FileName) then
  begin
    // load the style file
    TStyleManager.LoadFromFile(OpenDialog1.FileName);
    // refresh the list with the currently available styles
    StylesListRefresh;
    ShowMessage('New VCL Style has been loaded');
  end else
    ShowMessage('The file is not a valid VCL Style!');
  end;
end;
```

8. Just to give you an idea of how the different controls appear with the selected style, drag and drop some controls onto the right-hand side of the form.

9. Hit *F9* (or go to **Run** | **Run**), and play with your application, using and loading styles from the disk. The following screenshot shows my application with some styles loaded, some at design time and some from the disk:

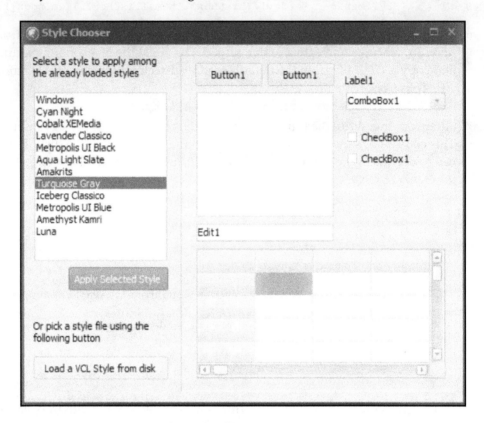

Figure 1.3: The Style Chooser form with a Turquoise Gray style loaded

How it works...

The `TStyleManager` class has all the methods we need to do the following:

- Inspect the loaded styles with `TStyleManager.StyleNames`
- Apply an already loaded style to the running application using the following code:

    ```
    TStyleManager.SetStyle('StyleName')
    ```

- Check whether a file has a valid style using the following code:

    ```
    TStyleManager.IsValidStyle('StylePathFileName')
    ```

- Load a style file from disk using the following code:

    ```
    TStyleManager.LoadFromFile('StylePathFileName')
    ```

After loading new styles from disk, the new styles are completely identical to the styles linked in the executable during the compile and link phases and can be used in the same way.

There's more...

Other things to consider are third-party controls. If your application uses third-party controls, take care with their style support (some third-party controls are not style-aware). If your external components do not support styles, you will end up with some styled controls (the originals included in Delphi) and some that are not styled (your external third-party controls).

Go to **Tools** | **Bitmap Style Designer**. Using a custom VCL Style, we can also:

- Change application colors, such as `ButtonNormal`, `ButtonPressed`, `ButtonFocused`, `ButtonHot`, and others.
- Override system colors, such as `clCaptionText`, `clBtnFace`, `clActiveCaption`, and so on.

- Font color and font name for particular controls should be familiar to ButtonTextNormal, ButtonTextPressed, ButtonTextFocused, ButtonTextHot, and many others:

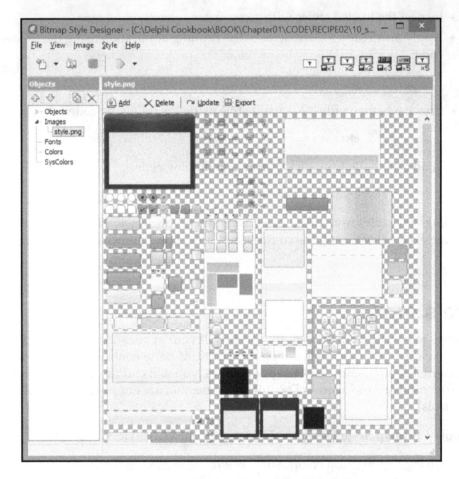

Figure 1.4: The Bitmap Style Designer while it is working on a custom style

Customizing TDBGrid

The adage *a picture is worth a thousand words* refers to the notion that a complex idea can be conveyed with just a single still image. Sometimes, even a simple concept is easier to understand and nicer to see if it is represented by images. In this recipe, we'll see how to customize **TDBGrid** to visualize a graphical representation of data.

Getting ready

Many VCL controls are able to delegate their drawing, or part of it, to user code. It means that we can use simple event handlers to draw standard components in different ways. It is not always simple, but **TDBGrid** is customizable in a really easy way. Let's say that we have a class of musicians that have to pass a set of exams. We want to show the percent of exams already passed with a progress bar and if the percent age is higher than 50, there should also be a check mark in another column. Moreover, after listening to the pieces played at the exams, each musician received votes from an external examination committee. The last column needs to show the mean of votes from this committee as a rating from zero to five.

How to do it...

We'll use a special in-memory table from the FireDAC library. FireDAC is a new data access library from Embarcadero included in RAD Studio since version XE5. If some of the code seems unclear at the moment, consider the in-memory table as a normal `TDataSet` descendant that holds its data only in memory. However, at the end of the section, there are some links to the FireDAC documentation, and I strongly suggest that you read it if you still don't understand FireDAC:

1. Create a brand new VCL application and drop a **TFDMemTable**, a **TDBGrid**, a **TDataSource**, and a **TDBNavigator** onto the form. Connect all the components in the usual way (**TDBGrid** connected to **TDataSource**, followed by **TFDMemTable**). Set the **TDBGrid** font size to **18**. This will create more space in the cell for our graphical representation.
2. Using the **TFDMemTable** fields editor, add the following fields and then activate the dataset by setting its **Active** property to **True**:

Field name	Field data type	Field type
FullName	String (size 50)	Data
TotalExams	Integer	Data
PassedExams	Integer	Data
Rating	Float	Data
PercPassedExams	Float	Calculated
MoreThan50Percent	Boolean	Calculated

3. Now, add all the columns to **TDBGrid** by right-clicking and selecting **Columns Editor...**. Then, again right-click and select **Add all fields** in the resultant window. Then, rearrange the columns as shown here and give it a nice title caption:

- `FullName`
- `TotalExams`
- `PassedExams`
- `PercPassedExams`
- `MoreThan50Percent`
- `Rating`

4. In a real application, we would load real data from some sort of database. However, for now, we'll use some custom data generated in code. We have to load this data into the dataset with the code that follows:

```
procedure TMainForm.FormCreate(Sender: TObject);
begin
  FDMemTable1.AppendRecord(['Ludwig van Beethoven', 30, 10,
4]);
  FDMemTable1.AppendRecord(['Johann Sebastian Bach', 24, 10,
2.5]);
  FDMemTable1.AppendRecord(['Wolfgang Amadeus Mozart', 30, 30,
5]);
  FDMemTable1.AppendRecord(['Giacomo Puccini', 25, 10, 2.2]);
  FDMemTable1.AppendRecord(['Antonio Vivaldi', 20, 20, 4.7]);
  FDMemTable1.AppendRecord(['Giuseppe Verdi', 30, 5, 5]);
  FDMemTable1.AppendRecord(['John Doe', 24, 5, 1.2]);
end;
```

5. Do you remember we have two calculated fields that need to be filled in some way? Calculated fields need a form of processing behind them to work. **TFDMemTable**, just like any other `TDataSet` descendant, has an event called **OnCalcFields** that allows the developer to do this. Create the **OnCalcFields** event handler for **TFDMemTable** and fill it with the following code:

```
procedure TMainForm.FDMemTable1CalcFields(DataSet: TDataSet);
var
  LPassedExams: Integer;
  LTotExams: Integer;
begin
  LPassedExams :=
FDMemTable1.FieldByName('PassedExams').AsInteger;
```

```
LTotExams := FDMemTable1.FieldByName('TotalExams').AsInteger;
if LTotExams = 0 then
  FDMemTable1.FieldByName('PercPassedExams').AsFloat := 0
else
  FDMemTable1.FieldByName('PercPassedExams').AsFloat :=
LPassedExams /
    LTotExams * 100;

FDMemTable1.FieldByName('MoreThan50Percent').AsBoolean :=
  FDMemTable1.FieldByName('PercPassedExams').AsFloat > 50;
```

```
end;
```

6. Run the application by hitting *F9* (or by going to **Run | Run**) and you will get the following screenshot:

Full Name	#Exams	#Passed Exams	% Passed Exams	More than 50%	Rating
Ludwig van Beethoven	30	10	33	False	4
Johann Sebastian Bach	24	10	42	False	2,5
Wolfgang Amadeus Mozart	30	30	100	True	5
Giacomo Puccini	25	10	40	False	2,2
Antonio Vivaldi	20	20	100	True	4,7
Giuseppe Verdi	30	5	17	False	5
▸John Doe	24	5	21	False	1,2

Figure 1.5: A normal form with some data

7. This is useful, but a bit boring. Let's start our customization. Close the application and return to the Delphi IDE.
8. Go to **Properties** of **TDBGrid** and set **Default Drawing** to **False**.
9. Now, we have to organize the resources used to draw the grid cells. Calculated fields will be drawn directly using code, but the Rating field will be drawn using a five-star rating image from 0 to 5. It starts with a 0.5 incremental step (0, 0.5, 1, 1.5, and so on). So, drop **TImageList** on the form, and set the **Height** as 32 and the **Width** as 160.

10. Select the **TImageList** component and open the image list's editor by right-clicking and then selecting **ImageList Editor....** You can find the required PNG images in the recipe's project folder (`ICONS\RATING_IMAGES`). Load the images in the correct order, as shown here:

- Index 0 as image `0_0_rating.png`
- Index 1 as image `0_5_rating.png`
- Index 2 as image `1_0_rating.png`
- Index 3 as image `1_5_rating.png`
- Index 4 as image `2_0_rating.png`

11. Go to the **TDBGrid** event and create the event handler for **OnDrawColumnCell**. All the customization code goes in this event. Include the `Vcl.GraphUtil` unit, and write the following code in the `DBGrid1DrawColumnCell` event:

```
procedure TMainForm.DBGrid1DrawColumnCell(Sender: TObject;
const Rect: TRect;
  DataCol: Integer; Column: TColumn; State: TGridDrawState);
var
  LRect: TRect;
  LGrid: TDBGrid;
  LText: string;
  LPerc: Extended;
  LTextWidth: TSize;
  LRating: Extended;
  LNeedOwnerDraw: Boolean;
  LImageIndex: Int64;
begin
  LGrid := TDBGrid(Sender);
  if [gdSelected, gdFocused] * State <> [] then
    LGrid.Canvas.Brush.Color := clHighlight;

  LNeedOwnerDraw := (Column.Field.FieldKind = fkCalculated) or
    Column.FieldName.Equals('Rating');

  // if doesn't need owner-draw, default draw is called
  if not LNeedOwnerDraw then
  begin
    LGrid.DefaultDrawColumnCell(Rect, DataCol, Column, State);
    exit;
  end;

  LRect := Rect;

  if Column.FieldName.Equals('PercPassedExams') then
```

```
begin
  LText := FormatFloat('##0', Column.Field.AsFloat) + ' %';
  LGrid.Canvas.Brush.Style := bsSolid;
  LGrid.Canvas.FillRect(LRect);
  LPerc := Column.Field.AsFloat / 100 * LRect.Width;
  LGrid.Canvas.Font.Size := LGrid.Font.Size - 1;
  LGrid.Canvas.Font.Color := clWhite;
  LGrid.Canvas.Brush.Color := clYellow;
  LGrid.Canvas.RoundRect(LRect.Left, LRect.Top,
Trunc(LRect.Left + LPerc),
    LRect.Bottom, 2, 2);
  LRect.Inflate(-1, -1);
  LGrid.Canvas.Pen.Style := psClear;
  LGrid.Canvas.Font.Color := clBlack;
  LGrid.Canvas.Brush.Style := bsClear;

  LTextWidth := LGrid.Canvas.TextExtent(LText);
  LGrid.Canvas.TextOut(LRect.Left + ((LRect.Width div 2) -
    (LTextWidth.cx div 2)), LRect.Top + ((LRect.Height div 2)
-
    (LTextWidth.cy div 2)), LText);
end
else if Column.FieldName.Equals('MoreThan50Percent') then
begin
  LGrid.Canvas.Brush.Style := bsSolid;
  LGrid.Canvas.Pen.Style := psClear;
  LGrid.Canvas.FillRect(LRect);
  if Column.Field.AsBoolean then
  begin
    LRect.Inflate(-4, -4);
    LGrid.Canvas.Pen.Color := clRed;
    LGrid.Canvas.Pen.Style := psSolid;
    DrawCheck(LGrid.Canvas, TPoint.Create(LRect.Left,
      LRect.Top + LRect.Height div 2), LRect.Height div 3);
  end;
end
else if Column.FieldName.Equals('Rating') then
begin
  LRating := Column.Field.AsFloat;
  if Frac(LRating) < 0.5 then
    LRating := Trunc(LRating)
  else
    LRating := Trunc(LRating) + 0.5;
  LText := LRating.ToString;
  LGrid.Canvas.Brush.Color := clWhite;
  LGrid.Canvas.Brush.Style := bsSolid;
  LGrid.Canvas.Pen.Style := psClear;
  LGrid.Canvas.FillRect(LRect);
```

```
        Inc(LRect.Left);
        LImageIndex := Trunc(LRating) * 2;
        if Frac(LRating) >= 0.5 then
          Inc(LImageIndex);
        ImageList1.Draw(LGrid.Canvas, LRect.CenterPoint.X -
          (ImageList1.Width div 2), LRect.CenterPoint.Y -
    (ImageList1.Height div 2),
          LImageIndex);
      end;

    end;
```

12. That's all folks! Press *F9* (or go to **Run** | **Run**), and we now have a nicer grid with more direct information about our data:

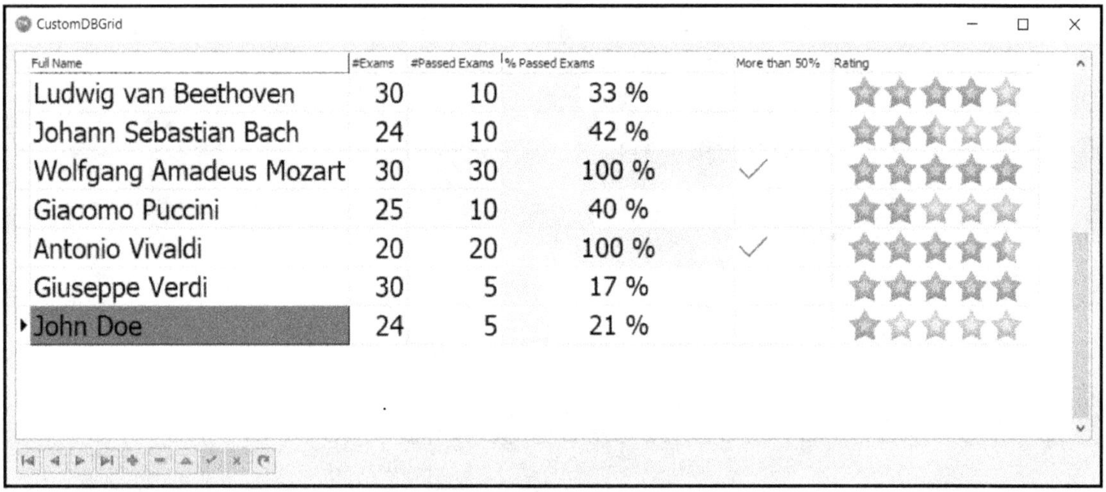

Full Name	#Exams	#Passed Exams	% Passed Exams	More than 50%	Rating
Ludwig van Beethoven	30	10	33 %		★★★★☆
Johann Sebastian Bach	24	10	42 %		★★☆☆☆
Wolfgang Amadeus Mozart	30	30	100 %	✓	★★★★★
Giacomo Puccini	25	10	40 %		★★☆☆☆
Antonio Vivaldi	20	20	100 %	✓	★★★★☆
Giuseppe Verdi	30	5	17 %		★★★★★
▸John Doe	24	5	21 %		★☆☆☆☆

Figure 1.6: The same grid with a bit of customization

How it works...

By setting the **TDBGrid** property **Default Drawing** to **False**, we told the grid that we wanted to manually draw all the data into every cell. OnDrawColumnCell allows us to actually draw using standard Delphi code. For each cell we are about to draw, the event handler was called with a list of useful parameters to know which cell we're about to draw and what data we have to read, considering the column we are currently drawing. In this case, we want to draw only the calculated columns and the Rating field in a custom way. This is not a rule, but this can be done to manipulate all cells. We can draw any cell in the way we like. For the cells where we don't want to do custom drawing, a simple call method, DefaultDrawColumnCell, passes the same parameters we got from the event and the VCL code will draw the current cell as usual.

Among the event parameters, there is a Rect object (of type **TRect**) that represents the specific area we're about to draw. There is a column object (of type **TColumn**), which is a reference to the current column of the grid, and a State (of type **TGridDrawState**), which is a set of the grid cell states (for example, Selected, Focused, HotTrack, and many more). If our drawing code ignores the State parameter, all the cells will be drawn in the same way, and users cannot see which cell or row is selected.

The event handler uses a **Pascal Sets Intersect** to know whether the current cell should be drawn as a Selected or Focused cell. Refer to the following code for better clarity:

```
if [gdSelected, gdFocused] * State <> [] then
   Grid.Canvas.Brush.Color := clHighlight;
```

Remember that if your dataset has 100 records and 20 fields, **OnDrawColumnCell** will potentially be called 2,000 times! So, the event code must be fast; otherwise, the application will become less responsive.

There's more...

Owner drawing is a really large topic and can be simple or tremendously complex, involving much Canvas-related code. However, often, the kind of drawing you need will be relatively similar. So, if you need checks, arrows, color gradients, and so on, check the procedures into the Vcl.GraphUtil unit. Otherwise, if you need images, you could use **TImageList** to hold all the images needed by your grid, as we did in this recipe for the Rating field.

The good news is that the drawing code can be reused by different kinds of controls, so try to organize your code in a way that allows code reutilization by avoiding direct dependencies to the form where the control is.

The code in the drawing events should not contain business logic or presentation logic. If you need presentation logic, put it in a separate, testable function or class.

Using owner-draw combos and listboxes

Many things are organized in a list. **Lists** are useful when you have to show items or when your user has to choose from a set of possible options. Usually, standard lists are flat, but sometimes you need to transmit more information in addition to a list of items. Let's think about when you go to choose a font in an advanced text editor such as Microsoft Word or Apache OpenOffice. Having the name of the font drawn in the font style itself helps users make a faster and more reasoned choice. In this recipe, we'll see how to make listboxes more useful. The code is perfectly valid for **TComboBox** as well.

Getting ready

As we saw in the *Customizing TDBGrid* recipe, many VCL controls are able to delegate their drawing, or part of it, to user code. This means that we can use simple event handlers to draw standard components in different ways. Let's say that we have a list of products in our store and we have to set discounts for these products. As there are many products, we want to set up the processing so that our users can make a fast selection in terms of the available discount percentages using a *color code*.

How to do it...

Let's look at the following steps:

1. Create a brand new VCL application and drop a **TListBox** onto the form. Set the following properties:

Property	Value
Style	lbOwnerDrawFixed
Font.Size	14

2. In the `Items` listbox property, add seven levels of discount. For example, you can use no discount, 10% discount, 20% discount, 30% discount, 40% discount, 50% discount, 60% discount, and 70% discount.

3. Then, drop a **TImageList** component onto the form and set the following properties:

Property	Value
ColorDepth	cd32Bit
DrawingStyle	dsTransparent
Width	32
Height	32

4. **TImageList** is our image repository and will be used to draw an image by index. Load seven PNG images (size 32 x 32) into **TImageList**. You can find some nice PNG icons in the respective recipe project folder (`ICONS\PNG\32`).

5. Create an `OnDrawItem` event handler for `TListBox` and write the following code:

```
procedure TMainForm.ListBox1DrawItem(Control: TWinControl;
Index: Integer;
  Rect: TRect; State: TOwnerDrawState);
var
  LBox: TListBox;
  R: TRect;
  S: string;
  TextTopPos, TextLeftPos, TextHeight: Integer;
const
  IMAGE_TEXT_SPACE = 5;
begin
  LBox := Control as TListBox;
  R := Rect;
  LBox.Canvas.FillRect(R);
  ImageList1.Draw(LBox.Canvas, R.Left, R.Top, Index);
  S := LBox.Items[Index];
  TextHeight := LBox.Canvas.TextHeight(S);
  TextLeftPos := R.Left + ImageList1.Width + IMAGE_TEXT_SPACE;
  TextTopPos := R.Top + R.Height div 2 - TextHeight div 2;
  LBox.Canvas.TextOut(TextLeftPos, TextTopPos, S);
end;
```

6. Run the application by hitting *F9* (or by going to **Run | Run**) and you will see the following:

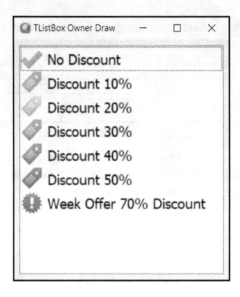

Figure 1.7: Our listbox with some custom icons read from TImageList

How it works...

The `TListBox.OnDrawItem` event handler allows us to customize the drawing of the listbox. In this recipe, we've used **TImageList** as the image repository for the listbox. Using the `Index` parameter, we've read the correspondent image in **TImageList** and drawn on the listbox `Canvas`. After this, all the other code is related to the alignment of image and text inside the listbox row.

Remember that this event handler will be called for each item in the list, so the code must be fast and should not do too much slow canvas writing. Otherwise, your GUI will be unresponsive. If you want to create complex graphics on the fly in the event, I strongly suggest that you prepare your images the first time you draw the item and then put them in a sort of cache memory (`TObjectList<TBitmap>` is enough).

There's more...

While you are in `OnDrawItem`, you can do whatever you want with the **TListBox** `Canvas`. Moreover, the `State` parameter (of type `TOwnerDrawState`) tells you which states the listbox item is in (for example, `Selected`, `Focused`, `HotTrack`, and so on). So, you can use a different kind of drawing, depending on the item state. Check out the *Customizing TDBGrid* recipe to find out about **TDBGrid** owner-drawing for an example of the `State` parameter.

If you want to make your code aware of the selected VCL Style, changing the color used according to the style, you can use `StyleServices.GetStyleColor()`, `StyleServices.GetStyleFontColor()`, and `StyleServices.GetSystemColor()` in the `Vcl.Themes` unit.

The icons used in this recipe are from the **ICOJAM** website (`http://www.icojam.com`). The specific set used is available at `http://www.icojam.com/blog/?p=259`.

Making an owner-draw control aware of the VCL styles

Owner-draw controls are powerful. They allow you to completely adapt your GUI to the needs of your users and potentially enable your application to display data in a more familiar way. In the end, owner-draw controls improve the user's experience with the application. However, owner-draw controls do not always fit in well with the VCL custom styles. Why? Because if you try to draw something by yourself, you could be tempted to use a fixed color, such `clRed` or `clYellow`, or you could be tempted to use the operating system color, such as `clBtnFace` or `clWindow`. In doing so, your owner-draw controls will not be style aware and will be drawn in the same way regardless of the current VCL style. In this recipe, you'll learn how to make custom graphics that remain relevant to the selected VCL style.

Getting ready

Let's say you are in charge of developing a control panel for a hotel's lighting system. You have a list of lamps to power on and you, using some hardware, have to power on some lamps by clicking on a button. Customers tell you that buttons should show some additional information about the lamp, for example:

- Served zone (corridor, hall, room number, and so on)
- State (on/off, using some fancy graphics)
- The time the lamp was powered on
- The time when electrical problems were detected, and a red icon to indicate that the lamp is off even when a current supplies the line, so the circuit is interrupted somewhere
- Other custom information not currently known, such as small graphs showing lamp state history during the last 24 hours

The question is how to implement this kind of UI. One of the possible ways is to use **TDrawGrid** and draw all the required details in each cell, using the cell as a button. Using **TDrawGrid**, you get a grid of buttons for free. You have also the greatest flexibility in terms of the information displayed because you are using the TCanvas method to custom draw each cell. This is quite a popular solution for this kind of non-standard UI.

However, when you deploy this application, the customers ask about the possibility of changing the style of the application to fit the needs of the current user. So you think about VCL styles, and you are right. However, the graphics drawn into the cells don't follow the currently selected VCL style, and your beautiful application becomes a bad mix of colors. In other words, when users change the selected VCL style, all the controls reflect the new style, but the owner-drawn grid, which is unaware to the selected style, doesn't look as nice as the rest of the UI. How do you solve this problem? How do you draw custom graphics adhering to the selected VCL style? In this recipe, you'll learn how to do it using the lamp control grid example.

How it works...

At design time, the form looks like the one shown in the following screenshot:

Figure 1.8: The form as it looks at design time

When the form is created, the list of available styles is loaded in the `Radio` group using code similar to the following:

```
RadioGroup1.Items.Clear;
RadioGroup1.Columns := Length(TStyleManager.StyleNames);
for LStyleName in TStyleManager.StyleNames do
  RadioGroup1.Items.Add(LStyleName);
RadioGroup1.ItemIndex := 0;
TStyleManager.SetStyle('Windows');
```

Then, a list of `TLampInfo` objects is created and initialized using the information contained in the `Zones` array. After that, the draw grid is initialized according to the `LAMPS_FOR_EACH_ROW` constant. Here's the relevant code:

```
FLamps := TObjectList<TLampInfo>.Create(True);
for I := 1 to LAMPS_FOR_EACH_ROW * 4 do
begin
  FLamps.Add(TLampInfo.Create(Zones[I]));
end;

DrawGrid1.DefaultColWidth := 128;
DrawGrid1.DefaultRowHeight := 64;
DrawGrid1.ColCount := LAMPS_FOR_EACH_ROW;
DrawGrid1.RowCount := FLamps.Count div LAMPS_FOR_EACH_ROW;
```

The `FormCreate` event handler initializes the styles list and the list of lamps (the model) on the form. Now, we'll see how the other event handlers will use them.

The `TDrawGrid OnSelectCell` event, as the name suggests, is used to address the current lamp from `FLamps` and to toggle its state. That's it. If the lamp is on, then the lamp will be powered down, otherwise the lamp will be powered on. After that, the code forces the grid to redraw using the `Invalidate` method:

```
procedure TMainForm.DrawGrid1SelectCell(Sender: TObject; ACol,
    ARow: Integer; var CanSelect: Boolean);
begin
  FLamps[ACol + ARow * LAMPS_FOR_EACH_ROW].ToggleState;
  DrawGrid1.Invalidate;
end;
```

Now, really interesting things happened in the `DrawThemed` method called inside the `TDrawGridOnDrawCell` event. This method receives information about the coordinates of the cell to draw, and then it draws a button on the canvas using the information contained in the corresponding `TLampInfo` instance. The code is quite long, but an interesting concept is that no specific colors are used. When it is necessary to draw something, the code asks `StyleService` to get the correct color according to the current style. This approach is also used for font color and for system colors. Here's a handy table that summarizes these concepts:

Method name	Description
`StyleServices.GetStyleColor(Color: TStyleColor)`	Returns the color defined in the style for the element specified by `Color`

`StyleServices.StyleFontColor(Font: TStyleFont)`	Returns the font color for the element specified by `Font`
`StyleServices.GetSystemColor(Color: TColor)`	Returns the system color defined in the current style

So, when we have to highlight the (pseudo) button if there are electrical problems on the power line, we use the following code:

```
if LLamp.ThereAreElectricalProblems then
    LCanvas.Brush.Color := StyleServices.GetStyleColor(scButtonHot)
else
    LCanvas.Brush.Color := StyleServices.GetStyleColor(scWindow);
LCanvas.FillRect(LRect);
```

When we've got to draw normal text, we use the following code:

```
LCanvas.Font.Color :=
    StyleServices.GetStyleFontColor(sfButtonTextNormal);
LCanvas.TextRect(LRect, LValue, [TTextFormats.tfCenter,
TTextFormats.tfVerticalCenter]);
```

It is clear that the paradigm is this:

1. Get the current color for the selected element of the UI according to the style
2. Draw the graphics using that color

Clicking on the **Simulate Problems** button, it is possible to see how the graphics are drawn in the case of problems on the power line. The images are drawn directly from the image list using the following code:

```
procedure TMainForm.DrawImageOnCanvas(ACanvas: TCanvas;
    var ARect: TRect; ImageIndex: Integer);
begin
    ImageList1.Draw(ACanvas, ARect.Left + 4,
        ARect.Top + ((ARect.Bottom - ARect.Top) div 2) - 16,
        ImageIndex);
end;
```

Using this approach, the application created in this recipe, which has a lot of custom graphics, behaves very well even for VCL styles. Here are some screenshots:

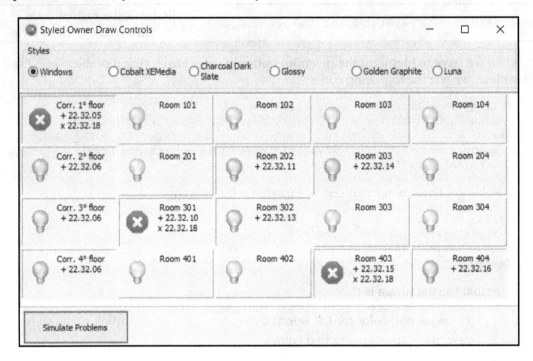

Figure 1.9: The application while it is using the Windows style

Figure 1.10: The application while it is using the Luna style

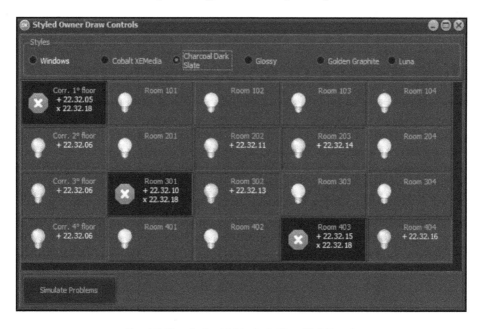

Figure 1.11: The application while it is using the Charcoal Dark Slate style

As you can see, the application correctly draws the owner-draw parts of the UI using the right colors from the selected style.

There's more...

The VCL style infrastructure is very powerful. In the case of `TWinControl` descendants, you can even define specific hooks for your components using `TStyleHook`. `TStyleHook` is a class that handles messages for controls acting as a wrapper for the hooked control. If you have a custom control that you want to be style enabled, inherit from `TStyleHook` and provide custom processing for that control. For examples, see `TEditStyleHook` and `TComboBoxStyleHook`. You need to register the style hook class with the style engine using the `RegisterStyleHook` method, as shown in the following code:

```
TCustomStyleEngine.RegisterStyleHook(TCustomEdit, TEditStyleHook);
```

Moreover, the `StyleServices` function returns an instance of `TCustomStyleServices`, which provides a lot of customization methods related to VCL styles. Check out the related documentation at `http://docwiki.embarcadero.com/Libraries/en/Vcl.Themes.TCustomStyleServices_Met hods` to see all the possibilities.

Creating a stack of embedded forms

Every modern browser has a tabbed interface. Also, many other kinds of multiple-view software have this kind of interface. Why? Because it's very useful. While you are reading one page, you can rapidly check another page and still come back to the first one at the same point you left some seconds ago. You don't have to redo a search or use a lot of mouse clicks to just go back to that particular point. You simply switch from one window to another window and come back to the first. I have seen too many business applications that are composed of a bunch of dialog windows. Every form is called with the `TForm.ShowModal` method. So, the user has to navigate into your application one form at time. This is simpler to handle for the programmer, but it's less user friendly for your customers. However, giving a switchable interface to your customer is not that difficult. In this recipe, we'll see a complete example of how to do it.

Getting ready

This recipe is a bit more complex than the previous recipes. So, I'll not explain all the code but only the fundamental parts. You can find the complete code in the book's code repository (`Chapter1/RECIPE06`).

Let's say we want to create a tabbed interface for our software that is used to manage product orders, sales, and invoices. All the forms must be usable at the same time, without us having to close the previous one. Before we begin, the following screenshot is what we want to create:

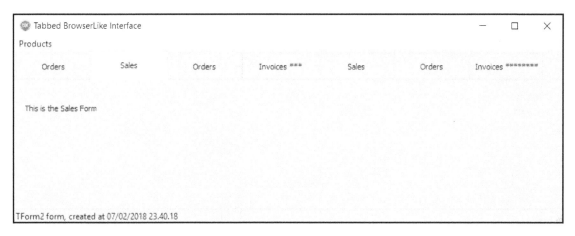

Figure 1.12: The main form containing seven embedded child forms

How it works...

The project is composed of a bunch of forms. The main form has `TTabControl`, which allows us to switch between the active forms. All embedded forms inherit from `EmbeddableForm`. The most important one is the `Show` method shown here:

```
procedure TEmbeddableForm.Show(AParent: TPanel);
begin
  Parent := AParent;
  BorderStyle := bsNone;
  BorderIcons := [];
  Align := alClient;
  Show;
end;
```

Note that all the forms apart from the main form have been removed from the **Auto-Create Form** list (you can access the list by going to **Project** | **Options** | **Forms**).

All the other forms descend from `EmbeddableForm` and are added to `TTabControl` on the main form, with a line of code similar to the following one:

```
procedure TMainForm.MenuOrdersClick(Sender: TObject);
begin
  AddForm(TForm1.Create(self));
end;
```

The `AddForm` method is in charge of adding an actual instance of a form into the tabs, keeping a reference to it. The following code shows how it is done:

```
// Add a form to the stack
procedure TMainForm.AddForm(AEmbeddableForm: TEmbeddableForm);
begin
  AEmbeddableForm.Show(Panel1);
  // each tab show the caption of the containing form and
  // hold the reference to it
  TabControl1.Tabs.AddObject(
     AEmbeddableForm.Caption, AEmbeddableForm);
  ResizeTabsWidth;
  ShowForm(AEmbeddableForm);
end;
```

Other methods are in charge of bringing an already created form to the front when a user clicks on the **Related** tab, and then to close a form when the **Related** tab is removed (check out the `ShowForm` and `WMEmbeddedFormClose` methods).

There is a bit of code, but the concepts are simple:

- When we need to create a new form, we add it in the `TabControl1.Tabs` property. The caption of the form is the caption of the tab, and the object is the form itself. This is what the `AddForm` method does with the following line:

  ```
  TabControl1.Tabs.AddObject(AEmbeddableForm.Caption,
  AEmbeddableForm);
  ```

- When a user clicks on a tab, we have to find the associated form by cycling through the `TabControl1.Tabs.Objects` list and bringing it to the front.

- When a form asks to be closed (sending a `WM_EMBEDDED_CLOSE` message), we have to set the `ParentWantClose` property and then call the `Close` method of the correspondent form.
- If the user wants to close a form by closing the corresponding tab (in the recipe code, there is `TPopMenu` connected to `TabControl`, which is used to close a form with a right-click), we have to call the `Close` method on the corresponding form.
- Every form frees itself in the `OnClose` event handler. This is done once for all the forms in the `TEmbeddableForm.CloseForm` event handler, using the `caFree` action.

There's more...

Embedding a form into another `TWinControl` is not difficult and allows us to create flexible GUIs without using `TPageControl` and `Frames`. For the end user, this multi-tabbed GUI is probably more familiar because all modern browsers use it, and your user may already know how to use a browser with different pages or screens open. From a developer's point of view, the multi-tabbed interface allows for much better programming patterns and practices. This technique can also be used for other scenarios where you have to embed one screen into another.

More flexible (and complex) solutions can be done involving the use of Observers, but in simple cases, this recipe's solution, based on Windows Messaging, is enough.

More information about the Observer design pattern can be found at `http://sourcemaking.com/design_patterns/observer/delphi`.

Other interesting solutions that don't rely on Windows Messaging and so are also cross-platform include the following:

- Solutions based on the `System.Messaging.TMessageManager` class. More information about `TMessageManager` can be obtained at `http://docwiki.embarcadero.com/Libraries/en/System.Messaging.TMessageManager`.
- **Delphi Event Bus** (**DEB**) is a publish/subscribe Event Bus framework for the Delphi platform. More information can be found at `https://github.com/spinettaro/delphi-event-bus` or you can take a look at `Chapter 5`, *The Thousand Faces of Multithreading*, in the *Communication made easy with Delphi Event Bus* recipe.

Code in this recipe can be used with every component that uses **TStringList** to show items (**TListBox**, **TComboBox**, and so on) and can be adapted easily for other scenarios.

In the recipe code, you'll also find a nice way to show status messages generated by the embedded forms and a centralized way to show application hints in the status bar.

Manipulating JSON

JavaScript Object Notation (**JSON**) is a lightweight data-interchange format. As the reference site says, *It is easy for humans to read and write. It is easy for machines to parse and generate*. It is based on a subset of the JavaScript programming language, but it is not limited to JavaScript in any way. Indeed, JSON is a text format that is completely language agnostic. These properties make JSON an ideal data-interchange language for many uses. In recent years, JSON has become on a par with XML in many applications, especially when data size matters, because of its intrinsic conciseness and simplicity.

Getting ready

JSON provides the following five datatypes—String, Number, Object, Array, Boolean, and Null.

This simplicity is an advantage when you have to read a JSON string into some kind of language-specific structure, because every modern language supports the JSON datatypes as simple types, or as a HashMap (in the case of JSON objects) or List (in the case of JSON arrays). So, it makes sense that a data format that is interchangeable with programming languages is also based on these types and structures.

Since version 2009, Delphi has provided built-in support for JSON. The System.JSON.pas unit contains all the JSON types with a nice object-oriented interface. In this recipe, you'll see how to generate, modify, and parse a JSON string.

How to do it...

Let's look at the following steps:

1. Create a new VCL application and drop in three **TButton** and a **TMemo**. Align all the buttons as a toolbar at the top of the form and the memo to all the remaining from client areas.

2. From left to right, name the buttons btnGenerateJSON, btnModifyJSON, and btnParseJSON.

3. We'll use static data as our data source. A simple matrix is enough for this recipe. Just after the start of the implementation section of the unit, write the following code:

```
type
  TCarInfo = (
    Manufacturer = 1,
    Name = 2,
    Currency = 3,
    Price = 4);

var
  Cars: array [1 .. 4] of
    array [Manufacturer .. Price] of string = (
    ('Ferrari','360 Modena','EUR', '250000'),
    ('Ford', 'Mustang', 'USD', '80000'),
    ('Lamborghini', 'Countach', 'EUR','300000'),
    ('Chevrolet', 'Corvette', 'USD', '100000')
  );
```

4. **TMemo** is used to show our JSON files and our data. To keep things clear, create a public property called JSON on the form and map its setter and getter to the Memo1.Lines.Text property. Use the following code:

```
//...other form methods declaration
private
  procedure SetJSON(const Value: String);
  function GetJSON: String;
public
  property JSON: String read GetJSON write SetJSON;
end;

//...then in the implementation section
function TMainForm.GetJSON: String;
begin
  Result := Memo1.Lines.Text;
end;

procedure TMainForm.SetJSON(const Value: String);
begin
  Memo1.Lines.Text := Value;
end;
```

5. Now, create event handlers for each button and write the code that follows. Pay attention to the event names:

```delphi
procedure TMainForm.btnGenerateJSONClick(Sender: TObject);
var
  i: Integer;
  JSONCars: TJSONArray;
  Car, Price: TJSONObject;
begin
  JSONCars := TJSONArray.Create;
  try
    for i := Low(Cars) to High(Cars) do
    begin
      Car := TJSONObject.Create;
      JSONCars.AddElement(Car);
      Car.AddPair('manufacturer',
          Cars[i][TCarInfo.Manufacturer]);
      Car.AddPair('name', Cars[i][TCarInfo.Name]);
      Price := TJSONObject.Create;
      Car.AddPair('price', Price);
      Price.AddPair('value',
          TJSONNumber.Create(
            Cars[i][TCarInfo.Price].ToInteger));
      Price.AddPair('currency',
          Cars[i][TCarInfo.Currency]);
    end;
    JSON := JSONCars.ToJSON;
  finally
    JSONCars.Free;
  end;
end;

procedure TMainForm.btnModifyJSONClick(Sender: TObject);
var
  JSONCars: TJSONArray;
  Car, Price: TJSONObject;
begin
  JSONCars := TJSONObject.ParseJSONValue(JSON)
as TJSONArray;
  try
    Car := TJSONObject.Create;
    JSONCars.AddElement(Car);
    Car.AddPair('manufacturer', 'Hennessey');
    Car.AddPair('name', 'Venom GT');
    Price := TJSONObject.Create;
    Car.AddPair('price', Price);
    Price.AddPair('value', TJSONNumber.Create(600000));
    Price.AddPair('currency', 'USD');
```

```
      JSON := JSONCars.ToJSON;
  finally
    JSONCars.Free;
  end;
end;

procedure TMainForm.btnParseJSONClick(Sender: TObject);
var
  JSONCars: TJSONArray;
  i: Integer;
  Car, JSONPrice: TJSONObject;
  CarPrice: Double;
  s, CarName, CarManufacturer, CarCurrencyType: string;
begin
  s := '';
  JSONCars := TJSONObject.ParseJSONValue(JSON)
                                         as TJSONArray;
  if not Assigned(JSONCars) then
    raise Exception.Create('Not a valid JSON');
  try
    for i := 0 to JSONCars.Count - 1 do
    begin
      Car := JSONCars.Items[i] as TJSONObject;
      CarName := Car.GetValue('name').Value;
      CarManufacturer :=
        Car.GetValue('manufacturer').Value;
      JSONPrice := Car.GetValue('price') as TJSONObject;
      CarPrice := (JSONPrice.GetValue('value') as
        TJSONNumber).AsDouble;
      CarCurrencyType := JSONPrice.GetValue('currency')
      .Value
      s := s + Format(
        'Name = %s' + sLineBreak +
        'Manufacturer = %s' + sLineBreak +
        'Price = %.0n%s' + sLineBreak +
        '-----' + sLineBreak,
        [CarName, CarManufacturer,
        CarPrice, CarCurrencyType]);
    end;
    JSON := s;
  finally
    JSONCars.Free;
  end;
end;
```

6. Run the application by hitting *F9* (or by going to **Run** | **Run**).
7. Click on the `btnGenerateJSON` button, and you should see a JSON array and some JSON objects in the memo.
8. Click on the `btnModifyJSON` button, and you should see one more JSON object inside the outer JSON array in the memo.
9. Click on the last button, and you should see the same data as before, but with a normal text representation.
10. After the third click, you should see something similar to the following screenshot:

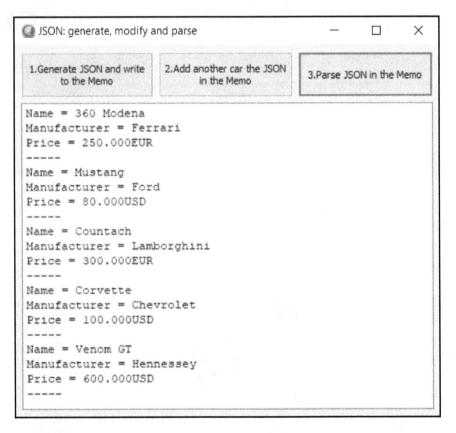

Figure 1.13: Text representation of the JSON data generated and modified

There's more...

In JSON objects, the `Owned` property determines whether the parent is responsible for the destruction of the object. This property by default is `True`, meaning all contained instances are owned by their parent. This is why, usually, if you have a combination of various JSON objects, you free only the last parent.

Although not the fastest or the most standards-compliant on the market, JSON usability is important because other Delphi technologies, such as **DataSnap**, use it. Luckily, there are a lot of alternative JSON parsers for Delphi, if you find you have trouble with the standard one.

Other notable JSON parsers are the following:

- **SuperObject** (`https://github.com/hgourvest/superobject`)
- The one included in **Delphi Web Script library** can be found at `https://bitbucket.org/egrange/dwscript/`
- A fast JSON parser from Andreas Hausladen can be found at `https://github.com/ahausladen/JsonDataObjects`

If your main concern is speed, then check out these alternative JSON parsers.

There are also a lot of serialization libraries that use JSON as a serialization format. In general, every parser has its own way to serialize an object to JSON. Find your favorite. Just as an example, in `Chapter 5`, *The Thousand Faces of Multithreading*, in the *Using tasks to make your customer happy* recipe, you will see an open source library containing a set of serialization helpers using the default Delphi JSON parser.

However, JSON is not the right tool for every interchange or data-representation job. XML has been creating other technologies that can help if you need to search, transform, and validate your data in a declarative way. In JSON land, there is no such level of standardization, apart from the format itself. However, over the years, there has been an effort to include at least the XML schema counterpart in JSON, and you can find more details at `http://json-schema.org/`.

 One of the reasons JSON was chosen over XML (in transfer protocol scenarios) is that JSON results in less data for the same amount of information.

Manipulating and transforming XML documents

XML stands for **Extensible Markup Language** (http://en.wikipedia.org/wiki/XML) and is designed to represent, transport, and store hierarchical data in a trees of nodes. You can use XML to communicate with different systems and store configuration files, complex entities, and so on. They all use a standard and powerful format. Delphi has had good support for XML for more than a decade now.

Getting ready

All the basic XML-related activities can be summarized with the following points:

- Generate XML data
- Parse XML data
- Parse XML data and modify it

In this recipe, you will see how to carry out all these activities.

How to do it...

Let's have a look at the following steps:

1. Create a new VCL application and drop three **TButton** and a **TMemo**. Align all the buttons as a toolbar at the top of the form and the memo on the remaining form client area.
2. From left to right, name the buttons btnGenerateXML, btnModifyXML, btnParseXML, and btnTransformXML.
3. The real work on the XML will be done by the **TXMLDocument** component. So, drop one instance of the form and set its DOMVendor property to Omni XML.
4. We will use static data as our data source. A simple matrix is enough for this recipe. Just after the implementation section of the unit, write the code that follows:

```
type
  TCarInfo = (
  Manufacturer = 1,
  Name = 2,
  Currency = 3,
```

```
    Price = 4);

var
  Cars: array [1 .. 4] of
    array [Manufacturer .. Price] of string = (
      (
        'Ferrari','360 Modena','EUR', '250,000'
      ),
      (
        'Ford', 'Mustang', 'USD', '80,000'
      ),
      (
        'Lamborghini', 'Countach', 'EUR','300,000'
      ),
      (
        'Chevrolet', 'Corvette', 'USD', '100,000'
      )
    );
```

5. We will use a `TMemo` to display the XML and the data. To keep things clear, create a public property called `Xml` on the form and map its setter and getter to the `Memo1.Lines.Text` property. Use the following code:

```
//...other form methods declaration
private
  procedure SetXML(const Value: String);
  function GetXML: String;
public
  property Xml: String read GetXML write SetXML;
end;

//...then in the implementation section
function TMainForm.GetXML: String;
begin
  Result := Memo1.Lines.Text;
end;

procedure TMainForm.SetXML(const Value: String);
begin
  Memo1.Lines.Text := Value;
end;
```

6. Now, create event handlers for each button. For `btnGenerateXML`, write the following code:

```
procedure TMainForm.btnGenerateXMLClick(Sender: TObject);
var
  RootNode, Car, CarPrice: IXMLNode;
  i: Integer;
  s: String;
begin
  XMLDocument1.Active := True;
  try
    XMLDocument1.Version := '1.0';
    RootNode := XMLDocument1.AddChild('cars');
    for i := Low(Cars) to High(Cars) do
    begin
      Car := XMLDocument1.CreateNode('car');
      Car.AddChild('manufacturer').Text :=
        Cars[i][TCarInfo.Manufacturer];
      Car.AddChild('name').Text :=
        Cars[i][TCarInfo.Name];
      CarPrice := Car.AddChild('price');
      CarPrice.Attributes['currency'] :=
        Cars[i][TCarInfo.Currency];
      CarPrice.Text := Cars[i][TCarInfo.Price];
      RootNode.ChildNodes.Add(Car);
    end;
    XMLDocument1.SaveToXML(s);
    Xml := s;
  finally
    XMLDocument1.Active := False;
  end;
end;
```

7. Now, we have to write the code to change the XML. In the `btnModifyXML` click event handler, write the following code:

```
procedure TMainForm.btnModifyXMLClick(Sender: TObject);
var
  Car, CarPrice: IXMLNode;
  s: string;
begin
  XMLDocument1.LoadFromXML(Xml);
  try
    Xml := '';
    Car := XMLDocument1.CreateNode('car');
    Car.AddChild('manufacturer').Text := 'Hennessey';
    Car.AddChild('name').Text := 'Venom GT';
    CarPrice := Car.AddChild('price');
```

```
      CarPrice.Attributes['currency'] := 'USD';
      CarPrice.Text := '600,000';
      XMLDocument1.DocumentElement.ChildNodes.Add(Car);
      XMLDocument1.SaveToXML(s);
      Xml := s;
    finally
      XMLDocument1.Active := False;
    end;
  end;
```

8. Write the following code under the `btnParseXML` click event handler:

```
procedure TMainForm.btnParseXMLClick(Sender: TObject);
var
  CarsList: IDOMNodeList;
  CurrNode: IDOMNode;
  childidx, i: Integer;
  CarName, CarManufacturer, CarPrice, CarCurrencyType:
      string;
begin
  XMLDocument1.LoadFromXML(Xml);
  try
    Xml := '';
    CarsList := XMLDocument1.
      DOMDocument.getElementsByTagName('car');
    for i := 0 to CarsList.length - 1 do
    begin
      CarName := '';  CarManufacturer := '';
      CarPrice := '';  CarCurrencyType := '';
      for childidx := 0 to
        CarsList[i].ChildNodes.length - 1 do
      begin
        CurrNode := CarsList[i].ChildNodes[childidx];
        if CurrNode.nodeName.Equals('name') then
          CarName := CurrNode.firstChild.nodeValue;
        if CurrNode.nodeName.Equals('manufacturer') then
          CarManufacturer := CurrNode.firstChild.nodeValue;
        if CurrNode.nodeName.Equals('price') then
        begin
          CarPrice := CurrNode.firstChild.nodeValue;
          CarCurrencyType :=
            CurrNode.Attributes.
            getNamedItem('currency').nodeValue;
        end;
      end;
      Xml := Xml +
          'Name = ' + CarName + sLineBreak +
          'Manufacturer = ' + CarManufacturer + sLineBreak +
```

```
                    'Price = ' +
                    CarPrice + CarCurrencyType + sLineBreak +
                    '-----' + sLineBreak;
           end;
        finally
           XMLDocument1.Active := False;
        end;
     end;
```

9. Write the following code under the `btnTransformXML` click event handler:

```
procedure TMainForm.btnTransformClick(Sender: TObject);
var
  LXML, LXSL: string;
  LOutput: string;
begin
  LXML := TFile.ReadAllText('..\..\..\cars.xml');
  LXSL := TFile.ReadAllText('..\..\..\cars.xslt');
  LOutput := Transform(LXML, LXSL);
  TFile.WriteAllText('..\..\..\cars.html', LOutput);
  ShellExecute(0, PChar('open'),
      PChar('file:///' +
      TPath.GetFullPath('..\..\..\cars.html')), nil,
      nil, SW_SHOW);
end;
```

10. Now, add the following function in your form implementation section:

```
function Transform(XMLData: string; XSLT: string): String;
var
  LXML, LXSL: IXMLDocument;
  LOutput: WideString;
begin
  LXML := LoadXMLData(XMLData);
  LXSL := LoadXMLData(XSLT);
  LXML.DocumentElement.TransformNode(LXSL.DocumentElement,
      LOutput);
  Result := String(LOutput);
end;
```

11. Run the application by hitting *F9* (or by going to **Run | Run**).
12. Click on the `btnGenerateXML` button, and you should see some XML data in the memo.
13. Click on the `btnModifyXML` button, and you should see some more XML in the memo.

14. Click on `btnParseXML`, and you should see the same data as before, but with normal text representation.

15. After the third click, you should see something similar to the following screenshot:

Figure 1.14: Text representation of the XML data generated and modified

16. Now, copy the `cars.xml` and `cars.xslt` files from the respective recipe folder to the parent folder of your project folder and click on the `btnTransformXML` button.

17. The system default browser should appear, showing something like the following screenshot:

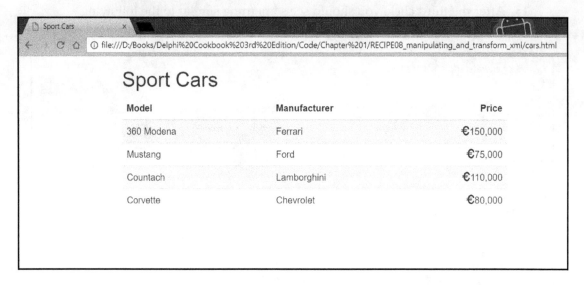

Fig. 1.15 XML data transformed into HTML using an XSLT transformation

How it works...

Let's look at these steps now:

1. The first button generates the XML representation of the data in our matrix. We've used some car information as sample data.

> Note that the prices of the cars are not real.

2. To create an XML attribute, there are three fundamental **TXMLDocument** methods:

   - XMLNode := XMLDocument1.CreateNode('node');
   - XMLNode.AddChild('childnode');
   - XMLNode.Attributes['attrname'] := 'attrvalue';

   There are other very useful methods, but these are the basics of XML generation.

3. The `btnModifyXML` button loads the XML into the memo and appends some other data (another car) to the list. Then, it updates the memo with the new updated XML. These are the most important lines to note:

```
//Create a node without adding it to the DOM
Car := XMLDocument1.CreateNode('car');

//fill Car XMLNode... and finally add it to the DOM
//as child of the root node
XMLDocument1.DocumentElement.ChildNodes.Add(Car);
```

4. The code under the `btnParseXMLClick` event handler allows us to read the display as normal text as the XML data navigating through XML tree.

5. The code under the `btnTransformXMLClick` event handler uses the XSLT transformation in `cars.xslt` and the data in `cars.xml` to generate a brand new HTML page. The XSLT code is as follows:

```
<?xml version="1.0" encoding="UTF-8"?>
<xsl:stylesheet version="1.0"
  xmlns:xsl="http://www.w3.org/1999/XSL/Transform">
    <xsl:output method="html" version="5.0"
      encoding="UTF-8" indent="yes"/>
  <xsl:template match="cars">
    <html>
      <head>
        <link href="https://maxcdn.bootstrapcdn.com/
          bootstrap/3.3.4/css/bootstrap.min.css"
          rel="stylesheet"/>
        <title>
          Sport Cars
        </title>
      </head>
      <body>
        <div class="container">
        <div class="row">
        <h1>Sport Cars</h1>
          <table class="table table-striped table-hover">
```

```
            <thead>
              <tr>
                <th>Model</th>
                <th>Manufacturer</th>
                <th class="text-right">Price</th>
              </tr>
            </thead>
            <tbody>
              <xsl:for-each select="car">
              <tr>
                <td>
                  <xsl:value-of select="name"/>
                </td>
                <td>
                  <xsl:value-of select="manufacturer"/>
                </td>
                <td class="text-right">
                  <span class="glyphicon glyphicon-euro">
                  </span>
                  <xsl:value-of select="price"/>
                </td>
              </tr>
              </xsl:for-each>
            </tbody>
          </table>
        </div>
        </div>
      </body>
    </html>
  </xsl:template>
</xsl:stylesheet>
```

There's more...

There are many things to say about the XML ecospace. There are XML engines that provide facilities to search data in an XML tree (XPath), to validate XML using other XML (XML Schema or DTD), to transform an XML into another kind of format using another XML (XSLT), and many others (`http://en.wikipedia.org/wiki/List_of_XML_markup_languages`). The good thing is that just like XML, the DOM object is also standardized. So, every library that is compliant with the standard has the same methods, from Delphi to JavaScript and from Python to C#.

TXMLDocument allows you to select the `DOMVendor` implementation. By default, there are three implementations available:

- **MSXML**:
  - From Microsoft; implemented as COM objects
  - Supports XML transformations
  - Available only on Windows (so no Android, iOS, or Mac OS X)
- **Omni XML**:
  - Much faster than ADOM and based on the Open Source Project.
  - It is cross-platform, so is available on all the supported Delphi platforms. If you plan to write XML handling code on mobile or Mac, this is the way to go.
- **ADOM XML**:
  - A quite old open source Delphi implementation
  - Does not support transformations
  - Available on all supported Delphi platforms
  - For backward compatibility, consider Omni XML instead in Delphi

TXMLDocument uses a Windows-only vendor by default. If you are designing a FireMonkey application that is intended to run on other platforms than Windows, select a cross-platform DOM vendor.

XSLT allows you to transform XML to something else, using other XML as a stylesheet. As we saw in this recipe, you can use an XML file and an XSLT file to generate an HTML page that shows the data contained in the XML, using XSLT to format the data.

The following function loads one XML and one XSLT document from two string variables. Then, we use the XSLT document to transform the XML document. The code that follows shows this in detail:

```
function Transform(XMLData: string; XSLT: string): String;
var
  LXML, LXSL: IXMLDocument;
  LOutput: WideString;
begin
  LXML := LoadXMLData(XMLData);
  LXSL := LoadXMLData(XSLT);
  LXML.DocumentElement.TransformNode(
    LXSL.DocumentElement, LOutput);
  Result := String(LOutput);
end;
```

This function doesn't know about the output format because it is defined by the XSLT document. The result could be XML, HTML, CSV, plain text, or whatever the XSLT defines, but the code does not change.

XSLT can be really useful. I recommend that you go and visit `http://www.w3schools.com/xml/xsl_languages.asp` for further details on the language.

I/O in the 21st century – knowing the streams

Many I/O-related activities handle streams of data. A stream is a sequence of data elements made available over time. Wikipedia says:

> *"A stream can be thought of as a conveyor belt that allows items to be processed one at a time rather than in large batches."*

At the lowest level, all streams are bytes, but using a high-level interface could obviously help the programmer handle their data. This is the reason why a stream object usually has methods such as `read`, `seek`, `write`, and so on, just to make handling a byte stream a bit simpler.

In this recipe, you'll see some stream utilization examples.

Getting ready

In the good old Pascal days, there was a set of functions to handle the I/O (`Assign`, `Reset`, `Rewrite`, `Close`, and many more). Now, we have a bunch of classes. All Delphi streams inherit from `TStream` and can be used as the internal stream of one of the adapter classes (by adapter, I mean an implementation of the `Adapter`, or `Wrapper`, design patterns from the *Gang of Four (GoF)* famous book about design patterns).

There are 10 fundamental types of streams:

Class	Use
`System.Classes.TBinaryWriter`	Writer for binary data
`System.Classes.TStreamWriter`	Writer for characters to a stream
`System.Classes.TStringWriter`	Writer for a string

`System.Classes.TTextWriter`	Writer of a sequence of characters; it is an abstract class
`System.Classes.TWriter`	Writes component data to an associated stream
`System.Classes.TReader`	Reads component data from an associated stream
`System.Classes.TStreamReader`	Reader for stream of characters
`System.Classes.TStringReader`	Reader for strings
`System.Classes.TTextReader`	Reader for sequence of characters; it is an abstract class
`System.Classes.TBinaryReader`	Reader for binary data

You can check out the complete list and their intended uses on the Embarcadero website at `http://docwiki.embarcadero.com/RADStudio/en/Streams,_Reader_and_Writers`.

As Joel Spolsky says, *You can no longer pretend that plain text is ASCII*. So, while we write streams, we have to pay attention to the encoding our text uses and the encoding our counterpart is waiting for.

One of the most frequent necessities is to efficiently read and write a text file using the correct encoding:

> *"The Single Most Important Fact About Encodings... It does not make sense to have a string without knowing what encoding it uses. You can no longer stick your head in the sand and pretend that 'plain' text is ASCII."*

> *– Joel Spolsky*

The point Joel is making is that the content of a string doesn't know about the type of character encoding it uses.

When you think about file handling, ask yourself—*Could this file become 10 MB? And 100 MB? And 1 GB? How will my program behave in that case?* Handling a file one line at time and not loading all the file contents in memory is usually good insurance for these cases. A stream of data is a good way to do this. In this recipe, you'll see the practical utilization of streams, stream writers, and stream readers.

How it works...

The project is not complex. All the interesting stuff happens in btnWriteFile and btnReadFile. To write the file, TStreamWriter is used. TStreamWriter (similar to its counterpart TStreamReader) is a wrapper for a TStream descendant and adds some useful high-level methods to write to the stream. There are a lot of overloaded methods (Write/WriteLine) to allow easy writing to the underlying stream. However, you can access the underlying stream using the BaseStream property of the wrapper. Just after having written the file, the memo reloads the file using the same encoding used to write it and displays it. This is only a fast check for this recipe; you don't need **TMemo** at all in your real project. The btnReadFile simply opens the file using a stream and passes the stream to TStreamReader which, using the right encoding, will read the file one line at a time (note that text is stored in the .pas file, which is in this case encoded as UTF-8, while by default Delphi .pas files use ASCII encoding).

Now, let's run some checks. Run the program and with the encoding set to ASCII, click on btnWriteFile. The memo will show garbage text, as shown in the following screenshot. This is because we are using the wrong encoding for the data we are writing in the file:

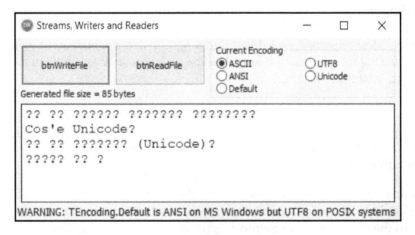

Figure 1.16: Garbage text written to the file using the wrong encoding

Now, select UTF8 from the radio group and retry it. By clicking on btnWriteFile, you will see the correct text in the memo. Try to change the current encoding using ASCII and click on btnReadFile. You will still get garbage text. Why? Because the file has been read with the wrong encoding. You have to know the encoding beforehand to safely read the file's contents. To read the text that we wrote, we have to use the same encoding. Play with other encodings to see the different behavior.

There's more...

Streams are very powerful and their uniform interface helps us write portable and generic code. With the help of streams and polymorphism, we can write code that uses TStream to do some work, without knowing what kind of stream it is!

Also, a well-known possibility is that if you ever need to write a program that needs to access the good old TD_INPUT, STD_OUTPUT, or STD_ERROR, you can use THandleStream to wrap these system handles with a nice TStream interface using the following code:

```
program StdInputOutputError;
//the following directive instructs the compiler to create a
//console application and not a GUI one, which is the default.
{$APPTYPE CONSOLE}
uses
  System.Classes, // required for Stream classes
  Winapi.Windows; // required to have access to the STD_* handles
var
  StdInput: TStreamReader;
  StdOutput, StrError: TStreamWriter;
begin
  StdInput := TStreamReader.Create(
THandleStream.Create(STD_INPUT_HANDLE));
  StdInput.OwnStream;
  StdOutput := TStreamWriter.Create(
THandleStream.Create(STD_OUTPUT_HANDLE));
  StdOutput.OwnStream;
  StdError := TStreamWriter.Create(
THandleStream.Create(STD_ERROR_HANDLE));
  StdError.OwnStream;
  { HERE WE CAN USE OURS STREAMS }
  // Let's copy a line of text from STD_IN to STD_OUT
  StdOutput.writeln(StdInput.ReadLine);
  { END - HERE WE CAN USE OURS STREAMS }
  StdError.Free;
  StdOutput.Free;
  StdInput.Free;
end;
```

Moreover, when you work with file-related streams, the TFile class (contained in System.IOUtils.pas) is very useful, and has some helper methods to write shorter and more readable code.

Creating a Windows Service

Some kinds of application need to be running 24/7. Usually, they are network servers or data transfer/monitoring applications. In these cases, you probably start with a normal GUI or console application. However, when the systems start to be used in production, you are faced with a lot of problems related to Windows session termination, reboots, user rights, and other issues related to the server environment.

Getting ready

The way to go, in the previous scenario, is to develop a Windows Service.

In this recipe, we'll see how to write a good Windows Service scaffold, and this can be the skeleton for many other services. So, feel free to use this code as a template to create all the services that you will need.

How it works...

The project has been created using the default project template accessible by going to **File | New | Other | Delphi Projects | Service Application**; it has been integrated with a set of functionalities to make it real.

All the low-level interfacing with the Windows Service Manager is done by the `TService` class. In `ServiceU.pas`, there is the actual descendant of `TService` that represents the Windows Service we are implementing. Its event handlers are used to communicate with the operating system.

Usually, a service needs to respond to Windows Service Controller commands independently of what it is doing. So, we need a background thread to do the actual work, while the `TService.OnExecute` event should not do any real work (this is not a must, but usually is the way to go). The unit named `WorkerThreadU.pas` contains the thread and the main service needed to hold a reference to the instance of this thread.

The background thread starts when the service is started (the `OnStart` event) and stops when the service is stopped (the `OnStop` event). The `OnExecute` event waits and handles `ServiceController` commands but doesn't do any actual functional work. This is done using `ServiceThread.ProcessRequests(false)` in a `while` loop.

Usually, the `OnExecute` event handler looks like this:

```
procedure TSampleService.ServiceExecute(Sender: TService);
begin
  while not Terminated do
  begin
    ServiceThread.ProcessRequests(false);
    TThread.Sleep(1000);
  end;
end;
```

The wait of 1,000 milliseconds is not a must, but consider that the wait time should not be too high because the service needs to be responsive to the Windows Service Controller messages. It should not be too low because otherwise the thread context switch may waste resources.

The background thread writes a line in a logfile once a second. While it is in a `Paused` state, the service stops writing. When the service continues, the thread will restart writing the log line. In the service event handlers is the logic to implement this change of state:

```
procedure TSampleService.ServiceContinue(Sender: TService;
    var Continued: Boolean);
begin
  FWorkerThread.Continue;
  Continued := True;
end;

procedure TSampleService.ServicePause(Sender: TService;
    var Paused: Boolean);
begin
  FWorkerThread.Pause;
  Paused := True;
end;
```

In the thread, there is actual logic to implement the `Paused` state, and in this case, it is fairly simple; we pause the writing of the logfile:

Here's an extract:

```
Log := TStreamWriter.Create(
  TFileStream.Create(LogFileName,
    fmCreate or fmShareDenyWrite));
try
  while not Terminated do
  begin
    if not FPaused then
    begin
```

```
                Log.WriteLine('Message from thread: ' + TimeToStr(now));
            end;
            TThread.Sleep(1000);
        end;
    finally
        Log.Free;
    end;
```

The `FPaused` Boolean instance variable can be considered thread safe for this use case.

Delphi services don't have a default description under the Windows Service Manager. If we want to give them a description, we have to write a specific key in the Windows registry. Usually, this is done in the **AfterInstall** event. In our service, this is the code to write in the **AfterInstall** event handler:

```
procedure TSampleService.ServiceAfterInstall(
Sender: TService);
var
  Reg: TRegistry; //declared in System.Win.Registry;
begin
  Reg := TRegistry.Create(KEY_READ or KEY_WRITE);
  try
    Reg.RootKey := HKEY_LOCAL_MACHINE;
    if Reg.OpenKey(
      'SYSTEM\CurrentControlSet\Services\' + name,
      False {do not create if not exists}) then
    begin
      Reg.WriteString('Description',
        'My Fantastic Windows Service');
      Reg.CloseKey;
    end;
  finally
    Reg.Free;
  end;
end;
```

It is not necessary to delete this key in the **AfterUnInstall** event because Windows deletes all the keys related to the service (under `HKEY_LOCAL_MACHINE\SYSTEM\CurrentControlSet\Services\<MyServiceName>`) when the service is actually uninstalled.

Let's try an installation. Build the project, open the Windows command prompt (with administrator-level privileges), and go to the folder where the project has been built. Then, run this command:

```
C:\<ExeProjectPath>\WindowsService.exe /install
```

If everything is okay, you should see this message:

Figure 1.17: The service installation is okay

Now, you can check under the Windows Services Console, and you should find the service installed. Click on **Start**, wait for the confirmation, and the service should start to write to its logfile.

Play with **Pause** and **Continue** and check the file activity.

> Some text editors could have a problem opening the logfile while the service is writing. I suggest that you use a Unix tail clone for Windows.

> There are many free choices. Here are some links:
>
> * http://sourceforge.net/projects/tailforwin32/
> * http://ophilipp.free.fr/op_tail.htm
> * http://www.baremetalsoft.com/baretail/

There's more...

Windows Services are very powerful. Using the abstractions that Delphi provides, you can also create an application that, reading a parameter on the command line, can act as a normal GUI application or as a Windows Service.

In the respective recipe folder, there is another recipe folder called 20_WindowsServiceOrGUI.

This application can be used as a normal Windows Service using the normal command-line switches used so far, but if launched with /GUI, it acts as a GUI application and can use the same application code (not TService). In our example, the GUI version uses the same worker thread as the service version. This can be very useful for debugging purposes also.

Run the application with the following command:

```
C:\<ExeProjectPath>\WindowsServiceOrGUI.exe /GUI
```

You will get a GUI version of the service, as shown here:

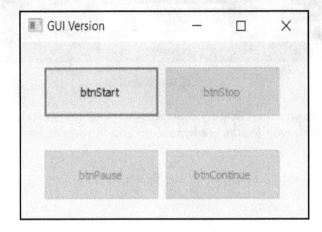

Figure 1.18: The GUI version of the Windows Service

Using the TService.LogMessage method

If something happens during the execution of the service that you want to log, and you want to log in to the system logger, you can use the `LogMessage` method to save a message. The message can be viewed later using the Windows built-in event viewer.

You can call the `LogMessage` method using an appropriate logging type, like this:

```
LogMessage('Your message goes here for SUCCESS',
    EVENTLOG_SUCCESS, 0, 1);
```

If you check the event in the **Event Viewer**, you will find a lot of garbage text that complains about the lack of description for the event.

If you really want to use the **Event Viewer** to view your log message (when I can, I use a logfile and don't concern myself with the Event Viewer, but there are scenarios where the **Event Viewer** log is needed), you have to use the **Microsoft © Message Compiler**.

The Microsoft © Message Compiler is a tool able to compile a file of messages into a set of RC files. Then, these files must be compiled by a resource compiler and linked into your executable.

More information on Microsoft © Message Compiler and the steps needed to provide the description for the log event can be found at `http://www.codeproject.com/Articles/4166/Using-MC-exe-message-resources-and-the -NT-event-lo`.

Associating a file extension with your application on Windows

In some cases, your fantastic application needs to be opened by just double-clicking on a file with an extension associated with it. This is the case with Microsoft Word, Microsoft Excel, and many other well-known pieces of software. If you have a file generated with a program, double-click on the file and the program that generated the file will come up, pointing to that file. So, if you click on `mywordfile.docx`, Microsoft Word will be opened and `mywordfile.docx` will be shown. This is what we'd like to do in this recipe. The association can be useful when you have multiple configurations for a program. Double-click on the `ConfigurationXYZ.myext` file and the program will start using that configuration.

Getting ready

The hard work is done by the operating system itself. We have to instruct Windows to provide the following information:

- The file extension to associate
- The description of the file type (it will be shown by Windows Explorer, describing the file type)
- The default icon for the file type (in this recipe, we'll use the application icon itself, but it is not mandatory)
- The application that we want to associate

Let's start!

How to do it...

Let's complete the following steps:

1. Create a new VCL application and drop two **TButton** components and a **TMemo** component. Align all the buttons as a toolbar at the top of the form and the memo to all the remaining form client area.

2. The button on the left-hand side will be used to register a file type, while the button on the right-hand side will be used to unregister the association (cleaning the registry).

3. We have to handle some features specific to Microsoft Windows, so we need some Windows-related units. Under the implementation section of the unit, write this uses clause:

```
uses System.Win.registry, Winapi.shlobj, System.IOUtils;
```

4. In the implementation section, we need two procedures to do the real work; so just after the uses clause, add this code:

```delphi
procedure UnregisterFileType(
  FileExt: String;
  OnlyForCurrentUser: boolean = true);
var
  R: TRegistry;
begin
  R := TRegistry.Create;
  try
    if OnlyForCurrentUser then
      R.RootKey := HKEY_CURRENT_USER
    else
      R.RootKey := HKEY_LOCAL_MACHINE;

    R.DeleteKey('\Software\Classes\.' + FileExt);
    R.DeleteKey('\Software\Classes\' + FileExt + 'File');
  finally
    R.Free;
  end;
  SHChangeNotify(SHCNE_ASSOCCHANGED, SHCNF_IDLIST, 0, 0);
end;

procedure RegisterFileType(
  FileExt: String;
  FileTypeDescription: String;
  ICONResourceFileFullPath: String;
  ApplicationFullPath: String;
  OnlyForCurrentUser: boolean = true);
```

```
var
  R: TRegistry;
begin
  R := TRegistry.Create;
  try
    if OnlyForCurrentUser then
      R.RootKey := HKEY_CURRENT_USER
    else
      R.RootKey := HKEY_LOCAL_MACHINE;

    if R.OpenKey('\Software\Classes\.' + FileExt,
      true) then begin
      R.WriteString('', FileExt + 'File');
      if R.OpenKey('\Software\Classes\' + FileExt + 'File',
        true) then begin
        R.WriteString('', FileTypeDescription);
        if R.OpenKey('\Software\Classes\' +
          FileExt + 'File\DefaultIcon', true) then
        begin
          R.WriteString('', ICONResourceFileFullPath);
          if R.OpenKey('\Software\Classes\' +
            FileExt + 'File\shell\open\command',
              true) then
          R.WriteString('',
            ApplicationFullPath + ' "%1"');
        end;
      end;
    end;
  finally
    R.Free;
  end;
  SHChangeNotify(SHCNE_ASSOCCHANGED, SHCNF_IDLIST, 0, 0);
end;
```

5. These two procedures allow us to register (and unregister) a file type, considering only the current user or all the machine users. Pay attention; if you want to register the association for every user, write your data to:

   HKEY_LOCAL_MACHINE\Software\Classes

6. If you want to register the association for the current user only, write your data to:

   HKEY_CURRENT_USER\Software\Classes

7. On the newest Windows versions, you need administrator rights to register a file type for all machine users. The last line of the procedures tells Explorer (the Microsoft Windows graphic interface) to refresh its settings to reflect the changes made to the file associations. As a result, the Explorer file list views will update.

8. We've almost finished. Change the left button name to btnRegister, the right button name to btnUnRegister, and put the following code into their **Onclick** event handlers:

```
procedure TMainForm.btnRegisterClick(Sender: TObject);
begin
  RegisterFileType(
    'secret',
    'This file is a secret',
    Application.ExeName,
    Application.ExeName,
    true);
  ShowMessage('File type registred');
end;

procedure TMainForm.btnUnRegisterClick(Sender: TObject);
begin
  UnregisterFileType('secret', true);
  ShowMessage('File type unregistered');
end;
```

9. Now, when our application is invoked after double-clicking, we'll get the file name as a parameter. It is possible to read a parameter passed by Windows Explorer (or the command line) using the ParamStr(1) function. Create a FormCreate event handler using the following code:

```
procedure TMainForm.FormCreate(Sender: TObject);
begin
  if TFile.Exists(ParamStr(1)) then
    Memo1.Lines.LoadFromFile(ParamStr(1))
  else begin
    Memo1.Lines.Text := 'No valid secret file type';
  end;
end;
```

10. Now, the application should be complete. However, nice integration with the operating system requires a nice icon. In the code, the associated file will get the same icon as the main program, so let's change our default icon by going to **Project** | **Options** | **Application dialog**, and choosing a nice icon. Click on the **Load Icon** button, choose an ICO file, and then select the third item from the resultant dialog:

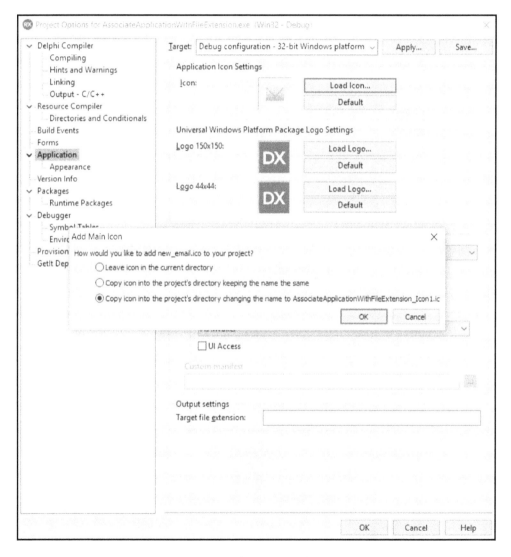

Figure 1.19: Changing the default application icon for our application

11. Now, create some text files with our registered extension, `.secret`.

12. These files will appear with the default Windows icons, but in a few seconds, they will have a brand new icon.

13. Run the application by hitting *F9* (or by going to **Run** | **Run**).

14. Click on the `btnRegister` button and close the application. Now, the files get new icons, as shown here:

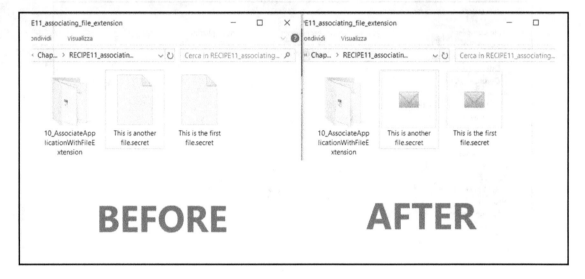

Figure 1.20: The files in Windows Explorer before and after having registered the SECRET extension

15. Now, with the application not running, double-click on a `.secret` file. Our program will be started by Windows itself, using the information stored in the registry about the `.secret` file, and we'll get this form (the text shown in the memo is the text contained in the file):

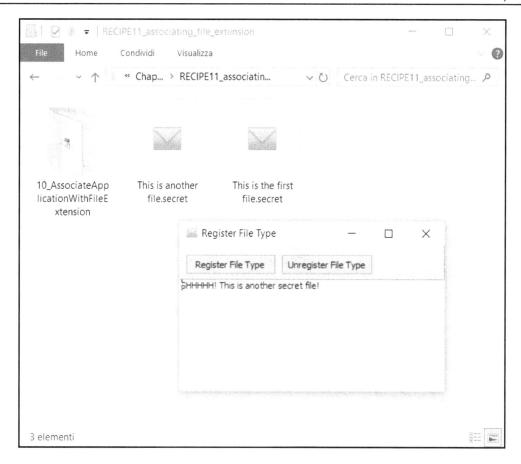

Figure 1.21: Our application, launched by the operating system, showing the contents of the file

There's more...

One application can register many file types. In some cases, I've used this technique to register some specific desktop database files to my application (Firebird SQL-embedded database files or SQLite database files). So, double-clicking actually creates a connection to that database.

Be coherent with the Windows look and feel using TTaskDialog

Version after version, the Windows OS changed its look and feel a lot from mid-2009 when the first Windows 95 came out. Also, the UX guidelines from Microsoft have changed a lot. Do you remember the **Multiple Document Interface** (**MDI**) paradigm? It was very popular in the 90s, but now is deprecated and an application using seems old, even if it has been released. Indeed, many Windows applications seem stuck in the past in terms of UI and UX.

What about dialog? Our beloved `ShowMessage` and `MessageDlg` have been there since Delphi 1, but now, the modern Windows versions use different dialogs to communicate to users. Many of the standard dialogs contain more than a question and a simple *Yes* and *No*. Some dialogs ask something and provide a list of choices using buttons; others have a nice progress bar; others have a nice button with an extended explanation of each choice just inside the button. How can our Delphi application benefit from the new dialogs offered by the OS? In other words, how can we create a coherent look and feel for our dialog windows so that our application does not look old? This recipe shows you how to use the **TTaskDialog** component.

Getting started

TTaskdialog is a dialog box, somewhat like the standard call to `Application.MessageBox` in the VCL, but much more powerful. The Task Dialog API has been available since Windows Vista and Windows Server 2008, and your application must be theme enabled to use it (go to **Project** | **Options** | **Application** | **Runtime Themes** | **Enable Runtime Themes**).

Besides the usual default set of buttons (**OK, Cancel, Yes, No, Retry**, and **Close**), you can define extra buttons and many other customizations. The following Windows APIs provides Task Dialog:

API Name	Description
`TaskDialog`	This creates, displays, and operates Task Dialog. The Task Dialog contains application-defined message text and title, icons, and any combination of predefined push buttons. This function does not support the registration of a callback function to receive notifications.
`TaskDialogCallbackProc`	This is an application-defined function used with the `TaskDialogIndirect` function. It receives messages from the Task Dialog when various events occur. `TaskDialogCallbackProc` is a placeholder for the application-defined function name.
`TaskDialogIndirect`	This creates, displays, and operates Task Dialog. The Task Dialog contains application-defined icons, messages, a title, a verification checkbox, command links, push buttons, and radio buttons. This function can register a callback function to receive notification messages.

More information about API utilization can be obtained from
`https://msdn.microsoft.com/en-`
`us/library/windows/desktop/bb787471(v=vs.85).aspx`.

While the API can be useful in some border cases, the VCL comes with a very nice component that does all the low-level stuff for us. Let's look at an example program and see how simple it is to create a modern application.

How it works...

Open the `TaskDialogs.dproj` project so we can examine how it works.

There are six buttons on the form. The first one shows a simple utilization of the Task Dialog API, while the other five show a different utilization of the **TTaskDialog** component, which wraps that API.

The first button uses the Windows API directly with the following code:

```
procedure TMainForm.btnAPIClick(Sender: TObject);
var
  LTDResult: Integer;
begin
  TaskDialog(0, HInstance,
    PChar('The Title'),
    PChar('These are the main instructions'),
    PChar('This is another content'),
    TDCBF_OK_BUTTON or TDCBF_CANCEL_BUTTON,
    TD_INFORMATION_ICON, @LTDResult);
  case LTDResult of
    IDOK:
      begin
        ShowMessage('Clicked OK');
      end;
    IDCANCEL:
      begin
        ShowMessage('Clicked Cancel');
      end;
  end;
end;
```

The `TaskDialog` function is declared inside the `Winapi.CommCtrl.pas` unit. So far, you could ask, *Why should I use a component for Task Dialogs?* Seems quite simple. Yes, it is, if you only want to mimic `MessageDlg`, but things get complicated very fast if you want to use all the features of the Task Dialog API. So, the second button uses the **TTaskDialog** component. Let's see the relevant properties configured at design time for the `tdSimple` component:

```
object tdSimple: TTaskDialog
  Caption = 'The question'
  CommonButtons = [tcbYes, tcbNo]
  DefaultButton = tcbYes
  ExpandButtonCaption = 'More information'
  ExpandedText =
      'Yes, you have to decide something about this question...' +
        ' but I cannot help you a lot'
  Flags = [tfUseHiconMain, tfUseHiconFooter,
      tfVerificationFlagChecked]
  FooterIcon = 4
  FooterText = 'This is an important question...'
  Text = 'To be or not to be, this is the question. To be?'
  Title = 'William ask:'
end
```

You can check the runtime appearance at design time by double-clicking on the component above your form, or by selecting **Test Dialog** from the menu above the component. You can access the menu by right-clicking on the component.

As you can see, only the minimum properties have been set, just to show the power of the component. This configuration shows a dialog with two buttons labeled **Yes** and **No**. The **TTaskDialog** component can be configured at design time using the **Object Inspector**, or can be configured at runtime by code. In the first example, the configuration is defined at design time so that at runtime we only have to call the `Execute` method and read the user response. Here's the code that actually uses the `tdSimple` instance:

```
procedure TMainForm.btnSimpleClick(Sender: TObject);
begin
  tdSimple.Execute; //show the taskdialog
  if tdSimple.ModalResult = mrYes then
    ShowMessage('yes')
  else
    ShowMessage('no')
end;
```

Even in this case, it is quite simple, but let's go deeper with the configuration. Let's say that we need a Task Dialog similar to the following screenshot:

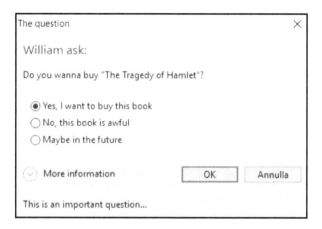

Fig. 1.22: The TTaskDialog component is configured to show three radio buttons

Using the plain API, this is not so simple. So, let's see how to configure the component:

```
object tdRadioButtons: TTaskDialog
  Caption = 'The question'
  DefaultButton = tcbYes
  ExpandButtonCaption = 'More information'
  ExpandedText =
    'Yes, you have to decide something about this question... ' +
      'but I cannot help you a lot'
  Flags = [tfUseHiconMain, tfUseHiconFooter,
    tfVerificationFlagChecked]
  FooterIcon = 4
  FooterText = 'This is an important question...'
  RadioButtons = <
    item
      Caption = 'Yes, I want to buy this book'
    end
    item
      Caption = 'No, this book is awful'
    end
    item
      Caption = 'Maybe in the future'
    end>
  Text = 'Do you wanna buy "The Tragedy of Hamlet"?'
  Title = 'William ask:'
end
```

The preceding block of code contains the definition for the three radio buttons. The following code shows the dialog and the retrieval of the result:

```
procedure TMainForm.btnRadioClick(Sender: TObject);
begin
  tdRadioButtons.Execute;
  if tdRadioButtons.ModalResult = mrOk then
    ShowMessage('Selected radio button ' +
        tdRadioButtons.RadioButton.ID.ToString);
end;
```

Even in this case, we have defined the properties at design time so that the runtime code is quite simple. Just note that the user's choice is stored in the `RadioButton.ID` property.

The `TTaskDialog.Flags` property can greatly change the behavior of the dialog. Here's the meaning of each element of its set:

Flag set element name	If set...
tfEnableHyperlinks	Content, footer, and expanded text can include hyperlinks
tfUseHiconMain	Uses the custom main icon
tfUseHiconFooter	Uses the custom footer icon
tfAllowDialogCancellation	Permits the Task Dialog to be closed in the absence of a **Cancel** button
tfUseCommandLinks	Buttons are displayed as command links using a standard dialog glyph
tfUseCommandLinksNoIcon	Buttons are displayed as command links without a glyph
tfExpandFooterArea	Displays expanded text in the footer
tfExpandedByDefault	Expanded text is displayed when the Task Dialog opens
tfVerificationFlagChecked	The verification checkbox is initially checked
tfShowProgressBar	Displays the progress bar
tfShowMarqueeProgressBar	Displays the marquee progress bar
tfCallbackTimer	Callback dialog will be called every 200 milliseconds
tfPositionRelativeToWindow	The Task Dialog is centered with respect to the parent window
tfRtlLayout	Text reads right to left
tfNoDefaultRadioButton	There is no default radio button
tfCanBeMinimized	The Task Dialog can be minimized

The real power of `TaskDialog` comes when you build your dialog at runtime. Let's check what the fourth button does under the hood:

```
procedure TMainForm.btnConfirmClick(Sender: TObject);
var
  LFileName: string;
  LGSearch: String;
const
  GOOGLE_SEARCH = 99;
begin
  LFileName := 'MyCoolProgram.exe';
  tdConfirm.Buttons.Clear;
  tdConfirm.Title := 'Confirm Removal';
  tdConfirm.Caption := 'My fantastic folder';
  tdConfirm.Text :=
    Format('Are you sure that you want to remove ' +
        'the file named "%s"?', [LFileName]);
  tdConfirm.CommonButtons := [];
  with TTaskDialogButtonItem(tdConfirm.Buttons.Add) do
  begin
    Caption := 'Remove';
    CommandLinkHint := Format('Delete file %s from the folder.',
        [LFileName]);
    ModalResult := mrYes;
  end;
  with TTaskDialogButtonItem(tdConfirm.Buttons.Add) do
  begin
    Caption := 'Keep';
    CommandLinkHint := 'Keep the file in the folder.';
    ModalResult := mrNo;
  end;

  if TPath.GetExtension(LFileName).ToLower.Equals('.exe') then
  begin
    with TTaskDialogButtonItem(tdConfirm.Buttons.Add) do
    begin
      Caption := 'Google search';
      CommandLinkHint := 'Let''s Google tell us what ' +
        'this program is.';
      ModalResult := GOOGLE_SEARCH;
    end;
  end;

  tdConfirm.Flags := [tfUseCommandLinks];
  tdConfirm.MainIcon := tdiInformation;

  if tdConfirm.Execute then
  begin
```

```
    case tdConfirm.ModalResult of
      mrYes:
        ShowMessage('Deleted');
      mrNo:
        ShowMessage(LFileName + ' has been preserved');
      GOOGLE_SEARCH:
        begin
          LGSearch := Format('https://www.google.it/#q=%s',
            [LFileName]);
          ShellExecute(0, 'open', PChar(LGSearch), nil, nil,
            SW_SHOWNORMAL);
        end;
    end; //case
  end; //if
end;
```

It seems like a lot of code, but it is simple and can be easily parameterized and reused inside your program. The resultant dialog is as shown:

Figure 1.23: The dialog customized by code

The third choice allows the user to search on Google for the program executable name. This is not a common choice in the `MessageDlg` dialog where buttons are predefined, but using **TTaskDialog,** you can even ask the user something strange (such as—*Do you want to ask Google about it?*).

To achieve a better apparent speed, progress bars are great! The Task Dialog API provides a simple way to use progress bars inside dialogs. The classic Delphi solution relays a custom form with a progress bar and some labels (just like the compiling dialog that you see when you compile a program within the Delphi IDE). However, in some cases, you need some simple stuff done and a Task Dialog is enough. If **TTaskDialog** has the tfCallbackTimer flag and tfShowProgressBar, the **OnTimer** event will be called every 200 milliseconds (five times a second), and the dialog will show a progress bar that you can update within the **OnTimer** event handler. However, the **OnTimer** event handler runs in the main thread so that all the related advice applies (if the UI becomes unresponsive, consider a proper background thread and a queue to send information to the main thread).

This is the design time configuration of tdProgress:

```
object tdProgress: TTaskDialog
  Caption = 'Please wait'
  CommonButtons = [tcbCancel]
  ExpandButtonCaption = 'More'
  ExpandedText =
    'A prime number (or a prime) is a natural number greater'+
      ' than 1 that has no positive divisors other than 1 ' +
      'and itself.'
  Flags = [tfAllowDialogCancellation, tfShowProgressBar,
tfCallbackTimer]
  FooterIcon = 3
  FooterText = 'Please wait while we are calculate prime numbers'
  Text = 'Let'#39's calculate prime numbers up to 1000'
  Title = 'Calculating prime numbers...'
  VerificationText = 'Remember my choice'
  OnButtonClicked = tdProgressButtonClicked
  OnTimer = tdProgressTimer
end
```

There are two event handlers, one to handle clicking on the Cancel button inside the dialog and one to handle the callback:

```
const
  MAX_NUMBERS = 1000;
  NUMBERS_IN_A_SINGLE_STEP = 50;

procedure TMainForm.tdProgressButtonClicked(Sender: TObject;
  ModalResult: TModalResult; var CanClose: Boolean);
begin
  if not FFinished then
  begin
    tdProgress.OnTimer := nil;
    ShowMessage('Calculation aborted by user');
```

```
      CanClose := True;
    end;
  end;


procedure TMainForm.tdProgressTimer(Sender: TObject;
    TickCount: Cardinal;
  var Reset: Boolean);
var
  I: Integer;
begin
  for I := 1 to NUMBERS_IN_A_SINGLE_STEP do
  begin
    if IsPrimeNumber(FCurrNumber) then
      Inc(FPrimeNumbersCount);
    tdProgress.ProgressBar.Position := FCurrNumber * 100
      div MAX_NUMBERS;
    Inc(FCurrNumber);
  end;

  FFinished := FCurrNumber >= MAX_NUMBERS;
  if FFinished then
  begin
    tdProgress.OnTimer := nil;
    tdProgress.ProgressBar.Position := 100;
    ShowMessage('There are ' + FPrimeNumbersCount.ToString +
        ' prime numbers up to ' + MAX_NUMBERS.ToString);
  end;
end;
```

To not block the `main` thread, the prime numbers are calculated a few at a time. When the calculation is ended, the callback is disabled by setting the **OnTimer** event handler to `nil`.

In other words, the real calculation is done in the `main` thread, so you should slice your process into smaller parts so that it can be executed one (small) piece at time.

The following code fires the progress Task Dialog:

```
procedure TMainForm.btnProgressClick(Sender: TObject);
begin
  FCurrNumber := 1;
  FFinished := False;
  FPrimeNumbersCount := 0;
  tdProgress.ProgressBar.Position := 0;
  tdProgress.OnTimer := tdProgressTimer;
  tdProgress.Execute;
end;
```

Here's the resultant dialog:

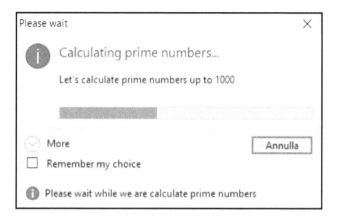

Figure 1.24: The Task Dialog with an embedded progress bar

There's more...

The new Task Dialog API can give your application a fresh look, but that comes with a cost because it works only on Vista or better, with enabled themes. So, how do you work around the problem if you need to run the application on Windows XP or on a machine without themes enabled? For button 6, there's simple code to check whether you can safely use the **TTaskDialog** component or whether you have to go back to the normal ShowMessage or MessageDlg. Here's the event handler for button 6:

```
procedure TMainForm.btnCheckWinVerClick(Sender: TObject);
var
  LTaskDialog: TTaskDialog;
begin
  if (Win32MajorVersion >= 6) and ThemeServices.ThemesEnabled then
  begin
    LTaskDialog := TTaskDialog.Create(Self);
    try
      LTaskDialog.Caption := 'MY Fantastic Application';
      LTaskDialog.Title := 'The Cook Task Dialog!';
      LTaskDialog.Text :=
        'This is a Task Dialog, so I''m on Vista ' +
          'or better with themes enabled';
      LTaskDialog.CommonButtons := [tcbOk];
      LTaskDialog.Execute;
    finally
      LTaskDialog.Free;
```

```
        end
      end
      else
      begin
        ShowMessage('This is an old and boring ShowMessage, ' +
            'here only to support old Microsoft Windows OS ' +
            '(XP and below)');
      end;
    end;
```

Try to disable the themes for your application and click on button 6.

Obviously, it is strongly suggested that you wrap this code in a function so that you do not have to write the same code repeatedly.

The amazing TFDTable – indices, aggregations, views, and SQL

Without question, the software industry is a data-driven environment. All of the IT industry runs on data (we are in the big data era, guys!)—customers, orders, purchases, billings; every day, our applications transform in data interactions between them. Undoubtedly, data is the wellspring of all IT businesses, so we must choose the best programs to interact with it, and fortunately with Delphi we are safe, we have FireDAC.

FireDAC is a unique set of Universal Data Access Components for developing multi-device database applications for Delphi and C++Builder. Here are some features that make this framework special:

- Cross-platform support
- You can use FireDAC on Windows, macOS, Android, iOS, and Linux applications
- Drivers for almost every major relational database, both commercial and open source

In the 90s, the catchphrase in software development was *developing database applications*. Delphi was the master, thanks to the way it was designed (TDataSet interface, Data Module, and ClientDataSet) and its frameworks (Borland Database Engine). In the spring of 2013, Embarcadero acquired AnyDAC and re-branded it as FireDAC. Now Delphi Database Developers have made an unrivaled framework available again.

Getting ready

The **TFDTable** component implements a dataset that works with a single database table. What makes this component amazing is a set of additional features—filtering, indexing, aggregation, cached updates, and persistence.

In this recipe, we'll see some of these at work: how to configure an **TFDTable**, how to manage the indexes to sort an associated grid, and how to collect new information via aggregates.

This recipe uses the `DELPHICOOKBOOK` database, an InterBase DB prepared for the last three recipes of this chapter. To speed up the mechanisms, I suggest adding it to the FireDAC connections in the **Data Explorer**:

1. Open Delphi.
2. Go to the **Data Explorer** tab.
3. Open the **FireDAC** section.
4. Open the **InterBase** section.
5. Right-click on it.
6. Click **Add New Connection**.
7. In the opened window, enter the name, `DELPHICOOKBOOK`.
8. Complete the configuration with this data:
   - Username: `sysdba`
   - Password: `masterkey`
   - Database: Choose the path of the database in your filesystem (the database is under the `data` folder)

Follow the same steps to register the `EMPLOYEE` database; you can find it at `C:\Users\Public\Documents\Embarcadero\Studio\19.0\Samples\data\employee.gdb`.

How to do it...

Let's look at the following steps:

1. Create a new VCL application by selecting **File** | **New** | **VCL Forms Application**.

2. Put a DBNavigator (aligned to the top), a DBGrid (aligned to the client), a DataSource, and a PopUpMenu into the form.

3. Set the DataSource property of DBGrid1 to DataSource1.

4. Select the EMPLOYEE connection in the Data Explorer and then drag and drop it on the form to generate the EmployeeConnection.

5. Put a TFDTable in the form and rename it to SalesTable.

6. The connection property of SalesTable is automatically set to EmployeeConnection.

7. Set the DataSet property of DataSource1 to SalesTable.

8. To choose the Table, you have to expand the Table property combobox and select SALES:

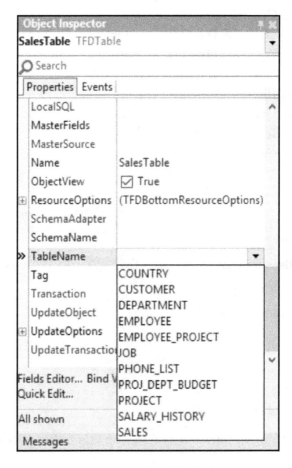

Figure 1.25: SalesTable in the Object Inspector

9. If you performed all the steps correctly, you should be in this situation:

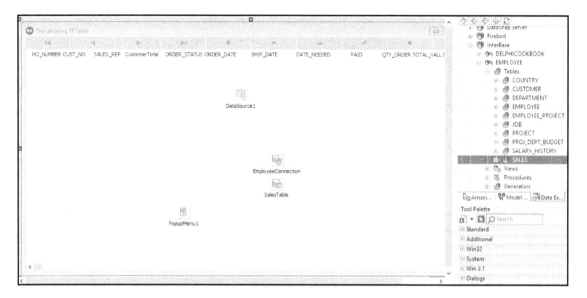

Figure 1.26: Form at design time

10. Declare the `CreateIndexes` procedure under the private section of the form and implement it with the following code:

```
procedure TMainForm.CreateIndexes;
var
  LCustNoIndex: TFDIndex;
begin
  LCustNoIndex := SalesTable.Indexes.Add;
  LCustNoIndex.Name := 'MyCustNoIdx';
  LCustNoIndex.Fields := 'Cust_No';
  LCustNoIndex.Active := true;
end;
```

11. Declare the `CreateAggregates` procedure under the private section of the form and implement it with the following code:

```
procedure TMainForm.CreateAggregates;
begin
  with SalesTable.Aggregates.Add do
  begin
    Name := 'CustomerTotal';
    Expression := 'SUM(TOTAL_VALUE)';
    GroupingLevel := 1;
```

```
      Active := true;
      IndexName := 'MyCustNoIdx';

  end;

  with SalesTable.Aggregates.Add do
  begin
    Name := 'CustomerMax';
    Expression := 'MAX(TOTAL_VALUE)';
    GroupingLevel := 1;
    Active := true;
    IndexName := 'MyCustNoIdx';

  end;

  with SalesTable.Aggregates.Add do
  begin
    Name := 'CustomerLastDate';
    Expression := 'MAX(ORDER_DATE)';
    GroupingLevel := 1;
    Active := true;
    IndexName := 'MyCustNoIdx';

  end;
end;
```

12. Now, we are able to set up the `SalesTable` component. So, implement the `OnCreate` event handler for the form and include this code:

```
procedure TMainForm.FormCreate(Sender: TObject);
begin
  SalesTable.Active := false;
  CreateIndexes;
  CreateAggregates;
  SalesTable.IndexName := 'MyCustNoIdx';
  // index activated
  SalesTable.IndexesActive := true;
  // aggregates activated
  SalesTable.AggregatesActive := true;
  SalesTable.Active := true;
end;
```

13. Now, we have to implement DBGrid1TitleClick to perform the right sorting method when the user clicks on a specific title:

```
procedure TMainForm.DBGrid1TitleClick(Column: TColumn);
begin

  // if reset the column caption of LastColumnClickIndex, because
index could be change...
  if FLastColumnClickIndex > 0 then
    DBGrid1.Columns[FLastColumnClickIndex].Title.Caption :=
      DBGrid1.Columns[FLastColumnClickIndex].FieldName;

  // if the order is descending set the IndexFieldNames to ''.
  if SalesTable.IndexFieldNames = (Column.Field.FieldName + ':D')
then
  begin
    Column.Title.Caption := Column.Field.FieldName;
    SalesTable.IndexFieldNames := '';
  end
  // if the order is ascending set it to descending
  else if SalesTable.IndexFieldNames = Column.Field.FieldName then
  begin
    SalesTable.IndexFieldNames := Column.Field.FieldName + ':D';
    Column.Title.Caption := Column.Field.FieldName + ' ▼';
  end
  // if no order is specified I'll use ascending one
  else
  begin
    SalesTable.IndexFieldNames := Column.Field.FieldName;
    Column.Title.Caption := Column.Field.FieldName + ' ▲';
  end;

  // set last column index
  FLastColumnClickIndex := Column.Index;

end;
```

14. It's time to insert the aggregates. The goal is to show some aggregated information through a simple ShowMessage procedure. Add a new menu item to PopupMenu1, rename it to **Customer Info**, and implement the **OnClick** event with the following code:

```
procedure TMainForm.CustomerInfoClick(Sender: TObject);
var
  LOldIndexFieldNames: string;
begin
```

```
    // i use LOldIndexFieldNames to reset the index to last user
choice
    LOldIndexFieldNames := SalesTable.IndexFieldNames;
    DBGrid1.Visible := false;
    // the right index for aggregate
    SalesTable.IndexName := 'MyCustNoIdx';

    // show some customer info
    ShowMessageFmt('The total value of order of this customer is %m.
' +
      'The max value order of this customer is %m. ' + 'Last order on
%s ',
      [StrToFloat(SalesTable.Aggregates[0].Value),
      StrToFloat(SalesTable.Aggregates[1].Value),
      DateTimeToStr(SalesTable.Aggregates[2].Value)]);

    SalesTable.IndexFieldNames := LOldIndexFieldNames;
    DBGrid1.Visible := true;
end;
```

15. Run the application by hitting *F9* (or by going to **Run** | **Run**):

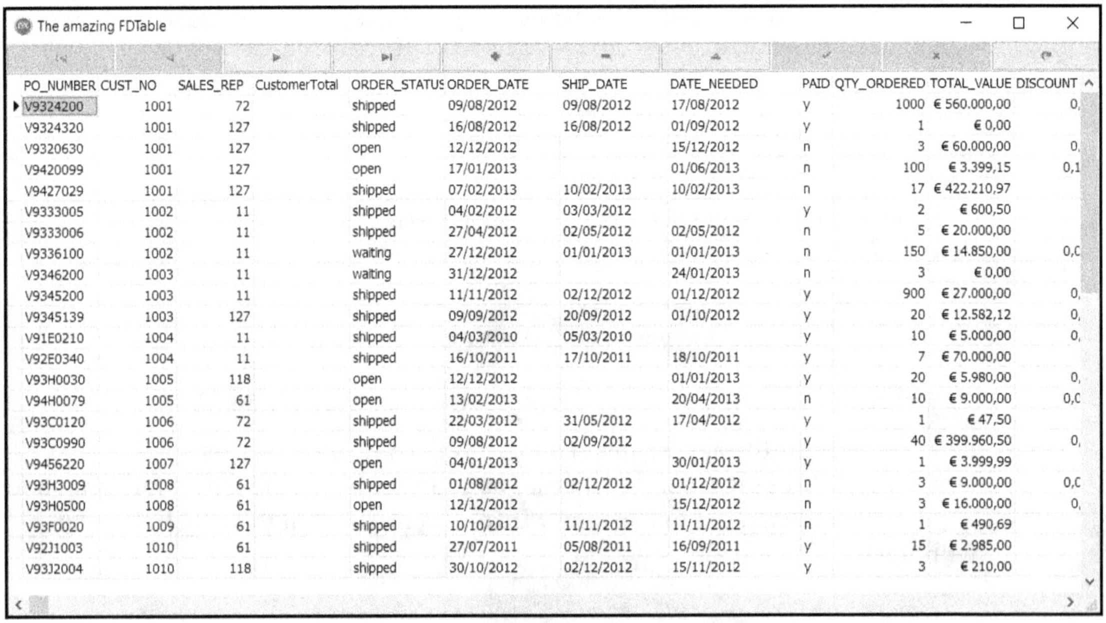

PO_NUMBER	CUST_NO	SALES_REP	CustomerTotal	ORDER_STATUS	ORDER_DATE	SHIP_DATE	DATE_NEEDED	PAID	QTY_ORDERED	TOTAL_VALUE	DISCOUNT
V9324200	1001	72		shipped	09/08/2012	09/08/2012	17/08/2012	y	1000	€ 560.000,00	0,
V9324320	1001	127		shipped	16/08/2012	16/08/2012	01/09/2012	y	1	€ 0,00	
V9320630	1001	127		open	12/12/2012		15/12/2012	n	3	€ 60.000,00	0,
V9420099	1001	127		open	17/01/2013		01/06/2013	n	100	€ 3.399,15	0,1
V9427029	1001	127		shipped	07/02/2013	10/02/2013	10/02/2013	n	17	€ 422.210,97	
V9333005	1002	11		shipped	04/02/2012	03/03/2012		y	2	€ 600,50	
V9333006	1002	11		shipped	27/04/2012	02/05/2012	02/05/2012	n	5	€ 20.000,00	
V9336100	1002	11		waiting	27/12/2012	01/01/2013	01/01/2013	n	150	€ 14.850,00	0,C
V9346200	1003	11		waiting	31/12/2012		24/01/2013	n	3	€ 0,00	
V9345200	1003	11		shipped	11/11/2012	02/12/2012	01/12/2012	y	900	€ 27.000,00	0,
V9345139	1003	127		shipped	09/09/2012	20/09/2012	01/10/2012	y	20	€ 12.582,12	0,
V91E0210	1004	11		shipped	04/03/2010	05/03/2010		y	10	€ 5.000,00	0,
V92E0340	1004	11		shipped	16/10/2011	17/10/2011	18/10/2011	y	7	€ 70.000,00	
V93H0030	1005	118		open	12/12/2012		01/01/2013	y	20	€ 5.980,00	0,
V94H0079	1005	61		open	13/02/2013		20/04/2013	n	10	€ 9.000,00	0,C
V93C0120	1006	72		shipped	22/03/2012	31/05/2012	17/04/2012	y	1	€ 47,50	
V93C0990	1006	72		shipped	09/08/2012	02/09/2012		y	40	€ 399.960,50	0,
V9456220	1007	127		open	04/01/2013		30/01/2013	y	1	€ 3.999,99	
V93H3009	1008	61		shipped	01/08/2012	02/12/2012	01/12/2012	n	3	€ 9.000,00	0,C
V93H0500	1008	61		open	12/12/2012		15/12/2012	n	3	€ 16.000,00	0,
V93F0020	1009	61		shipped	10/10/2012	11/11/2012	11/11/2012	n	1	€ 490,69	
V92J1003	1010	61		shipped	27/07/2011	05/08/2011	16/09/2011	y	15	€ 2.985,00	
V93J2004	1010	118		shipped	30/10/2012	02/12/2012	15/11/2012	y	3	€ 210,00	

Figure 1.27: Amazing FDTable at startup

16. Click on the `Total Value` column twice in the descending order:

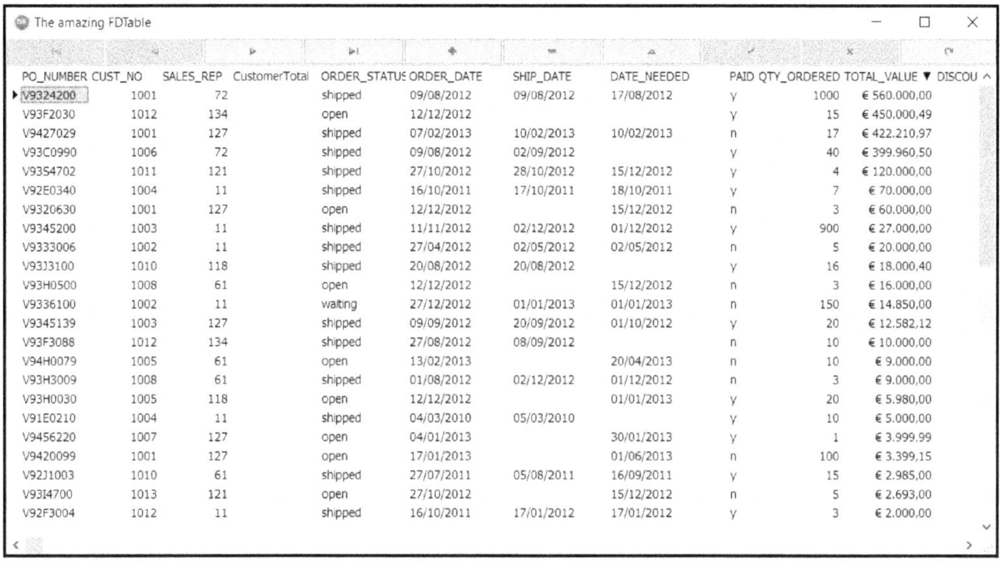

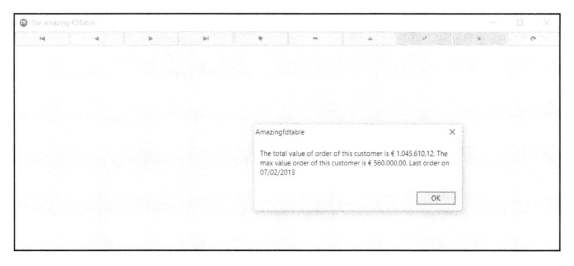

Figure 1.28: Descending order on total_value field

17. Right-click on the first record to bring up the pop-up menu, then click on `Customer Info`:

Figure 1.29: Aggregates in action

How it works...

The core concepts of this recipe are enclosed in the `DBGrid1TitleClick` and `CustomerInfoClick` functions.

In the first procedure, we used the `IndexFieldNames` property to generate a temporary index to perform sorting based on a field related to the `DBGrid` column clicked, and also applying a graphical change to the column to better understand the ordering.

 A temporary index accepts more than one field, so if you want to sort data by several fields you can do it by separating fields, with a semicolon. In addition, you can also specify the sort order, such as ascending or descending, adding the suffixes `:A` for ascending and `:D` for descending.

In the second procedure, we used **Aggregate** to report some customer info:

- `Total Value`: This represents the total amount of all orders
- `Max Value`: This represents the order with the maximum amount
- `Last Order`: This represents the last date order

Aggregate are created in the `CreateAggregates` procedure. Here is some more information about the properties used:

- `Expression` property: This defines the expression to be used to calculate the aggregate.
    - `GroupingLevel` property: This defines the number of indexed fields to use for grouping. By default, its value is set to `0` (no fields and no grouping; all records in a dataset).
- `IndexName` property: This defines the name of the index to use for grouping. If none is specified, it will use the `IndexName` property of `DataSet`.

There's more...

The expression engine provided by FireDAC is a powerful engine used for filtering, indexing, and calculated fields. For more information on how to write powerful expressions, go here: `http://docwiki.embarcadero.com/RADStudio/en/Writing_Expressions_(FireDAC)`.

More information about aggregate and calculated fields can be found here: `http://docwiki.embarcadero.com/RADStudio/en/Calculated_and_Aggregated_Fields_(FireDAC)`.

ETL made easy – TFDBatchMode

In computing, **extract**, **transform**, **load** (**ETL**) refers to a process where the following applies:

- The **Extract** process is where data is extracted from homogeneous or heterogeneous data sources
- The **Transform** process involves in a series of rules or functions applied to the extracted data in order to prepare it for the end target
- The **Load** process loads the data into the end target

Nowadays, these operations can be everyday operations because we can retrieve information from any source (IoT, big data) and we need to enter this heterogeneous data into our systems. We may simply need to transfer our data to a new and different data source system.

FireDAC provides a component to make these operations really easy: **TFDBatchMove**.

Getting ready

In this recipe, we will see how to import the old information distributed under heterogeneous sources, CSV and table, into our new data system. We will also be able to export the new data in CSV format.

As already mentioned, **TFDBatchMove** implements the engine to process the data movement between different types of data sources and destinations. This operation is made possible through reader and writer components. FireDAC provides three types of standard reader and writer:

Component	Use
TFDBatchMoveTextReader	Reader for text file
TFDBatchMoveTextWriter	Writer for text file
TFDBatchMoveDataSetReader	Reader for TDataSet
TFDBatchMoveDataSetWriter	Writer for TDataSet
TFDBatchMoveSQLReader	Reader for SQL
TFDBatchMoveSQLWriter	Writer for SQL

Ensure that you have followed the instructions in the *The Amazing FDTable* recipe on database preparation. If you don't, then go to it and set up your environment.

How to do it...

Let's look at the following steps:

1. Create a new VCL application and drop these components (every time you add a component, align it to the top)—**TComboBox**, **TButton**, **TPanel**, **TDBGrid**, **TPanel**, and **TDBGrid** (this time, align the component to the client).
2. Ensure you perform caption refactoring, adjust the component size, and so on to make your form look like this:

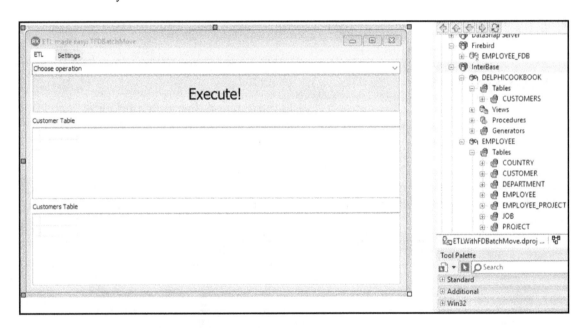

Figure 1.30: Form layout at design time

3. If you have followed the instructions of the *The Amazing FDTable* recipe on database preparation, you should see the database connections, as in *Figure 14.1*, in the **Data Explorer** tab under the `InterBase` entry. Select the `DELPHICOOKBOOK` and `EMPLOYEE` connections, and drag and drop the `CUSTOMERS` table from `DELPHICOOKBOOK` and the `CUSTOMER` table from `EMPLOYEE` onto the form.

4. This operation generates four components:
    *   `DelphiCookbookConnection`: The `FDConnection` to `DELPHICOOKBOOK`
    *   `CustomersTable`: The `TFDQuery` component relating to the `CUSTOMERS` table
    *   `EmployeeConnection`: The `FDConnection` to `Employee`
    *   `CustomerTable`: The `TFDQuery` component relating to the `CUSTOMER` table

5. Set these SQL statements to `TFDQuery` components into the form:
    *   `CustomerTable`: `select CUST_NO as ID, CONTACT_FIRST as FIRSTNAME, CONTACT_LAST as LASTNAME from {id CUSTOMER}`
    *   `CustomersTable`: `select * from {id CUSTOMERS}`

6. Put the `TFDBatchMove` component, and two `TDataSource` components:
    *   Rename `TDataSource` to `dsCustomer`, set the `DataSet` property to `CustomerTable`, and assign it to the `DataSource` property of the first `DBGrid`
    *   Rename the second `TDataSource` to `dsCustomers`, set the `DataSet` property to `CustomersTable`, and assign it to the `DataSource` property of the second `DBGrid`

7. We'll use the **TCombobox** component to allow the user to choose the operation to be performed, so set the **Items** property as follows:
    *   CSV to Table
    *   Table to Table
    *   Table to CSV

8. Declare the `CloseDataSets` procedure in the private section of the form and use the following code:

```
procedure TMainForm.CloseDataSets;
begin
  CustomersTable.Close;
end;
```

9. Declare the `OpenDataSets` procedure in private section of the form and use the following code:

```
procedure TMainForm.OpenDataSets;
begin
  CustomersTable.Close;
  CustomersTable.Open;
  CustomerTable.Close;
  CustomerTable.Open;
end;
```

10. Declare the `SetUpReader` procedure in the private section of the form and use the following code:

```
procedure TMainForm.SetUpReader;
var
  LTextReader: TFDBatchMoveTextReader;
  LDataSetReader: TFDBatchMoveDataSetReader;
begin
  case ComboBox1.ItemIndex of
    0:
      begin
        // Create text reader
        // FDBatchMove will automatically manage the reader
instance.
        LTextReader := TFDBatchMoveTextReader.Create(FDBatchMove);
        // Set source text data file name
        // data.txt provided with demo
        LTextReader.FileName :=
ExtractFilePath(Application.ExeName) +
          '..\..\data\data.txt';
        // Setup file format
        LTextReader.DataDef.Separator := ';';
        // to estabilish if first row is definition row (it is this
case)
        LTextReader.DataDef.WithFieldNames := True;
      end;
    1:
      begin
        // Create text reader
        // FDBatchMove will automatically manage the reader
instance.
        LDataSetReader :=
TFDBatchMoveDataSetReader.Create(FDBatchMove);
        // Set source dataset
        LDataSetReader.DataSet := CustomerTable;
        LDataSetReader.Optimise := False;
      end;
```

```
    2:
      begin
        LDataSetReader :=
TFDBatchMoveDataSetReader.Create(FDBatchMove);
        // set dataset source
        LDataSetReader.DataSet := CustomersTable;
        // because dataset will be show on ui
        LDataSetReader.Optimise := False;
      end;
  end;
end;
```

11. Declare the `SetUpWriter` procedure in the private section of the form and use the following code:

```
procedure TMainForm.SetUpWriter;
var
  LDataSetWriter: TFDBatchMoveDataSetWriter;
  LTextWriter: TFDBatchMoveTextWriter;
begin
  case ComboBox1.ItemIndex of
    0:
      begin
        // Create dataset writer and set FDBatchMode as owner. Then
        // FDBatchMove will automatically manage the writer
instance.
        LDataSetWriter :=
TFDBatchMoveDataSetWriter.Create(FDBatchMove);
        // Set destination dataset
        LDataSetWriter.DataSet := CustomersTable;
        // because dataset will be show on ui
        LDataSetWriter.Optimise := False;
      end;
    1:
      begin
        // Create dataset writer and set FDBatchMode as owner. Then
        // FDBatchMove will automatically manage the writer
instance.
        LDataSetWriter :=
TFDBatchMoveDataSetWriter.Create(FDBatchMove);
        // Set destination dataset
        LDataSetWriter.DataSet := CustomersTable;
        // because dataset will be show on ui
        LDataSetWriter.Optimise := False;
      end;
    2:
      begin
        LTextWriter := TFDBatchMoveTextWriter.Create(FDBatchMove);
```

```
                    // set destination file
                    LTextWriter.FileName :=
            ExtractFilePath(Application.ExeName) +
                    'DataOut.txt';
                    // ensure to write on empty file
                    if TFile.Exists(LTextWriter.FileName) then
                        TFile.Delete(LTextWriter.FileName);
                end;
            end;
        end;
```

12. Now, create event handlers for the **Execute** button and write the code that follows:

```
procedure TMainForm.Button1Click(Sender: TObject);
begin
  // ensure user make a choice
  if ComboBox1.ItemIndex = -1 then
  begin
    ShowMessage('You have to make a choice');
    exit;
  end;

  CloseDataSets;

  // SetUp reader
  SetUpReader;

  // SetUp writer
  SetUpWriter;

  // Analyze source text file structure
  FDBatchMove.GuessFormat;
  FDBatchMove.Execute;

  // show data
  OpenDataSets;

end;
```

13. Run the application by hitting *F9* (or by going to **Run** | **Run**).

14. In the order they are shown, select the item of **TComboBox** and click on the **ExecuteButton** to perform the operation.

15. After the third click, you should see something similar to the following screenshot:

Figure 1.31: Customers table after batchmove operations

16. In addition, at the same level as the executable file, you should find the `DataOut.txt` file as follows:

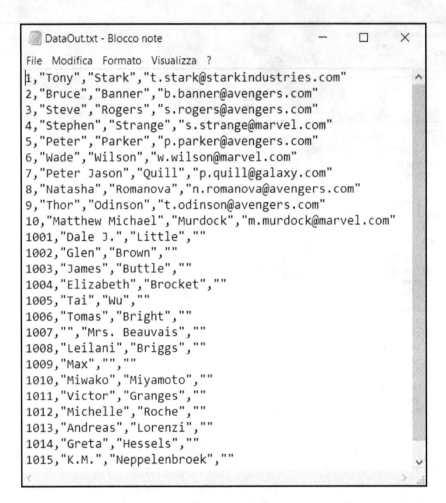

Figure 1.32: Output file generated

How it works...

This recipe allowed you to do the following operations:

- CSV to Table
- Table to Table
- Table to CSV

Depending on the chosen operation, specific readers and writers are created and hooked to the `FDBatchMove` component, to allow it to perform the `BatchMove` operation.

All the important stuff contained in this recipe resides under these operations—`SetUpReader`, `SetUpWriter`, and `FDBatchMove.Execute`.

`FDBatchMove.Execute` moves data from the data source to a data destination, but to do it we need to set up the reader and writer to tell `FDBatchMove` how to perform these operations.

In `SetUpReader`, we create the reader that will be used to read source data. If it is a `Text` source (CSV), we need to set the `FileName` and specify the separator. If it is a `DataSet` source (DB table), we need only to set the `DataSet` property only.

In `SetUpWriter`, we create the writer which well be used to write destination data. If it is a `Text` destination (CSV), we need to set the `FileName` to specify the output file path. If it is a `DataSet` destination (DB table), we need to set the `DataSet` property only.

Once the readers and writers have been prepared, it is possible to call the execute function that will perform the operations according to the specified instructions. Ensure you use the `GuessFormat` method to automatically recognize the data source format.

There's more...

You can use the `Mappings` collection property if you need different fields mapped from source to destination.

You can use the `LogFileAction` and the `LogFileName` properties, provided by the `TFDBatchMove` component, to log data movement.

You can use the `ReadCount`, `WriteCount` (or `InsertCount`, `UpdateCount`, `DeleteCount`), and `ErrorCount` properties to get the batch moving statistic.

Here are some Embarcadero documents about TFDBatchMove: http://docwiki.
embarcadero.com/Libraries/en/FireDAC.Comp.BatchMove.TFDBatchMove.

Data integration made easy – TFDLocalSQL

As Wikipedia says:

> *"Data integration involves combining data residing in different sources and providing users with a unified view of them."*

Traditionally, information must be stored in a single database with a single schema, but many organizations store information on multiple databases, so they need a way to retrieve data from different sources and assemble it in a unified way.

FireDAC provides a component that permits you to execute SQL statements against any dataset: TFDLocalSQL.

Getting ready

Let's imagine that a company wants to gain some business intelligence on their data. The marketing department, to allow special customers to take advantage of a special promotion, wants a list of customers who have spent at least a certain sum in at least one order.

The problem is that customers are provided in XML format and sales are stored in a database table. We want to achieve the aim of executing heterogeneous queries—XML and database tables. Let's go!

 Ensure you have followed the instructions in the *The Amazing FDTable* recipe on database preparation. If you haven't, go to it and set up your environment.

How to do it...

Let's look at the following steps:

1. Create a new VCL application by selecting **File | New | VCL Forms Application**.
2. Put a **TFDConnection** on the form and set its DriverName to SQLite (because SQL Local uses SQLLite in its engine).

3. Place on the form a **DBEdit** (aligned to the top), a **TButton** (aligned to the top), a **DBNavigator** (aligned to the top), a **DBGrid** (aligned to the client), and a **DataSource**. Set the **DataSource** property of **DBNavigator1** and DBGrid1 to DataSource1.

4. From **DataExplorer**, drag and drop onto the form the SALES table from the DELPHICOOKBOOK connection under the InterBase voice.

5. Now, put on the form one **TFDQuery**, one **TFDLocalSQL**, and one **TClientDataSet**.

6. It's time to rename components:

Old	New
FDQuery1	LocalQuery
DataSource1	dsLocalQuery
ClientDataSet1	CustomersCDS

7. If you performed all the steps correctly, you should be in this situation:

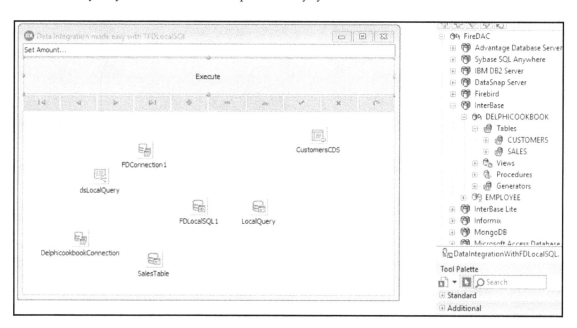

Figure 1.33: Form layout at design time

8. Set the FDLocalSQL1 connection to FDConnection1.

9. Select the DataSets property of FDLocalSQL1 and click the ellipsis button (...) to enter the editor.

10. Click the **Add New** button on the editor twice to add two datasets to the `DataSets` collection.

11. Select the first dataset in the collection and set the `DataSet` property to `SalesTable`; set the `Name` property to `Sales` in order to use the Sales identifier in SQL to refer to this dataset.

12. Select the second dataset in the collection and set the `DataSet` property to `CustomersCDS`; set the `Name` property to `Customers` in order to use the customers identifier in SQL to refer to this dataset:

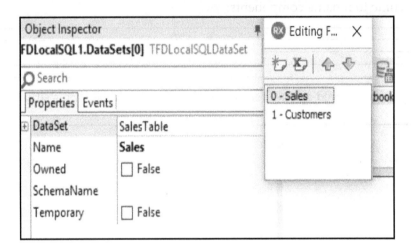

Fig 1.34: FDLocalSQL DataSets editor collection

13. In the private section of the form, declare a procedure named `OpenDataSets` and put in the following code:

```
procedure TMainForm.OpenDataSets;
begin
  SalesTable.Open();
  CustomersCDS.Active := True;
end;
```

14. In the private section of the form, declare a procedure named `PrepareDataSets` and put in the following code:

```
procedure TMainForm.PrepareDataSets;
begin
  CustomersCDS.FileName :=
  'C:\Users\Public\Documents\Embarcadero\Studio\19.0\Samples\Data\cus
  tomer.xml';
  LocalQuery.SQL.Text := 'select distinct c.* from Customers c ' +
```

```
                ' JOIN Sales s on cast (s.CUST_NO as integer) = c.CustNo ' +
                ' where s.total_value > :v order by c.CustNo ';
        end;
```

15. Generate a `FormCreate` event handler and put in this code:

```
procedure TMainForm.FormCreate(Sender: TObject);
begin
    PrepareDataSets;
end;
```

16. We have almost finished; now, we need to put everything together. Generate the `Button1 Click` event handler and put in this code:

```
procedure TMainForm.btnExecuteClick(Sender: TObject);
var
    LAmount: Integer;
begin

    // ensure amount is an integer
    if not TryStrToInt(Edit1.Text, LAmount) then
    begin
        ShowMessage('Amount must be integer...');
        exit;
    end;

    LocalQuery.Close;
    OpenDataSets;
    // apply user data
    LocalQuery.ParamByName('v').AsInteger := LAmount;
    // Execute the query through eterogeneous sources
    LocalQuery.Open;
end;
```

17. Run the application by hitting *F9* (or by going to **Run** | **Run**).

18. Try different amounts to filter the different customers:

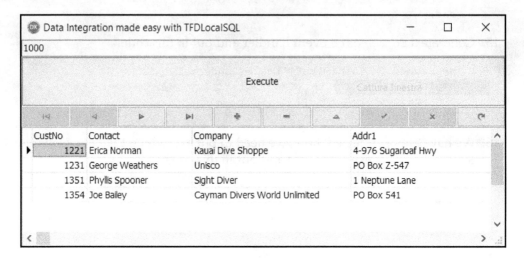

Figure 1.35: Data integration in action

Following image is an example showing different amounts to filter the different customers:

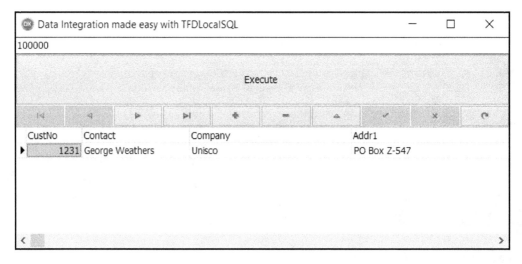

Figure 1.36: Another example of Data integration in action

How it works...

The code of this recipe is quite simple but I want to explain it anyway.

Our data is stored in two different dataset—SalesTable, which refers to a database table, and CustomerCDS, which refers to an XML file. By setting the FDConnection1, FDLocalSQL1, and LocalQuery components as explained in the previous *How to do it...* section, it is possible to have an FDQuery component (LocalQuery) where we write the query using different heterogeneous sources:

```
select distinct c.* from Customers c
JOIN Sales s on cast (s.CUST_NO as integer) = c.CustNo
where s.total_value > :v order by c.CustNo
```

When you click on the **Execute** button, preliminary checks are carried out on the validity of the data entered, then the query in LocalQuery is performed and the LocalQuery dataset is populated with data... from heterogeneous sources! This is a really great feature!

There's more...

The Local SQL is based on the SQLite database and supports most of the SQLite SQL dialect.

All the read and write operations are performed through the TDataSet API with some extensions, which means that FireDAC performs the operations by converting SQL into dataset calls. This is the reason why you can execute SQL statements against any dataset—FDQuery, IBQuery, ClientDataSet, third-party components, and so on.

The possible applications of Local SQL are (from Embarcadero DocWiki):

- **Heterogeneous queries**: Queryable datasets have result sets from different DBs
- **In-memory database**: TFDMemTables serve the datasets
- **Advanced offline mode**: In this case, although the main DB is not accessible, an application is still able to perform SQL queries
- **Advanced DataSnap client**: The data delivered by the DataSnap driver to the client can be queried locally
- **Simplified migration**: A developer can use the third-party TDataSet objects in an application, and can use a FireDAC API to work with these data sources

Here some important notes (from Embarcadero DocWiki):

- The Local SQL engine does not support datasets with multiple result sets.
- The Local SQL engine supports the INSERT/UPDATE/DELETE SQL commands as transactions and savepoints. Also, it transforms the corresponding SQL commands intoTDataSet API calls.
- The Local SQL engine supports INSERT/REPLACE, but uses only primary key fields to find a record to replace when a primary or unique key constraint is violated. Additionally, when only several fields are specified in INSERT/REPLACE INTO tab (<field list>), the fields that are not specified get null values on updating.
- The Local SQL engine uses the TDataSet API with some extensions provided by the IFDPhysLocalQueryAdapter interface. FireDAC datasets implement this interface. Optionally, for non-FireDAC datasets, a developer can create a class implementing the interface and assign its instance to the TFDLocalSQL.DataSets[..].Adapter property.

See also

- For more information, take a look at http://docwiki.embarcadero.com/ RADStudio/en/Local_SQL_(FireDAC)

Becoming a Delphi Language Ninja

2

In this chapter, we will cover the following recipes:

- Fun with anonymous methods—using higher-order functions
- Writing enumerable types
- Using enumerable types to create new language features
- RTTI to the rescue—configuring your class at runtime
- Duck typing using RTTI
- BOs validation using RTTI attributes
- Creating helpers for your classes

Introduction

This chapter will explain some of the not-so-obvious features of the language and RTL that every Delphi programmer should know. Anonymous methods, enumerable types, extended RTTI, and class helpers are powerful tools for every Delphi developer, but usually they are not mastered as they should be. In this chapter, there are ready-to-use recipes that use these concepts to implement something really useful.

Fun with anonymous methods – using higher-order functions

Since version 2009, the Delphi language (or better, its Object Pascal dialect) supports **anonymous methods**. What's an anonymous method? Not surprisingly, an anonymous method is a procedure or a function that does not have an associated name. An anonymous method treats a block of code just like a value so that it can be assigned to a variable, used as a parameter to a method, or returned by a function as its result value. In addition, an anonymous method can refer to variables and bind values to them in the context scope in which the anonymous method is defined. Anonymous methods are similar to closures defined in other languages, such as JavaScript or C#. An anonymous method type is declared as a reference to a function:

```
type
  TFuncOfString = reference to function(S: String): String;
```

Anonymous methods (or anonymous functions) are convenient for passing as an argument to a **higher-order function**. What's a higher-order function?

Wikipedia (http://en.wikipedia.org/wiki/Higher-order_function) gives the following explanation:

In mathematics and computer science, a higher-order function (also functional form, functional, or functor) is a function that does at least one of the following:

- Takes one or more functions as an input
- Outputs a function

All other functions are first-order functions.

Getting ready

In this recipe, you'll see how to use Delphi's anonymous methods with some of the most popular and useful higher-order functions:

- Map: This is available in many functional programming languages. This takes as arguments a func function and a list of elements list, and returns a new list with func applied to each element of list.
  - Reduce: This is also known as Fold. This requires a combining function, a starting point for a data structure, and possibly some default values to be used under certain conditions. The Reduce function proceeds to combine elements of the data structure using the injected function. This is used to perform operations on a set of values to get only one result (or a smaller set of values) that represents the reduction of that initial data. For example, the values 1, 2, and 3 can be reduced to the single value 6 using the SUM.
- Filter: This requires a data structure and a filter condition. This returns all the elements in the structure that match the filter condition.

How to do it...

1. For the HigherOrderFunctions.dproj project, the actual high-order functions are implemented in the HigherOrderFunctionsU.pas unit as generic class functions, as shown here:

```
type
  HigherOrder = class sealed
    class function Map<T>(InputArray: TArray<T>;
      MapFunction: TFunc<T, T>): TArray<T>;
    class function Reduce<T: record>(InputArray: TArray<T>;
      ReduceFunction: TFunc<T, T, T>; InitValue: T): T;
    class function Filter<T>(InputArray: TArray<T>;
      FilterFunction: TFunc<T, boolean>): TArray<T>;
  end;
```

2. Let's analyze each of these functions. The Map function requires a list of T parameters as its input data structure, and an anonymous method that accepts and returns the same type of data T. For each element of the input data structure, the MapFunction is called, and another data list is built to contain all its results. This is the body of the Map function:

```
class function HigherOrder.Map<T>(InputArray: TArray<T>;
    MapFunction: TFunc<T, T>): TArray<T>;
var
  I: Integer;
begin
  SetLength(Result, length(InputArray));
  for I := 0 to length(InputArray) - 1 do
    Result[I] := MapFunction(InputArray[I]);
end;
```

3. The main form uses the Map function in the following way:

```
procedure TMainForm.btnMapCapitalizeClick(Sender: TObject);
var
  InputData, OutputData: TArray<string>;
begin
  //let's generate some sample data
  InputData := GetStringArrayOfData;
  //call the map function on an array of string
  OutputData := HigherOrder.Map<string>(
    InputData,
    function(Item: String): String
    begin
      //this is the "map" criteria that will be applied to each
      //item to capitalize the first word in the item
      Result := String(Item.Chars[0]).ToUpper + Item.Substring(1);
    end);

  //fill the related listbox with the results
  FillList(OutputData, lbMap.Items);
end;
```

4. The `Reduce` function requires a list of `T` as its input data structure, and an anonymous method that accepts two parameters of type `T` and returns a value of type `T`. It can also be passed as a default for each element of the input data structure; `ReduceFunction` is called by passing the intermediate result calculated so far, and the current element of the list. After the last call, the result is returned to the caller function. This is the body of the `Reduce` function:

```
class function HigherOrder.Reduce<T>(InputArray: TArray<T>;
  ReduceFunction: TFunc<T, T, T>; InitValue: T): T;
var
  I: T;
begin
  Result := InitValue;
  for I in InputArray do
  begin
    Result := ReduceFunction(Result, I);
  end;
end;
```

5. The main form uses the `Reduce` function in the following way:

```
procedure TMainForm.btnReduceSumClick(Sender: TObject);
var
  InputData: TArray<Integer>;
  OutputData: Integer;
begin
  InputData := GetIntArrayOfData;
  //sum the input data using as starting value 0
  OutputData := HigherOrder.Reduce<Integer>(InputData,
    function(Item1, Item2: Integer): Integer
    begin
      Result := Item1 + Item2;
    end, 0);
  lbReduce.Items.Add('SUM: ' + OutputData.ToString);
end;
```

6. The last implemented function is `Filter`. The `Filter` function requires a list of T as its input data structure, and an anonymous method that accepts a single parameter of type T and returns a Boolean value. This anonymous method represents the filter criteria that will be applied to the input data. For each element of the input data structure, the `FilterFunction` is called; and if it returns `true`, then the current element will be in the returning list, but not otherwise. After the last call, the filtered list is returned to the caller function. Here is the body of the `FilterFunction`:

```
class function HigherOrder.Filter<T>(InputArray: TArray<T>;
    FilterFunction: TFunc<T, boolean>): TArray<T>;
var
  I: Integer;
  List: TList<T>;
begin
  List := TList<T>.Create;
  try
    for I := 0 to length(InputArray) - 1 do
      if FilterFunction(InputArray[I]) then
        List.Add(InputArray[I]);
    Result := List.ToArray;
  finally
    List.Free;
  end;
end;
```

7. The main form uses the `Filter` function to filter only even numbers. The code is as follows:

```
procedure TMainForm.btnFilterEvenClick(Sender: TObject);
var
  InputData, OutputData: TArray<Integer>;
begin
  InputData := GetIntArrayOfData;
  OutputData := HigherOrder.Filter<Integer>(InputData,
    function(Item: Integer): boolean
    begin
      Result := Item mod 2 = 0; //gets only the even numbers
    end);
  FillList(OutputData, lbFilter.Items);
end;
```

In the recipe's code, there are other utilization samples related to higher-order functions.

There's more...

Higher-order functions are a vast and interesting topic, so in this recipe we only scratched the surface. One of the main concepts is the abstraction of the internal loop over the data structure. Consider this by abstracting the concept of looping, you can implement looping; any way you want, including implementing it in a way that scales nicely with extra hardware. A good sample of what can be done using functional programming is the parallel extension of the `OmniThreadLibrary` (a nice library that simplifies multithreading programming), written by Primož Gabrijelčič (http://www.thedelphigeek.com/). This is a simple code sample that executes a parallel function for defining a single iteration with an anonymous method and runs it using multiple threads:

```
Parallel.ForEach(1, 100000).Execute(
    procedure (Const elem: integer)
begin
    //check if the current element is
    //a prime number (can be slow)
  if IsPrime(elem) then
      MyOutputList.Add(elem);
    end);
```

Writing enumerable types

When the `for...in` loop was introduced in Delphi 2005, the concept of enumerable types was also introduced into the Delphi language.

 A type is defined as **enumerable** if it can be iterated. That is, it can be the target of the `for...in` loop.

As you know, there are some built-in enumerable types; however, you can create your own enumerable types using a very simple pattern.

To make your container enumerable, implement a single method called `GetEnumerator`, which must return a reference to an object, interface, or record that implements the following two methods and one property (in the example, the element to enumerate is `TFoo`):

```
function GetCurrent: TFoo;
function MoveNext: Boolean;
property Current: TFoo read GetCurrent;
```

There are a lot of samples related to standard enumerable types, so in this recipe you'll look at some not-so-common utilizations.

Getting ready

In this recipe, you'll see a file enumerable function as it exists in other, mostly dynamic, languages. The goal is to enumerate all the rows in a text file without explicitly opening, reading, or closing the file (this is done automatically in the background), as shown in the following code:

```
var
  row: String;
begin
  for row in EachRows('....myfile.txt') do
    WriteLn(row);
end;
```

Nice, isn't it? Let's start.

How to do it...

1. We have to create an enumerable function result. The function simply returns the actual enumerable type. This type is not freed automatically by the compiler, so you've got to use a value type or an interfaced type. For the sake of simplicity, let's code to return a record type:

```
function EachRows(const AFileName: String): TFileEnumerable;
begin
  Result := TFileEnumerable.Create(AFileName);
end;
```

2. The `TFileEnumerable` type is defined as follows:

```
type
  TFileEnumerable = record
  private
    FFileName: string;
  public
    constructor Create(AFileName: String);
    function GetEnumerator: TEnumerator<String>;
  end;
. . .
constructor TFileEnumerable.Create(AFileName: String);
```

```
begin
  FFileName := AFileName;
end;

function TFileEnumerable.GetEnumerator: TEnumerator<String>;
begin
  Result := TFileEnumerator.Create(FFileName);
end;
```

3. This record is required only because you need a type that has a GetEnumerator method defined. This method is called automatically by the compiler when the type is used on the right side of the for...in loop.

4. The TFileEnumerator type is the actual enumerator, and is declared in the implementation section of the unit. Remember, this object is automatically freed by the compiler, because it is the return of the GetEnumerator call:

```
type
  TFileEnumerator = class(TEnumerator<String>)
  private
    FCurrent: String;
    FFile: TStreamReader;
  protected
    constructor Create(AFileName: String);
    destructor Destroy; override;
    function DoGetCurrent: String; override;
    function DoMoveNext: Boolean; override;
  end;

{ TFileEnumerator }

constructor TFileEnumerator.Create(AFileName: String);
begin
  inherited Create;
  FFile := TFile.OpenText(AFileName);
end;

destructor TFileEnumerator.Destroy;
begin
  FFile.Free;
  inherited;
end;

function TFileEnumerator.DoGetCurrent: String;
begin
  Result := FCurrent;
end;
```

```
function TFileEnumerator.DoMoveNext: Boolean;
begin
  Result := not FFile.EndOfStream;
  if Result then
    FCurrent := FFile.ReadLine;
end;
```

5. The enumerator inherits from `TEnumerator<String>` because each row of the file is represented as a string. This class also gives a mechanism to implement the required methods. The `DoGetCurrent` method (called internally by the `TEnumerator<T>.GetCurrent` method) returns the current line. The `DoMoveNext` method (called internally by the `TEnumerator<T>.MoveNext` method) returns `true` or `false`, depending on whether there are more lines to read in the file or not. Remember that this method is called before the first call to the `GetCurrent` method. After the first call to the `DoMoveNext` method, `FCurrent` is properly set to the first row of the file.

6. The compiler generates a piece of code similar to the following pseudo-code:

```
it = typetoenumerate.GetEnumerator;
while it.MoveNext do
begin
  S := it.Current;
  //do something useful with string S
end
it.free;
```

There's more...

Enumerable types are really powerful and help you write less, and less error-prone, code. There are some shortcuts to iterate over in-place data without even creating an actual container.

If you have a bunch, or integers, or if you want to create a non-homogeneous `for` loop over some kind of data type, you can use the new `TArray<T>` type, as shown here:

```
for i in TArray<Integer>.Create(2, 4, 8, 16) do
  WriteLn(i);
//write 2 4 8 16
```

The `TArray<T>` type is a generic type, so the same technique works also for strings:

```
for s in TArray<String>.Create('Hello','Delphi','World') do
  WriteLn(s);
```

It can also be used for **Plain Old Delphi Objects** (**PODOs**) or controls:

```
for btn in TArray<TButton>.Create(btn1, btn31,btn2) do
  btn.Enabled := false;
```

See also

- Here is a link to the Embarcadero documentation, which provides a detailed introduction to enumerable types:
  `http://docwiki.embarcadero.com/RADStudio/en/Declarations_and_Statements#Iteration_Over_Containers_Using_For_statements`.
- Don't start from zero! If you have to implement your own enumerable type, Delphi provides you with classes for a better start:
- `TEnumerable`: An abstract class that provides the `DoGetEnumerator` method to give your container enumerating functionality (`http://docwiki.embarcadero.com/Libraries/en/System.Generics.Collections.TEnumerable`)
- `TEnumerator`: An abstract class that provides the basic framework for enumerating a container (`http://docwiki.embarcadero.com/Libraries/en/System.Generics.Collections.TEnumerator`)

Using enumerable types to create new language features

Enumerable types, thanks to their versatility, are one of the most amazing language features in Delphi. In the previous recipe, we saw what they are and how they are used; now I want to introduce them in another way. They make the impossible possible as enumerables iterate over infinite ranges you need!

Did I stimulate your curiosity enough? Well, let's proceed!

Getting ready

Suppose you have to solve a **Project Euler Problem 10** (`https://projecteuler.net/problem=10`); that is, to find the sum of all the primes below 2 million. How can you iterate through an infinite container like prime numbers?

How to do it...

Perform the following steps:

1. First, declare an enumerable that returns, by using the `GetEnumerator` function, an enumerator able to iterate over prime numbers. In addition, we have to declare `GetPrimes` and `IsPrime` functions. The first to returns the enumerable (`TPrimesEnumerable`) over which we can iterate and the second to check if a given number is prime:

```
unit PrimeEnumU;

interface

uses
  System.Classes, System.Generics.Collections;

type
  TPrimesEnumerable = record
  private
    FStart: Integer;
  public
    constructor Create(AStart: Integer);
    function GetEnumerator: TEnumerator<Integer>;
  end;

function GetPrimes(const AStart: Integer): TPrimesEnumerable;

function IsPrime(const ANumber: Integer): Boolean;
```

2. In the `implementation` section we define `TPrimesEnumerator`. This type, returned by `GetEnumerator` function of `TPrimesEnumerable`, implements the two methods `GetCurrent`, `MoveNext`, and the `Current` property to make your container enumerable. In this case we inherit from `TEnumerator<Integer>`, so to respect the `Enumerator` interface we have only to override `DoGetCurrent` and `DoMoveNext` (take a look at `System.Generics.Collections.pas` to understand it):

```
implementation

uses
  System.IOUtils;

type
  TPrimesEnumerator = class(TEnumerator<Integer>)
```

```
private
  FCurrent: Integer;
protected
  constructor Create(const AStart: Integer);
  function DoGetCurrent: Integer; override;
  function DoMoveNext: Boolean; override;
end;
```

3. Now we can proceed with implementation:

```
constructor TPrimesEnumerable.Create(AStart: Integer);
begin
  FStart := AStart;
end;

function TPrimesEnumerable.GetEnumerator: TEnumerator<Integer>;
begin
  Result := TPrimesEnumerator.Create(FStart);
end;

{ TPrimesEnumerator }

constructor TPrimesEnumerator.Create(const AStart: Integer);
begin
  inherited Create;
  FCurrent := AStart;
end;

function TPrimesEnumerator.DoGetCurrent: Integer;
begin
  Result := FCurrent;
end;

function TPrimesEnumerator.DoMoveNext: Boolean;
begin
  // in the while true loop inc FCurrent while FCurrent is not a
prime number,
  // when it's a prime number, this is only the case you can exit
from function
  while True do
  begin
    try
      if IsPrime(FCurrent) then
        Exit(True);
    finally
      Inc(FCurrent);
    end;
  end;
end;
```

```
end;

// the function GetPrimes returns directly the Enumerable
function GetPrimes(const AStart: Integer): TPrimesEnumerable;
begin
  Result := TPrimesEnumerable.Create(AStart);
end;

// determines if the number passed as a parameter is a prime number
function IsPrime(const ANumber: Integer): Boolean;
var
  count: Int64;
  I: Integer;
begin

  if ANumber <= 1 then
    Exit(false);

  if ANumber = 2 then
    Exit(True);

  if (ANumber mod 2) = 0 then
    Exit(false);

  count := Trunc(Sqrt(ANumber) + 1);

  for I := 3 to count do
    if (ANumber Mod I) = 0 then
    begin
      Exit(false)
    end

  Result := True;
end;

end.
```

4. Then, write the code to iterate over prime numbers and add it to the sum, if it is below 2 million:

```
procedure Main;
var
  LTotal, LNextPrime: Integer;
begin
  LTotal := 2;
  for LNextPrime in GetPrimes(3) do
    if LNextPrime < 2000000 then
      LTotal := LTotal + LNextPrime
```

```
        else
        begin
            WriteLn(LTotal.ToString);
            break;
        end;
    end;
```

5. Finally, if you run the code, you'll see the Project Euler Problem 10 solved!

How it works...

Let's start by analyzing the main procedure. The first function we encounter is `GetPrimes`, which directly returns the enumerable (`TPrimesEnumerable`) over which you can iterate. The rest of the code is used to define the interface relative to the enumerable types, as seen in the previous recipe.

The other interesting part of the code is related to the `DoMoveNext` function of the enumerator—the `while True` loop is executed until a prime number is found, each time incrementing `FCurrent` by `1`.

What's the magic part? Glad you asked!

With a classic iteration approach, you should start with a set already defined, and scroll through the elements one by one. By an enumerable mechanism instead, it is possible to calculate on the fly the current value with considerable benefits.

It is not necessary to have a list (or, more generic, a container) already full of values available, so in the case of very large or even infinite sets, there would be a problem with memory management. In this case, only the value of the current iteration is kept in memory, and is only calculated each time.

The power of enumerable types lies in the fact that you can simply return the next value instead of all the values at once. It doesn't need to create a list at all. No list, no memory issues.

There's more...

I would also like to bring to your attention another aspect that needs clarification. This approach may seem trivial and surmountable in other ways, if we look at the specific example of the recipe; we could have iterated with a for loop and a counter, checked if the number is prime, and so on. We would have had the same results. But what happens if the container, as we have supposed, very large or infinite is not of numbers? This approach continues to work and gives the same advantages regardless of the type of data container.

RTTI to the rescue – configuring your class at runtime

Since Delphi 2010, the Delphi RTTI has been greatly expanded. Now, it is comparable to what is called **reflection** in other languages, such as C# or Java. A much-improved RTTI can dramatically change the way you write or even think about your code and your architecture. Now, it is possible to write highly flexible code without too much effort.

Getting ready

What we want to do in this recipe is to dynamically create a class by looking for it by name among the classes that have been linked in the executable (or loaded from dynamic packages). The goal is to change the behavior of the program using only an external file, without relying on a lot of parameters and complex configuration code; just create the right class. Wonderful! Let's say you've developed a program to do orders. Your program allows only one-line orders, so you cannot buy different things in the same order (this is an example). The form is shown in the following:

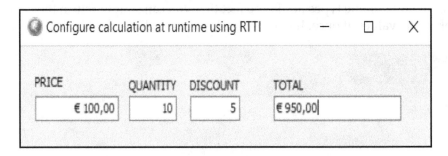

Figure 2.1 : The main form

There is a dataset field connected to each **TDBEdit** in the form. The **TOTAL** field is a calculated field, and its value is calculated in the **OnCalcFields** dataset.

The calculation is simple:

$$total = price * quantity * (1 - discount / 100)$$

The customer is happy, and you are happy as well.

Now, a new, big customer, the City Mall, wants a customization—*If the total is greater than $1,000, apply another 10% discount*. Okay, you can create the customized version easily. So far so good, but now you have two different versions to maintain.

Now, another customer, the Country Road Shop, says, *If there are more than 10 pieces, the discount must be at least 50%*. Another customer, Spark Industries, specifies—*Only at the weekend, all the calculated prices will be cut by 50%*.

Argh! Four customers and four different versions of your software to maintain because of customization! You get the point in the beginning things are simple, but when you start to customize something, complexity (and bugs) can arise. Let's fix this problem in this recipe.

How to do it...

The sample customization is simple; however, the difficulty begins when you have to handle which customization you have to choose from among those available. Sure, you can define some sort of parameters; but your code will get a lot of if just to understand which calculation to apply. And, even worse, a change in one of your criteria could break something in another. Bad approach! We can configure our software without if statements using RTTI. In this recipe, all the calculus engines are implemented in four different classes in four different units. You can also define all the criteria in only one unit, but it is not mandatory.

In the following table, we see a summary of the customers and the customizations implemented:

Customer	Unit/class name	Calculation criteria
Default (no customization)	CalculationCustomerDefaultU TCalculationCustomerDefault	Result := (Price * Quantity) * (1 - Discount / 100);
City Mall	CalculationCustomer_CityMall TCalculationCustomer_CityMall	Result := (Price * Quantity) * (1 - Discount / 100); if Result > 1000 then Result := Result * 0.90;

Customer	Unit/class name	Calculation criteria

Country Road Shop	CalculationCustomer_CountryRoad
	TCalculationCustomer_CountryRoad
		`if Quantity > 10 then`
		`  if Discount < 50 then`
		`    Discount := 50;`
		`Result := (Price * Quantity) * (1 - Disco`
		`unt / 100);`

Spark Industries	CalculationCustomer_Spark
	TCalculationCustomer_Spark
		`Result := (Price * Quantity) * (1 - Discount /`
		`100);`
		`if DayOfTheWeek(Date) in [1, 7] then`
		`  Result :=Result * 0.50;`

When the program starts, it looks for a configuration file. In the first line of the file, there is a fully qualified class name (`UnitName.ClassName`) that implements the needed calculus criteria. That string is used to create the related class, and the instance will be used to calculate the total price when needed. The interesting code is as follows:

```
procedure TMainForm.LoadCalculationEngine;
var
  TheClassName: string;
  CalcEngineType: TRttiType;
const
  CONFIG_FILENAME = '..\..\calculation.config.txt';
begin
  if not TFile.Exists(CONFIG_FILENAME) then
    TheClassName := 'CalculationCustomerDefaultU.' +
      'TCalculationCustomerDefault'
  else
    TheClassName := TFile.ReadAllLines(CONFIG_FILENAME)[0];

  CalcEngineType := FCTX.FindType(TheClassName.Trim);
  if not assigned(CalcEngineType) then
    raise Exception.CreateFmt('Class %s not found', [TheClassName]);
  if not CalcEngineType.GetMethod('Create').IsConstructor then
    raise Exception.CreateFmt('Cannot find Create in %s', [TheClassName]);

  FCalcEngineObj := CalcEngineType.GetMethod('Create')
    .Invoke(CalcEngineType.AsInstance.MetaclassType, []).AsObject;
  FCalcEngineMethod := CalcEngineType.GetMethod('GetTotal');
  Label5.Caption := 'Current Calc Engine: ' + TheClassName;
end;
```

`FCalcEngineObj` is a `TObject` reference that holds your actual calculation engine, while `FCalcEngineMethod` is an RTTI object that keeps a reference to the method to call when the calculus is needed.

Now, in the dataset **OnCalcFields** event handler, there is this code:

```
procedure TMainForm.ClientDataSet1CalcFields(DataSet: TDataSet);
begin
ClientDataSet1TOTAL.Value :=
FCalcEngineMethod.Invoke(FCalcEngineObj,
    [ClientDataSet1PRICE.Value,
    ClientDataSet1QUANTITY.Value,
    ClientDataSet1DISCOUNT.Value]).AsCurrency;
end;
```

Run the program and check which calculus engine is loaded. Then stop the program, open the configuration file, and write another `QualifiedClassName` (unit name plus class name), choosing from all those available. Run the program. As you can see, the correct engine is selected and the customization is applied without changing the working code.

On writing the `CalculationCustomer_CityMall.TCalculationCustomer_CityMall` class in the file, you will get the following behavior:

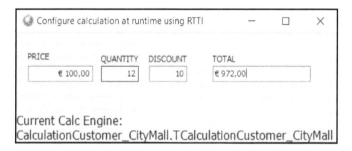

Figure 2.2: The main form using the customized calculus engine specified in the configuration file

There's more...

RTTI is a really vast topic. There are endless possibilities for using it in smart ways.

Remember, however, that, if the Delphi linker sees that your class is not used in the actual code (because it is used only in RTTI calls), it could eliminate the class from the executable. So, to be sure that your class will be included in the final executable, write a line of code, even a useless one, referring to the class. In this recipe, I've included a line of code similar to the following one in every initialization section of the different calculus classes:

```
//. . . other code before
```

```
initialization

//Linker will not remove the class from the final executable
//because now it is used somewhere

TCalculationCustomer_CityMall.ClassName;

end.
```

See also

- The documentation from Embarcadero gives more information about extended RTTI:
  http://docwiki.embarcadero.com/RADStudio/en/Working_with_RTTI_Index.

Duck typing using RTTI

"When I see a bird that walks like a duck and swims like a duck and quacks like a duck, I call that bird a duck."

– James Whitcomb Riley

Clear, isn't it? What may not be so clear is that this approach can also be used in computer programming. Yes, even without an actual duck!

 For design patterns lovers, duck typing is the smartest way to achieve the Visitor pattern goal.

Getting ready

Referring to duck typing, Wikipedia (http://en.wikipedia.org/wiki/Duck_typing) gives the following explanation :

"In computer programming with object-oriented programming languages, duck typing is a style of typing in which an object's methods and properties determine the valid semantics, rather than its inheritance from a particular class or the implementation of an explicit interface."

How can all these concepts be used in everyday programming? This is the question that this recipe aims to answer.

Let's say that you have a form, and you want to inform the user that something bad happened by changing all the colorable components to clRed. You don't know what the property Color means for any control that has that property, you only want to set all the properties named Color to clRed. How can you achieve this? The naive approach could be to cycle the Components property, check whether the current control is a control that you know has a Color property, and then cast that control reference to an actual **TEdit** (or **TComboBox**, **TListBox**, or whatever) reference, and change the Color property to clRed; however, what if tomorrow you need to color another kind of control as well? Or you have to change the Color property on **TPanel**, but the Font.Color property on **TEdit**? You get the point, I think. Using the naive approach can raise the complexity of your code. A programmer should hate complexity. More complexity means more time dealing with code, and more time means more money to spend. As usual, the **KISS** approach is the best one—**Keep It Simple, Stupid!**

How to do it...

The code in this recipe allows you to write code like the following snippets. In this following snippet, the Color property for all control in the form will be set to clRed. I don't know which kinds of controls there are on the form, but if they have a property named Color, that property will be set to clRed:

```
Duck.Apply(Self, 'Color', clRed);
```

In this snippet, the Caption property of the controls is in the array; if it exists, it will be set to 'Hello There':

```
Duck.Apply(
  TArray<TObject>.Create(Button1, Button2, Edit1),
  'Caption',
  'Hello There');
```

The following code disables all the **TDataSource** on the form, preventing data editing:

```
Duck.Apply(Self, 'Enabled', False,
  function(Item: TObject): boolean
  begin
    Result := Item is TDataSource;
  end);
```

The following code sets the font name to `Courier New` for some controls:

```
Duck.Apply(TArray<TObject>.Create(Edit1, Edit2, Button2),
    'Font.Name', 'Courier New');
```

This code works for every kind of control. If you change the **TButton** to **TSpeedButton**, it continues to work. If you replace a **TListBox** with a **TComboBox**, the code still works. The concept is simple—if you have a property *X*, then set that property independently of the actual object type.

Let's see the code that actually does the job.

The main `Duck` class is a mere method container (this is the reason its name is `Duck`, and not `TDuck`; it is not a real type) declared, as shown, in the following code:

```
type
  Duck = class sealed
    class procedure Apply(ArrayOf: TArray<TObject>; PropName: string;
      PropValue: TValue;
      AcceptFunction: TFunc<TObject, boolean> = nil); overload;
    class procedure Apply(AContainer: TComponent; PropName: string;
      PropValue: TValue;
      AcceptFunction: TFunc<TObject, boolean> = nil); overload;
  end;
```

Methods are very similar, and the second one adds a helper to work with `TComponent`; the real job is done by the first one:

```
class procedure Duck.Apply(ArrayOf: TArray<TObject>; PropName: string;
  PropValue: TValue; AcceptFunction: TFunc<TObject, boolean>);
var
  CTX: TRttiContext;
  Item, PropObj: TObject;
  RttiType: TRttiType;
  Prop: TRttiProperty;
  PropertyPath: TArray<string>;
  i: Integer;
begin
  CTX := TRttiContext.Create;
  try
    for Item in ArrayOf do
    begin
      if (not Assigned(AcceptFunction)) or (AcceptFunction(Item)) then
      begin
        RttiType := CTX.GetType(Item.ClassType);
        if Assigned(RttiType) then
        begin
```

```
      PropertyPath := PropName.Split(['.']);
      Prop := RttiType.GetProperty(PropertyPath[0]);
      if not Assigned(Prop) then
        Continue;
      PropObj := Item;
      if Prop.GetValue(PropObj).isObject then
      begin
        PropObj := Prop.GetValue(Item).AsObject;
        for i := 1 to Length(PropertyPath) - 1 do
        begin
          RttiType := CTX.GetType(PropObj.ClassType);
          Prop := RttiType.GetProperty(PropertyPath[i]);
          if not Assigned(Prop) then
            break;
          if Prop.GetValue(PropObj).isObject then
            PropObj := Prop.GetValue(PropObj).AsObject
          else
            break;
        end;
      end;
      if Assigned(Prop) and (Prop.IsWritable) then
        Prop.SetValue(PropObj, PropValue);
      end;
    end;
  end;
  finally
    CTX.Free;
  end;
end;
```

This is not very simple, I know, but you can see all the pieces we've already talked about. Obviously, we use RTTI to get the names and set the values of the properties.

The main loop cycles over the array parameter, and asks AcceptFunction whether the object must be inspected or not. AcceptFunction is optional, so the value can be nil. In this case, all the objects are inspected. To allow a syntax such as Font.Name, there is a small parser that splits the strings and walks through each piece to check whether there is a property with that name. If the last piece (or the only one) is found, then check whether that property is writable; if it is writable, set the property to the passed value. In this way, you can write code that walks through a complex object graph with a simple syntax:

```
Duck.Apply(TArray<TObject>.Create(
DataSource1, DataSource2, Button2), 'DataSet.Active', true);
```

There's more...

Duck typing is a very broad topic and allows you to do wonderful things with a few lines of code. In this recipe's code, there is a bonus recipe project called `DuckTypingUsingRTTIExtended.dproj`, which contains an advanced version of the base recipe. It uses a fluent interface, allows you to select the components that you want to change, and defines what type of change to make on those components; something similar to the following code snippets. Set all the `Caption` properties of the components on the form to `On All Captions`:

```
Duck(Self).All.SetProperty('Caption').ToValue('On All Captions');
```

Set all the `Text` properties to `'Hello There'` for the components with a name starting with `'Edit'`, using an anonymous method as a filter to select the components:

```
Duck(Self)
.Where(function(C: TComponent): boolean
begin
        Result := String(C.Name).StartsWith('Edit');
 end)
.SetProperty('Text')
.ToValue('Hello There');
```

Set the `Color` property to `clRed` for all the **TEdit** components on the form. Use an anonymous method to define what to do on the components:

```
Duck(Self).Where(TEdit).Apply(
procedure(C: TComponent)
begin
     TEdit(C).Color := clRed;
end);
```

In the bonus recipe, there are more examples. Feel free to experiment and expand on them.

BOs validation using RTTI attributes

What does **Business Object (BO)** mean?

> *"A business object is an actor within the business layer of a layered object-oriented computer program that represents a part of a business or an item within it.... It represents business entities such as an invoice, a transaction or a person."*

> *- Techopedia*

If we fine-tune our applications, we are dealing with BOs—TCustomer, TOrder, and TInvoice are classic examples. In the end we always go to meet the same necessity: their validation; that is, to verify their validity through appropriate procedures. Regardless of whether you use an ORM, a data access framework such as FireDAC, or another, it is always a good idea to develop the BO layer to remain independent of them and respect **separation of concerns (SoC)**.

In this recipe, we'll learn about RTTI Attributes, and we'll see how to use them to solve this problem elegantly.

Getting ready

RTTI attributes are a Delphi language feature that allows you to add metadata information to types and type members (comparable to C# attributes or Java annotations, to be clear). By attributes, you can add metadata information in a declarative way. An attribute is depicted by square brackets ([ ]) placed above the element it is used for.

Attributes do not modify the behavior of types or members by themselves, but, after it is associated, the attribute can be queried at runtime by using RTTI techniques (this topic has already been covered in the *RTTI to the Rescue* recipe), and you take appropriate actions when you think they are required.

Imagine the following scenario: we must proceed to register a client through a classic user registration form! How can we ensure, through the RTTI attributes, fields are correctly compiled?

Let's find out!

How to do it...

Let's look at the following steps:

1. Perform the necessary steps to obtain a form like this (design time):

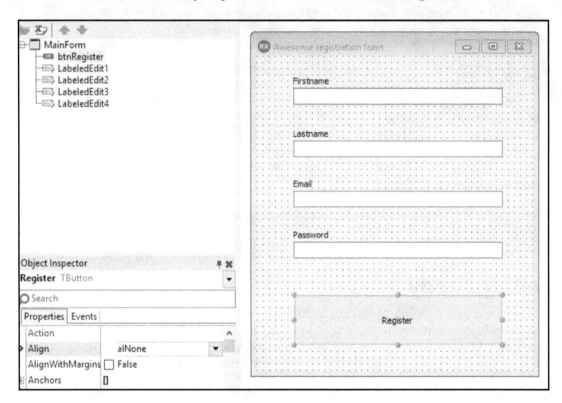

Figure 2.3: Form at design time

First of all, we have to give a correct representation (in terms of OOP) of the fields on the form.

2. Declare the BO:

```
TCustomer = class(TObject)
private
   FEmail: String;
   FLastname: String;
   FPassword: String;
   FFirstname: String;
```

```
    procedure SetEmail(const Value: String);
    procedure SetFirstname(const Value: String);
    procedure SetLastname(const Value: String);
    procedure SetPassword(const Value: String);
  public
    property Firstname: String read FFirstname write SetFirstname;
    property Lastname: String read FLastname write SetLastname;
    property Email: String read FEmail write SetEmail;
    property Password: String read FPassword write SetPassword;
  end;
```

3. Now, we have to declare the attributes that define our validation rules. In Delphi, to declare your own custom attribute, you must derive it from a special predefined class, `System.TCustomAttribute`. In the following table, there is a summary of the attributes and their purpose:

Attributes	Purpose
ValidationAttribute	Abstract base class to provide common stuff
RequiredValidationAttribute	Note that the field is mandatory, it is not possible to leave it empty
MaxLengthValidationAttribute	Note that the field must respect a maximum length
MinLengthValidationAttribute	Note that the field must respect a minimum length
RegexValidationAttribute	Note that the field must respect the validation set by a regular expression

 If it's not clear to you how regular expressions work, here is a dedicated recipe, *Check strings with regular expressions.*

We proceed to the declaration and definition of the attributes just described:

```
type

  ValidationAttribute = class abstract(TCustomAttribute)
  protected
    FErrorMessage: string;
    function DoValidate(AValue: string): Boolean; virtual;
abstract;
  public
    constructor Create(AErrorMessage: string);
    function Validate(AValue: string): TValidationResult;
  end;

  RequiredValidationAttribute = class(ValidationAttribute)
  public
    function DoValidate(AValue: string): Boolean; override;
  end;

  MaxLengthValidationAttribute = class(ValidationAttribute)
```

```delphi
  protected
    FLength: Integer;
  public
    constructor Create(AErrorMessage: string; ALength: Integer);
    function DoValidate(AValue: string): Boolean; override;
  end;

  MinLengthValidationAttribute =
class(MaxLengthValidationAttribute)
  public
    function DoValidate(AValue: string): Boolean; override;
  end;

  RegexValidationAttribute = class(ValidationAttribute)
  private
    FRegex: string;
  public
    constructor Create(AErrorMessage: string; ARegex: string);
    function DoValidate(AValue: string): Boolean; override;
  end;

implementation

uses
  System.SysUtils, System.RegularExpressions;

{ ValidationAttribute }

constructor ValidationAttribute.Create(AErrorMessage: string);
begin
  FErrorMessage := AErrorMessage;
end;

function ValidationAttribute.Validate(AValue: string):
TValidationResult;
begin
  if not DoValidate(AValue) then
    Result.ErrorMessage := FErrorMessage;
end;

{ RequiredValidationAttribute }

function RequiredValidationAttribute.DoValidate(AValue: string):
Boolean;
begin
  // value cannot be empty
  Result := not AValue.IsEmpty;
end;
```

```
{ MaxLengthValidationAttribute }

constructor MaxLengthValidationAttribute.Create(AErrorMessage:
string;
  ALength: Integer);
begin
  inherited Create(AErrorMessage);
  FLength := ALength;
end;


function MaxLengthValidationAttribute.DoValidate(AValue: string):
Boolean;
begin
  // length of value must be less than or equal to length provided
in attributes
  Result := Length(AValue) <= FLength;
end;


{ MinLengthValidationAttribute }

function MinLengthValidationAttribute.DoValidate(AValue: string):
Boolean;
begin
  // length of value must be greather than or equal to length
provided in attributes
  Result := Length(AValue) >= FLength;
end;


{ RegexValidationAttribute }

constructor RegexValidationAttribute.Create(AErrorMessage, ARegex:
string);
begin
  inherited Create(AErrorMessage);
  FRegex := ARegex;
end;


function RegexValidationAttribute.DoValidate(AValue: string):
Boolean;
begin
  // Match the value of regular expression
  Result := TRegEx.IsMatch(AValue, FRegex);
end;
```

The following code shows the attributes applied to the properties of `TCustomer`:

```
[RequiredValidation('Firstname cannot be blank...')]
[MinLengthValidationAttribute
  ('Firstname must have a length between 4 and 15 characters', 4)]
[MaxLengthValidationAttribute
  ('Firstname must have a length between 4 and 15 characters', 15)]
property Firstname: String read FFirstname write SetFirstname;

[RequiredValidation('Lastname cannot be blank...')]
[MinLengthValidationAttribute
  ('Lastname must have a length between 4 and 15 characters', 4)]
[MaxLengthValidationAttribute
  ('Lastname must have a length between 4 and 15 characters', 15)]
property Lastname: String read FLastname write SetLastname;

[RequiredValidation('Email cannot be blank...')]
[RegexValidation('The value does not seem to be a correct
email...',
'^(([^<>()\[\]\\.,;:\s@"]+(\.[^<>()\[\]\\.,;:\s@"]+)*)|(".+"))@((\[[0-9
]{1,3}\.[0-9]{1,3}\.[0-9]{1,3}\.[0-9]{1,3}])|(([a-zA-Z\-0-9]+\.)+[a-zA-
Z]{2,}))$')
  ]
property Email: String read FEmail write SetEmail;

[RequiredValidation('Password cannot be blank...')]
[RegexValidation('The value does not respect password criteria...',
  '^(?=.*[A-Za-z])(?=.*\d)[A-Za-z\d]{8,}$')]
property Password: String read FPassword write SetPassword;
```

4. We have almost finished, missing only the code related to RTTI techniques to understand if and how to applyattributes validation rules:

```
TValidator = class(TObject)
protected
  class var FContext: TRTTIContext;
public
  class constructor Create;
  class function Validate(AObject: TObject;
    out AErrors: TArray<string>): Boolean;
end;

{ TValidator }

class constructor TValidator.Create;
begin
  FContext := TRTTIContext.Create;
end;
```

```pascal
class function TValidator.Validate(AObject: TObject;
  out AErrors: TArray<string>): Boolean;
var
  rt: TRttiType;
  a: TCustomAttribute;
  p: TRttiProperty;
  m: TRttiMethod;
  LResult: TValidationResult;
begin
  AErrors := [];
  rt := FContext.GetType(AObject.ClassType);
  // iterate over object properties
  for p in rt.GetProperties do
    // iterate over attributes of this property
    for a in p.GetAttributes do
    begin
      // ensure that is a ValidationAttribute
      if not(a is ValidationAttribute) then
        continue;
      // reference to Validate method
      m := FContext.GetType(a.ClassType).GetMethod('Validate');
      if m = nil then
        continue;
      // invoke method
      LResult := m.Invoke(a, [p.GetValue(AObject).AsString])
        .AsType<TValidationResult>;
      // eventually add error to errors array
      if not LResult.Validation then
        AErrors := AErrors + [LResult.ErrorMessage];
    end;
  Result := Length(AErrors) = 0;
end;
```

 As previously mentioned, it's possible to add attributes to types and type members. In this specific scenario, the attributes are applied only to properties.

How it works...

On startup, the application is as shown in the following screenshot:

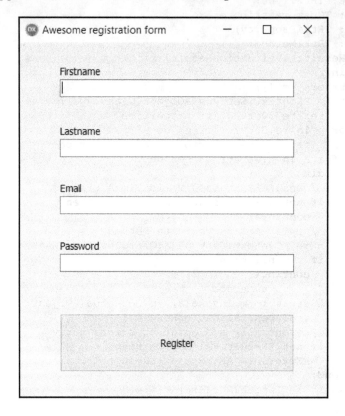

Figure 2.4: Validation using RTTI attributes application on startup

When the **Register** button is clicked, the validation mechanism is triggered through the following steps:

1. Retrieving a `TCustomer` instance based on the form data.
2. Using the `Validate` class method of the `TValidator` class validates the instance just retrieved.
3. If the validation is successful, a success message is shown; otherwise, errors are shown:

```
procedure TMainForm.btnRegisterClick(Sender: TObject);
var
```

```
    LCustomer: TCustomer;
    LErrors: TArray<string>;
begin
    LCustomer := GetCustomer;
    try
        LErrors := [];
        if TValidator.Validate(LCustomer, LErrors) then
            ShowMessage('Well done! You''re now registered...')
        else
            ShowErrors(LErrors);
    finally
        LCustomer.Free;
    end;
end;
```

The concepts of querying, consideration, and application are contained in this code snippet:

```
// iterate over object properties
for p in rt.GetProperties do
    // iterate over attributes of this property
    for a in p.GetAttributes do
    begin
        // ensure that is a ValidationAttribute
        if not (a is ValidationAttribute) then
            continue;
        // reference to Validate method
        m := FContext.GetType(a.ClassType).GetMethod('Validate');
        if m = nil then
            continue;
        // invoke method
        LResult := m.Invoke(a, [p.GetValue(AObject).AsString])
            .AsType<TValidationResult>;
        // eventually add error to errors array
        if not LResult.Validation then
            AErrors := AErrors + [LResult.ErrorMessage];
    end;
```

Specifically, in the validated method of `TValidator`, the properties of the BOs are iterated to understand whether the validation attributes have been applied. If yes, the `Validate` method of the specific attribute is applied to the current value of the object property. If the property value does not pass validation, the error message specified in the attribute declaration phase is added to the errors array.

In the following screenshot, you can see the mechanism in place:

Figure 2.5: Validation in action

There's more...

Attributes are a very powerful mechanism with which to add information in a declarative way. In fact, attributes can extend the normal object-oriented model with aspect-oriented elements (you can find more info here: `https://en.wikipedia.org/wiki/Aspect-oriented_programming`).

Attributes are useful when building general-purpose frameworks: to figure out what attributes and RTTI are capable of doing, try taking a look at the `RESTAdapter` of the DMVC framework (`https://github.com/danieleteti/delphimvcframework` or `https://danieleteti.gitbooks.io/delphimvcframework/content/chapterstaterest_adapter_md.html`).

In some cases, you might write an incorrect attribute name, or not import the unit containing that attribute; in this case, the compiler generates a warning as following—**Unsupported language feature: 'custom attribute'**. Watch out! As I wrote, it generates only a warning and the written code will not work as desired (if you have done some consideration on that attribute); if you want to generate an error instead, you have to set the **Unsupported language feature** to error. Go to **Project** | **Options** | **Delphi Compiler** | **Hint and warnings** | **Unsupported Language Feature** and set the value to error. From now on, every **Unsupported language feature** generates a compiler error.

Bonus recipe – Validation using RTTI attributes

Another benefit from structuring code in this way is that it is able to respond to different contexts of use. I want to show you a scenario to make you understand better: what if we have to validate the same BO form in different contexts? For example, an evolution of the application could be to add a login form. Since the customer is registered, we now have to authenticate it in the system, so we should add to the application a login form always based on the same BO—TCustomer. A solution could be to create another class (BO) with other attributes for the validation of the new login context, following the same steps as the previous one. What if another context based on TCustomer comes out? Another BO... and another one? Another BO... At the end, there will be *n * BO* representing the same entity, but in different contexts. So, this solution is neither very elegant nor an efficient one, because we are dealing with what is called a **classes explosion**, with all the negative consequences that derive from it.

> Analyze the *Bonus recipe - Validation using RTTI attributes* recipe to understand the context mechanism and solve class explosions.

A better idea is to add a simple context mechanism within the already implemented validation and attributes system to support different validation contexts, allowing you to always use the same BO. Further to this, I have supplied a bonus recipe that I highly recommend you analyze—Validation using RTTI attributes.

See also...

- The documentation from Embarcadero gives more information about RTTI Attributes: http://docwiki.embarcadero.com/RADStudio/en/Attributes_ (RTTI)

Creating helpers for your classes

As you know (and if you don't know, you can read the documentation about it), a **class helper** is a type that can be associated to a class. When a class helper is associated with another class, all the methods and properties defined in the helper are also available in the other class and in its descendants. Helpers are a way to extend a class without using inheritance; however, it is not the same thing as inheritance. In other words, if the TFooHelper helper is in the same scope as TFoo, the compiler's resolution scope then becomes the original type (TFoo), plus the helper (TFooHelper). So, if the TFoo class defines the DoSomething method and the TFooHelper (the TFoo class helper) defines DoAnotherThing, when TFoo is used in the same scope as the TFooHelper, the TFoo instances, and all its descendants, also have the DoAnotherThing method.

Getting ready

In this recipe, you'll see how to use class helpers to add iterators (or a sort of iterator) to the TDataSet class, so that any other TDataSet descendants, even from another vendor, can automatically support this kind of iterator. Moreover, you'll also add a SaveToCSV method, so that any TDataSet can be saved in a CSV file with only one line of code.

How to do it...

For the DataSetClassHelpers.dproj project, let's start to talk about the simpler helper—the SaveToCSV method.

The current compiler implementation of class helpers allows only one helper active at a time. So, if you need to add two or more helpers at the same time, you have to merge all the methods and properties in a single helper class. Your TDataSet helper is contained in the DataSetHelpersU.pas unit, and is defined as follows:

```
TDataSetHelper = class helper for TDataSet
  public
    procedure SaveToCSVFile(AFileName: string);
    function GetEnumerator: TDataSetEnumerator;
  end;
```

To use this helper with your **TDataSet** instances, you have to add the `DataSetHelpersU`
unit in the `uses` clause of the unit where you want to use the helper. The helper adds the
following features to all the **TDataSet** descendants:

Method name	Description
SaveToCSV	This allows any dataset to be saved as a CSV file. The first row contains all the fieldnames. All string values are correctly quoted, while the numeric values aren't. The resultant CSV file is compatible with MS Excel, and can be opened directly into it.
GetEnumerator	This enables the dataset to be used as an enumerable type in the `for...in` loops. This removes the necessity to cycle the dataset using the usual `while` loop, so you cannot forget the `DataSet.Next` call at the end of the loop. The dataset is correctly cycled from the current position to the end, and for each record the `for` loop is executed. The enumerator item type is a wrapper type called `TDSIterator`, and is able to access the individual values of the current record using a simplified interface.

To get an idea about what the helpers can do, check the following code:

```
//all the interface section before

implementationuses
  DataSetHelpersU; // add the TDataSet helper to the compiler scope

procedure TClassHelpersForm.btnSaveToCSVClick(Sender: TObject);
begin
  // use the SaveToCSVFile helper method
  FDMemTable1.SaveToCSVFile('mydata.csv');
  ListBox1.Items.LoadFromFile('mydata.csv');
end;

procedure TClassHelpersForm.btnIterateClick(Sender: TObject);
var
  it: TDSIterator;
begin
  ListBox1.Clear;
  ListBox1.Items.Add(Format('%-10s %-10s %8s', ['FirstName', 'LastName',
    'EmpNo']));
  ListBox1.Items.Add(StringOfChar('-', 30));

  for it in FDMemTable1 do
  begin
    ListBox1.Items.Add(Format('%-10s %-10s %8d',
      [it.Value['FirstName'].AsString, it.S['LastName'], it.I['EmpNo']]));
  end;
end;
```

Useful, isn't it? The following screenshot shows the status of the demo application after the **SaveToCSV** button was clicked. The demo application is seen as running:

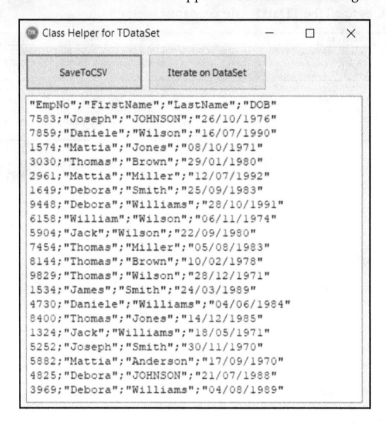

Figure 2.6: The form after the Save To CSV button is clicked

The following screenshot shows the output of the dataset iteration using the helper:

Figure 2.7: The form after the Iterate on DataSet button is clicked; the iteration is used to show dataset data in the listbox

Let's see the implementation details.

The `SaveToCSV` method has been implemented as shown here:

```
procedure TDataSetHelper.SaveToCSVFile(AFileName: string);
var
  Fields: TArray<string>;
  CSVWriter: TStreamWriter;
  I: Integer;
  CurrPos: TArray<Byte>;
begin
  // save the current dataset position
  CurrPos := Self.Bookmark;
  Self.DisableControls;
  try
    Self.First;
```

```delphi
      // create a TStreamWriter to write the CSV file
      CSVWriter := TStreamWriter.Create(AFileName);
      try
        SetLength(Fields, Self.Fields.Count);
        for I := 0 to Self.Fields.Count - 1 do
        begin
          Fields[I] := Self.Fields[I].FieldName.QuotedString('"');
        end;

        // Write the headers line joining the fieldnames with a ";"
        CSVWriter.WriteLine(string.Join(';', Fields));

        // Cycle the dataset
        while not Self.Eof do
        begin
          for I := 0 to Self.Fields.Count - 1 do
          begin
            // DoubleQuote the string values
            case Self.Fields[I].DataType of
              ftInteger, ftWord, ftSmallint, ftShortInt, ftLargeint,
ftBoolean,
                ftFloat, ftSingle:
                begin
                  CSVWriter.Write(Self.Fields[I].AsString);
                end;
            else
              CSVWriter.Write(Self.Fields[I].AsString.QuotedString('"'));
            end;
            // if at the last columns, newline, otherwise ";"
            if I < Self.FieldCount - 1 then
              CSVWriter.Write(';')
            else
              CSVWriter.WriteLine;
          end;
          // next record
          Self.Next;
        end;
      finally
        CSVWriter.Free;
      end;
    finally
      Self.EnableControls;
    end;
    // return to the position where the dataset was before
    if Self.BookmarkValid(CurrPos) then
      Self.Bookmark := CurrPos;
  end;
```

The other helper is a bit more complex, but all the concepts have been already introduced earlier on in this chapter, in the *Writing enumerable types* recipe, so this should not be too complex to understand.

The method in the class helper simply returns TDataSetEnumerator by passing the current dataset to the constructor:

```
function TDataSetHelper.GetEnumerator: TDataSetEnumerator;
begin
  Self.First;
  Result := TDataSetEnumerator.Create(Self);
end;
```

Now, some magic happens in TDataSetEnumerator! Methods to access the current record are encapsulated in a TDSIterator instance. This class allows you to access field values using a limited and simpler interface (compared to the TDataSet one).

The following is the declaration of the enumerator and the iterator:

```
TDataSetEnumerator = class(TEnumerator<TDSIterator>)
private
  FDataSet: TDataSet; // the current dataset
  FDSIterator: TDSIterator; // the current "position"
  FFirstTime: Boolean;
public
  constructor Create(ADataSet: TDataSet);
  destructor Destroy; override;
protected
  // methods to override to support the for..in loop
  function DoGetCurrent: TDSIterator; override;
  function DoMoveNext: Boolean; override;
end;

// This is the actual iterator
TDSIterator = class
private
  FDataSet: TDataSet;
  function GetValue(const FieldName: string): TField;
  function GetValueAsString(const FieldName: string): string;
  function GetValueAsInteger(const FieldName: string): Integer;
public
  constructor Create(ADataSet: TDataSet);
  // properties to access the current record values using the fieldname
  property Value[const FieldName: string]: TField read GetValue;
  property S[const FieldName: string]: string read GetValueAsString;
  property I[const FieldName: string]: Integer read GetValueAsInteger;
end;
```

The `TDataSetEnumerator` handles the mechanism needed by the enumerable type; however, instead of implementing all the needed methods directly (as you saw in the *Write enumerable types* recipe), you've inherited from the `TEnumerator<T>`, so the code to implement is shorter and simpler. The following is the implementation:

```
{ TDataSetEnumerator }

constructor TDataSetEnumerator.Create(ADataSet: TDataSet);
begin
  inherited Create;
  FFirstTime := True;
  FDataSet := ADataSet;
  FDSIterator := TDSIterator.Create(ADataSet);
end;

destructor TDataSetEnumerator.Destroy;
begin
  FDSIterator.Free;
  inherited;
end;

function TDataSetEnumerator.DoGetCurrent: TDSIterator;
begin
  Result := FDSIterator;
end;

function TDataSetEnumerator.DoMoveNext: Boolean;
begin
  if not FFirstTime then
    FDataSet.Next;
  FFirstTime := False;
  Result := not FDataSet.Eof;
end;
```

It is clear that the current record is encapsulated by a `TDSIterator` instance that uses the current dataset. This class is in charge of handling real data access to the underlying dataset fields. Here's the implementation:

```
constructor TDSIterator.Create(ADataSet: TDataSet);
begin
  inherited Create;
  FDataSet := ADataSet;
end;

function TDSIterator.GetValue(const FieldName: String): TField;
begin
  Result := FDataSet.FieldByName(FieldName);
```

```
end;

function TDSIterator.GetValueAsInteger(const FieldName: String): Integer;
begin
  Result := GetValue(FieldName).AsInteger;
end;

function TDSIterator.GetValueAsString(const FieldName: String): String;
begin
  Result := GetValue(FieldName).AsString;
end;
```

Let's summarize the relationship between the three classes involved. The `helper` class adds a method `GetEnumerator` to the `TDataSet` instance, which returns the `TDataSetEnumerator`. The `TDataSetEnumerator` method uses the underlying dataset to handle the enumerable mechanism. The current element returned by the `DataSetEnumerator` is a `TDSIterator` that encapsulates the dataset's current position, allowing the user code to iterate the dataset using the `for...in` loop.

There's more...

What we discussed for class helpers is valid for record helpers as well. If you find the content of this chapter too difficult, you can refresh your understanding about helpers by (re)reading the `Class` and record helpers section in the Embarcadero DocWiki website (`http://docwiki.embarcadero.com/RADStudio/en/Declarations_and_Statements#Iteration_Over_Containers_Using_For_statements`), and perhaps trying it yourself.

Usually, when I talk about class and record helpers during my live training, just before showing the samples, the attendants ask, *I understand the concepts, but in which cases should I use them?* Now, you've seen some nice use cases; however, if you need some others too, read this interesting thread on Stack OverFlows (`http://stackoverflow.com/questions/253399/what-are-good-uses-for-class-helpers`).

Knowing Your Friends – The Delphi RTL

3

In this chapter, we will cover the following recipes:

- Checking strings with regular expressions
- Consuming RESTful services using native HTTP(S) client libraries
- Coping with the encoded internet world using System.NetEncodings
- Saving space using System.Zip
- Decoupling your code by using a cross-platform publish/subscribe mechanism

Introduction

Don't reinvent the wheel! Many programmers ignore what the big Delphi RTL can offer them. Some old-time Delphi lovers continue to write code in the same way they wrote it years ago. New language features and better attention to the community let Embarcadero add many useful classes in the Delphi RTL, and if you don't know them well, you risk reinventing the wheel or simply writing inefficient code because you don't have the time to write it correctly. But often, the correct way to write code is just in the RTL! To minimize such risks, in this chapter we'll introduce some new or lesser known RTL classes.

Checking strings with regular expressions

A **regular expression** (**RegEx**) is a sequence of characters that form a search pattern where some characters have a special meaning. It's mainly used to match patterns on strings. A simple case is something like this: check whether string A matches the criteria defined in string B. Regular expressions follow a specific language to define the criteria. Regular expressions are not only present in Delphi. Many languages have a regular expression library in their standard built-in library. So, if you don't know what a regular expression is, you can read the general documentation at
`http://en.wikipedia.org/wiki/Regular_expression` and then check the Delphi-specific built-in implementation at
`http://docwiki.embarcadero.com/RADStudio/en/Regular_Expressions.`

With regular expressions, perhaps you'll need an external tool to test the most complex ones (just like you want to test a complex query using a database tool instead of changing the SQL in your code over and over again). There are a lot of sites offering this type of tool. One of the most complete websites offering such tools is `http://regex101.com.`

Getting ready

This recipe is a small, complete project with specific objectives. It contains a list of checks that could be daunting to code from scratch, but are trivial using regular expressions. There's just one thing to remember: you always require a RegEx string and an input string that you must check, and the RegEx library gives back the result of the match. In this case, the result is true or false.

Here are some samples of very simple regular expressions with some input strings as a test. In the last column, you can see the result of the match (using the `IsMatch` method). RegEx can be used to perform smart string replaces as well in order to find another strings and so on, but the concept is the same as the check. You only need to call the right method, such as `IsMatch`, `Split`, `Matches`, and so on, to give the right meaning to the RegEx:

RegEx	RegEx description	Input string	Result

`rocks`	Contains `rocks`	`delphi rocks`
		`rocks`
		`rocks of the mountain`
			True
			True
			True

`^rocks`	Starts with `rocks`	`delphi rocks`
		`rocks`
		`rocks of the mountain`
			False
			True
			True

`rocks$`	Ends with `rocks`	`delphi rocks`
		`rocks`
		`rocks of the mountain`
			True
			True
			False

RegEx	RegEx description	Input string	Result
^[ABC]3	Starts with A, B, or C and then is followed by a 3. Anything after the 3 matches.	A3	True
		B3	True
		C33	True
		F3	False
		A2	False
^[ABC][01]$	Starts with A, B, or C and then is followed by 0 or 1. Then the input ends. No more characters are allowed.	A0	True
		A1	True
		A2	False
		B1	True
		AA0	False
		C3	False
^\d{2}\.\d{1,3}$	Starts with 2 digits, then a point, then 1, 2, or 3 digits at the end of the string. \d matches a digit. It's a shortcut for [0-9].	12.3	True
		123.4	False
		89.123	True
		8.3	False
		34.23	True

How to do it...

The test application is shown in the following screenshot:

Figure 3.1: The RegEx recipe main form with some checks on it

Each button checks the value written to the edit at its left. The checks don't test the real validity of the data inserted. They only check the format validity (for example, if the email address is formally valid, the check returns `true`, even if the address doesn't really exist).

Open the recipe project called `RegEx.dproj` in the IDE and check the code for the form.

In Delphi, the necessary classes and records to work with regular expressions are contained in the `System.RegularExpressions.pas` unit and follow the standard of regular expressions as handled by the Perl language (one of the first languages that started to use RegEx). The unit is included in the implementation section of the form. I suggest putting all your validation code in a separate unit in some testable validator types. However, in this recipe, the validation code is in the form under the event handler (please do not do this in your production software!).

Let's start from the IP check. Under the `btnCheckIP` button, you will see the following code:

```
procedure TRegExForm.btnCheckIPClick(Sender: TObject);
begin
  if TRegEx.IsMatch(EditIP.Text,
     '^[0-9]{1,3}\.[0-9]{1,3}\.[0-9]{1,3}\.[0-9]{1,3}$') then
    ShowMessage('IPv4 address is valid')
  else
    ShowMessage('IPv4 address is not valid');
end;
```

The code is really simple; only the RegEx needs some more explanation. The regular expression checks a string that starts with 1, 2, or 3 with numbers from 0 to 9 (`[0-9]{1,3}`), and then expects a point. Consider that the point character in the regex syntax means any character, so if you simply want to check a point, you have to escape the character. This is the reason why in regular expressions there is a \ before the point (\ is used to escape the successive character).

RegEx continues with the same pattern repeated four times (for the four octets contained in the IPv4 address). The last pattern doesn't expect a point.

In this case, this regex is enough, but it also considers an IP such as 999.999.999.999 valid, which is not a valid one. For our needs, it is okay, but for a complete RegEx to check IPv4 addresses, read http://stackoverflow.com/questions/4890789/regex-for-an-ip-address/30023010#30023010.

Using the static `TRegEx.IsMatch` method, you can easily check whether a string matches a RegEx.

The second check is on the email address. The code used is shown as follows:

```
procedure TRegExForm.btnCheckEmailClick(Sender: TObject);
begin
  // Email RegEx from
  // http://www.regular-expressions.info/email.html

  if TRegEx.IsMatch(
        EditEMail.Text,
        '^[A-Z0-9._%+-]+@[A-Z0-9.-]+\.[A-Z]{2,4}$',
        [roIgnoreCase]) then
    ShowMessage('EMail address is valid')
  else
    ShowMessage('EMail address is not valid');
end;
```

In this case, the RegEx is a little bit more complicated. The string must start with at least a letter from A to Z, with a number from 0 to 9, or with a permitted character (., _, %, +, -). The + sign after the square brackets stands for at least one of. Then, there should be an @ sign. After the @ sign, the RegEx checks for letters, numbers, dots, and the minus sign (the domain part of the address) and, finally checks and looks for two, three, or four letters (.com, .it, .net, and so on). The RegEx syntax is case-sensitive, but an email address validity check must be case-insensitive, so I've put the roIgnoreCase modifier on the IsMatch to make the RegEx case-insensitive ([A-Z] is considered as [A-Z/a-z]).

As you can see, if you can read the RegEx syntax, you can easily understand what the RegEx checks. Obviously, there are really complex RegExes, so before you can use them, make sure that you are confident with what you are using.

The last button checks the Italian tax code. I put this example in because the criterion is not very complex and it is good to understand RegEx's flexibility.

In Italy, there is a tax code called **Codice Fiscale** that is assigned to all citizens when they reach a certain age. The criteria is as follows:

- Three letters
- Three letters
- Two numbers
- One letter
- Two numbers
- One letter
- Three numbers
- One letter

So, for instance, this is a formally valid Italian tax code: RSSMRA79S04H501V. As you can see, it is not complex; however, by checking it using plain code, Delphi can be boring and error-prone. Let's build the RegEx together and check it.

Start with six letters:

```
^[A-Z]{6}
```

Then, two numbers:

```
^[A-Z]{6}[0-9]{2}
```

Then, one letter and two numbers:

```
^[A-Z]{6}[0-9]{2}[A-Z][0-9]{2}
```

Then, one letter, three numbers, and one letter. Then, the code must terminate:

```
^[A-Z]{6}[0-9]{2}[A-Z][0-9]{2}[A-Z][0-9]{3}[A-Z]$
```

Now, the check is really simple:

```
procedure TRegExForm.btnCheckItalianTaxCodeClick(
                                    Sender: TObject);
begin
  if TRegEx.IsMatch(EditTaxCodeIT.Text,
        '^[A-Z]{6}[0-9]{2}[A-Z][0-9]{2}[A-Z][0-9]{3}[A-Z]$',
        [roIgnoreCase]) then
    ShowMessage('This italian tax code is valid')
else
    ShowMessage('This italian tax code is not valid');
end;
```

After some exercises, you will be able to master the RegEx syntax and you will find it really useful to check and manipulate strings and texts.

The following screenshot shows the sample application while it is checking an invalid email address:

Figure 3.2: The RegEx sample application while is checking an invalid email address

There's more...

RegEx can be used to perform many string-related tasks. You can match strings, search for strings in another string, split a string using a RegEx as a separator, and so on.

If you work with strings (who doesn't?), do yourself a favor and study regular expressions very well. Just as an example, suppose you have a string with a list of names separated by , or and. Something like: `Daniele, Bruce, Mark, and Scott`. Now, you only want to retrieve the names. How you can do this?

Here's the regular expression that does the job (in this case, it matches all valid delimiters that can appear between the names): `[ ]*,[ ]*|[ ]+and[ ]+`.

To do the actual split, we have to use the following code:

```
procedure MyCoolSplitter;
var
  LNames: TArray<string>;
  LInputString, LName, LRegEx: string;
begin
  //this is the input string, it is also badly formatted...
```

```
    LInputString := 'Daniele  , Bruce,  Mark     and   Scott';
    LRegEx := '[ ]*,[ ]*|[ ]+and[ ]+'; //regex to do the splitting
    LNames := TRegEx.Split(LInputString, LRegEx, []);
    for LName in LNames do
    begin
      ShowMessage(LName); //show each name
    end;
  end;
```

Remember to check the Delphi documentation about the built-in RegEx engine syntax at http://docwiki.embarcadero.com/RADStudio/en/Regular_Expressions.

Some nice RegEx samples (not Delphi-related) can be found at http://www.regular-expressions.info/examples.html.

As a bonus recipe, there is a RegEx tester called RegExTester.dproj in the attached code that helps you exploit all the functionalities. Play with it and become a RegEx ninja!

Consuming RESTful services using native HTTP(S) client libraries

We live in an interconnected world! A lot of applications nowadays have to exchange data with remote systems. Some of the most commonly used and powerful mechanisms used to define a communication interface between software over the internet are RESTful web services (more information about REST and RESTful interfaces will be provided in Chapter 6, *Putting Delphi on the Server*, in the recipe *Implementing a RESTful interface using WebBroker*). Usually, in Delphi, you can use the INDY suite to access HTTP servers. When dealing with HTTPS, INDY produces some headaches because it doesn't use the same SSL layer of the operating systems, but relies on OpenSSL libraries, so you have to provide a specific version of OpenSSL for each different OS your application supports. You cannot benefit from the security updates from OSes vendor. This was a *just-to-keep-in-mind* problem up until April 7, 2014, when the Heartbleed security bug was disclosed in the OpenSSL cryptography library.

Here's what Wikipedia has to say about it:

> *"At the time of disclosure, some 17% (around half a million) of the Internet's secure web servers certified by trusted authorities were believed to be vulnerable to the attack, allowing theft of the servers' private keys and users' session cookies and passwords."*

Client applications that use OpenSSL have been affected by this bug. The bug was fixed on the same day by the OpenSSL team, but the problem was still there for all deployed applications. Let's think about the problems that this situation produced! Think about your customer, who calls you because some security expert told him that there is a catastrophic bug in the OpenSSL system. Now, he wants an immediate update for all his systems! Then, there's a call from a second customer, then a third, and so on... Argh!

So, to overcome this bad situation, Embarcadero developed the native HTTP client library, which is not based on INDY or OpenSSL; it simply relies on the OS API to implement the HTTP protocol. So, when Microsoft, Apple, or Google releases a new security patch, your application is already updated. Great! You simply rely on the OS security infrastructure and don't depend on the OpenSSL DLLs anymore!

In this recipe, we'll see how to consume a RESTful interface provided by a sample server using the new native HTTP library that was introduced in RAD Studio XE8. The server is not HTTPS, but the same concepts apply.

 A new update is one its way with Delphi Tokyo 10.2.2: regarding support for security protocols (which is currently only implemented for Windows). It wasn't possible to specify the required security protocols, so a new property has been added—SecureProtocols. This property is based on the THTTPSecureProtocol enumeration, with the values SSL2, SSL3, TLS1, TLS11, and TLS12.

Getting ready

This is a showcase recipe. We'll see how to issue HTTP requests to a RESTful server using the new THTTPClient class. There is also the TNetHTTPClient component, which wraps the THTTPClient functionalities in a non-visual component (we don't need it now), and the raw THTTPClient class, which is a quite high level to be used directly. However, if you want to use the component, the interface is quite similar (almost identical).

Some HTTP considerations

While GET and POST are by far the most common methods that are used to access information provided by a web server, the **Hypertext Transfer Protocol** (**HTTP**) allows several other (and somewhat lesser-known) methods. The current standard is HTTP 1.1 and is defined by RFC 2616. RFC 2616 defines the following eight methods:

- HEAD
- GET
- POST
- PUT
- DELETE
- TRACE
- OPTIONS
- CONNECT

The THTTPClient class supports all of these verbs except for CONNECT, with specific methods that map each verb. To get an idea about the methods provided by this class, see the *There's more...* section.

How it works...

We will use a sample server called PeopleManagerServer.dproj, which is located in the Chapter03\RECIPE02\Server folder. Ensure that TCP port 8080 is available on your system and that your local InterBase instance is running. Run the server project. The server is not in the scope of this recipe, so we'll just use it as a black box; however, it handles CRUD and some kind of business logic and searches over a database table called PEOPLE, which is contained in a sample database called SAMPLES.IB. This server will be developed in Chapter 6, *Putting Delphi on the Server* in the recipe *Implementing a RESTful interface using WebBroker*, so if you are curious about its details, read that chapter and then come back here.

Now, open the client project located in Chapter03\RECIPE02\ClientPeopleManagerClient.dproj, which is the VCL client for the RESTful service that's provided by the PeopleManagerServer.

In the `FormCreate` event, the `FHTTPClient` is initialized, while in `FormDestroy`, it is destroyed:

```
procedure TMainForm.FormCreate(Sender: TObject);
begin
  FHTTPClient := THTTPClient.Create;
  pcMain.ActivePageIndex := 0;
end;

procedure TMainForm.FormDestroy(Sender: TObject);
begin
  FHTTPClient.Free;
end;
```

Note that, to use the `THTTPClient` class, you have to include `System.Net.HttpClient` (which contains the class definition itself) and the unit `System.Net.URLClient`, which holds common functionality that's relative to a generic URL Client (HTTP, FTP, and so on).

Run the client, go to the first tab, click on the **Open** button, and you should see something similar to the following screenshot:

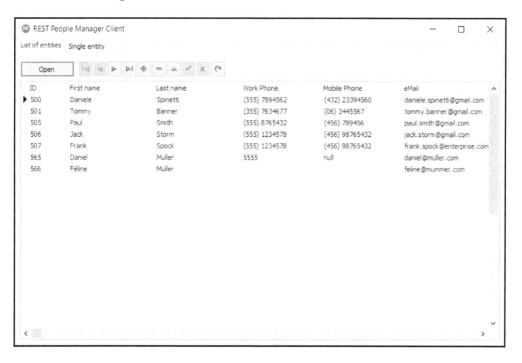

Figure 3.3: The VCL RESTful HTTP client for the PeopleManagerServer showing a list of people

So, let's look under the hood. Open the main form in the RAD Studio form designer and check the `btnOpenClick` event handler:

```
procedure TMainForm.btnOpenClick(Sender: TObject);
begin
  dsPeople.Close;
  dsPeople.BeforePost := nil;
  try
    dsPeople.Open;
    dsPeople.First;
  finally
    dsPeople.BeforePost := dsPeopleBeforePost;
  end;
end;
```

The `dsPeople` object is a `TFDMemTable` (an in-memory table from the FireDAC components suite provided with RAD Studio), so where does the data come from? Let's check the relevant properties of this `TFDMemTable`:

```
object dsPeople: TFDMemTable
  AfterOpen = dsPeopleAfterOpen
  BeforePost = dsPeopleBeforePost
  BeforeDelete = dsPeopleBeforeDelete
  object dsPeopleID: TStringField
    FieldName = 'ID'
    Size = 255
  end
  object dsPeopleFIRST_NAME: TStringField
    DisplayLabel = 'First name'
    DisplayWidth = 50
    FieldName = 'FIRST_NAME'
    Size = 255
  end
  object dsPeopleLAST_NAME: TStringField
    DisplayLabel = 'Last name'
    DisplayWidth = 50
    FieldName = 'LAST_NAME'
    Size = 255
  end
  object dsPeopleWORK_PHONE_NUMBER: TStringField
    DisplayLabel = 'Work Phone'
    DisplayWidth = 50
    FieldName = 'WORK_PHONE_NUMBER'
    Size = 255
  end
  object dsPeopleMOBILE_PHONE_NUMBER: TStringField
    DisplayLabel = 'Mobile Phone'
    FieldName = 'MOBILE_PHONE_NUMBER'
```

```
      Size = 50
    end
    object dsPeopleEMAIL: TStringField
      DisplayLabel = 'eMail'
      DisplayWidth = 50
      FieldName = 'EMAIL'
      Size = 255
    end
  end
end
```

As you can see, on `dsPeople` some event handlers and some persistent fields have been defined. The persistent fields are used to do the mapping between the JSON objects and the dataset structure (each field has a corresponding property in the JSON objects) and we'll talk about them in a moment. The event handlers allow us to use `MemTable` as though it was a normal database table, but it consumes the data from a RESTful web service instead of a normal database table. Let's see the code.

Just after the dataset is opened, in the `AfterOpen` event handler, we'll issue the request to the server and get the data. The JSON that's retrieved is converted and loaded into the dataset using the helpers of the `TDataSet` class. These helpers are provided by the unit `MVCFramework.DataSet.Utils.pas`, which is contained in the Open Source project `DelphiMVCFramework` (see `https://github.com/danieleteti/delphimvcframework`. More information regarding this is in `Chapter 6`, *Putting Delphi on the Server*, in the recipe *Implementing a RESTful interface using WebBroker,* so if you are curious about its details, you can read that chapter and then come back here):

```
procedure TMainForm.dsPeopleAfterOpen(DataSet: TDataSet);
var
  LResponse: IHTTPResponse;
begin
  // FHTTPClient is an instance of THTTPClient created in the
  // FormCreate event.
  // Here we are sending a GET request passing the URL, and one
  // ACCEPT header
  LResponse := FHTTPClient.Get(BASEURL + '/people', nil,
    [TNameValuePair.Create('accept', 'application/json')]);

  if LResponse.StatusCode = HTTP_STATUS.OK then
  begin
    // Load JSON data from the body request to the dataset using
    // the TDataSet class helpers provided by the open source
    // library DelphiMVCFramework project
    DataSet.AppendFromJSONArrayString(LResponse.ContentAsString);
  end
  else
  begin
```

```
    raise Exception.CreateFmt(ERROR_FORMAT_STRING,
      [LResponse.StatusCode, LResponse.StatusText]);
  end;
end;
```

Now, when we need to delete a record, just before deleting it we have to ask the server about the deletion. If some exceptions are raised in the server, the `delete` method also fails on the client and the record remains in the dataset:

```
procedure TMainForm.dsPeopleBeforeDelete(DataSet: TDataSet);
begin
  DeleteRecordOnServer(DataSet);
end;
```

The following is the code for the `DeleteRecordOnServer` method:

```
procedure TMainForm.DeleteRecordOnServer(ADataSet: TDataSet);
var
  LResponse: IHTTPResponse;
begin
  LResponse := FHTTPClient.Delete(BASEURL + '/people/' +
    ADataSet.FieldByName('ID').AsString);
  if LResponse.StatusCode <> HTTP_STATUS.NoContent then
    raise Exception.CreateFmt(ERROR_FORMAT_STRING,
      [LResponse.StatusCode, LResponse.StatusText]);
end;
```

When the user needs to update, or insert a new entity in, `BeforePost`, the event handler executes the following code:

```
procedure TMainForm.dsPeopleBeforePost(DataSet: TDataSet);
var
  LNewlyCreatedResourceURI: string;
begin
  case DataSet.State of
    dsInsert:
      begin
        LNewlyCreatedResourceURI := CreateRecordOnServer(DataSet);
        UpdateDataSetFromURL(LNewlyCreatedResourceURI, DataSet);
      end;
    dsEdit:
      UpdateRecordOnServer(DataSet);
  else
    raise Exception.Create('Invalid state');
  end;
end;
```

This is the code for the methods used in this event:

```
function TMainForm.CreateRecordOnServer(ADataSet:
  TDataSet): string;
var
  LPOSTRequest: IHTTPRequest;
  LResponse: IHTTPResponse;
  LBody: TStringStream;
begin
  LPOSTRequest := FHTTPClient.GetRequest('POST',
    BASEURL + '/people');
  LPOSTRequest.AddHeader('content-type', 'application/json');
  LBody := TStringStream.Create(ADataSet.asJSONObjectString,
TEncoding.UTF8);
  try
    LPOSTRequest.SourceStream := LBody;
    LResponse := FHTTPClient.Execute(LPOSTRequest);
  finally
    LBody.Free;
  end;
  if LResponse.StatusCode <> HTTP_STATUS.Created then
    raise Exception.CreateFmt(ERROR_FORMAT_STRING,
      [LResponse.StatusCode, LResponse.StatusText]);

  // the server returned the newly created resource
  // in the LOCATION header
  Result := LResponse.HeaderValue['location'];
end;

procedure TMainForm.UpdateDataSetFromURL(AURL: string;
  ADataSet: TDataSet);
var
  LResponse: IHTTPResponse;
begin
  LResponse := FHTTPClient.Get(BASEURL + AURL, nil,
    [TNameValuePair.Create('accept', 'application/json')]);
  if LResponse.StatusCode <> HTTP_STATUS.OK then
    raise Exception.CreateFmt(ERROR_FORMAT_STRING,
      [LResponse.StatusCode, LResponse.StatusText]);

  //load the JSON response body into the current record of the
  //dataset (using the class helpers from the DelphiMVCFramework)
  ADataSet.LoadFromJSONObjectString(LResponse.ContentAsString);
end;

procedure TMainForm.UpdateRecordOnServer(ADataSet: TDataSet);
var
  LPUTRequest: IHTTPRequest;
```

```
  LResponse: IHTTPResponse;
  LBody: TStringStream;
begin
  LPUTRequest := FHTTPClient.GetRequest('PUT',
    BASEURL + '/people/' +
    ADataSet.FieldByName('ID').AsString);
  LPUTRequest.AddHeader('content-type', 'application/json');
  LBody := TStringStream.Create(ADataSet.asJSONObjectString,
TEncoding.UTF8);
  try
    LPUTRequest.SourceStream := LBody;
    LResponse := FHTTPClient.Execute(LPUTRequest);
  finally
    LBody.Free;
  end;
  if LResponse.StatusCode <> HTTP_STATUS.OK then
    raise Exception.CreateFmt(ERROR_FORMAT_STRING,
      [LResponse.StatusCode, LResponse.StatusText]);
end;
```

As you can see, for simple HTTP requests, you can use shortcut methods such as the following:

```
//A simple GET request
LResp := FHTTPClient.Get('http://www.myserver.com/api/customers');

//A simple DELETE request
LResp :=
  FHTTPClient.Delete('http://www.myserver.com/api/customers/1');
```

There are methods that map each HTTP VERB defined in the HTTP 1.1 protocol (see the *There's more...* section for more detail).

If you need to send a more complex request, or if you want to prepare your request and execute it later, you can use the GetRequest method, as shown in the CreateRecordOnServer method. The following code is a smaller example of the GetRequest use case:

```
procedure TMyForm.MyComplexRequest;
var
  LPOSTRequest: IHTTPRequest;
  LResponse: IHTTPResponse;
  LBody: TStringStream;
begin
  //get the POST request
  LPOSTRequest := FHTTPClient.GetRequest(
    'POST', 'http://localhost/people');
```

```
//now we can customize the request with headers...
LPOSTRequest.AddHeader('content-type', 'application/json');
LPOSTRequest.AddHeader('accept', 'application/json');

//...and a request body
LBody := TStringStream
  .Create('{"firstname":"Daniele","lastname":"Teti"');
try
  LPOSTRequest.SourceStream := LBody;
  //now, execute the request
  LResponse := FHTTPClient.Execute(LPOSTRequest);
finally
  LBody.Free;
end;
//after the request, we can use the response object (which is an
//interface instance, so it is reference counted)
if LResponse.StatusCode <> 201 then
  raise Exception.Create(
    'Invalid response: ' + LResponse.StatusText);
end;
```

Let's go back to the recipe application. The second tab of TPageControl contains all the controls used to do a simple search by ID on the server. Let's write a valid ID in the **EditSearch** button; edit and click on the **Get by ID** button. The code under this button is quite simple to understand now:

```
procedure TMainForm.btnGetPersonClick(Sender: TObject);
var
  LResponse: IHTTPResponse;
begin
  dsPerson.Close;
  LResponse := FHTTPClient.Get(BASEURL + '/people/' +
    TNetEncoding.URL.Encode(EditSearch.Text));
  if LResponse.StatusCode = HTTP_STATUS.OK then
  begin
    //check if the response is in a supported format (JSON)
    if LResponse
        .HeaderValue['Content-Type']
        .StartsWith('application/json') then
    begin
      dsPerson.Open;
      //unhook the BeforePost event handler because the
      //LoadFromJSONObjectString call could fire it.
      dsPerson.BeforePost := nil;
      dsPerson.Insert;
      //Load the JSON string containing a JSON object into the
      //dataset which has the same structure
      dsPerson.LoadFromJSONObjectString(LResponse.ContentAsString);
```

```
        //hook the event
        dsPerson.BeforePost := dsPersonBeforePost;
    end
    else
    begin
        //the response content-type is not supported
        ShowMessageFmt('Invalid response format ' +
          '(expected application/json, actual %s',
          [LResponse.HeaderValue['Content-Type']]);
    end;
  end
  else
  begin
    ShowMessageFmt(ERROR_FORMAT_STRING,
      [LResponse.StatusCode, LResponse.StatusText]);
  end;
end;
```

Clicking on the button, you should see that the data has been loaded into `TDBEdits`. Now, you can change the data in the controls and click on the **post** button in the small `TDBNavigator` at the bottom. In the `BeforePost` event of the hooked dataset, called `dsPerson`, there is this code:

```
procedure TMainForm.dsPersonBeforePost(DataSet: TDataSet);
begin
  // only updates allowed here
  if DataSet.State <> dsEdit then
    raise Exception.Create('Invalid dataset state');
  //here we can use the same method used for
  //the dsPeople dataset
  UpdateRecordOnServer(DataSet);
end;
```

There's more...

There's a lot of stuff in this recipe! Let's give you some more detail on the proposed code.

In the source code, you will have seen some reference to a strange `HTTP_STATUS` complex variable. It seems strange, but in Delphi `System.Net.*` there is no list of valid HTTP status codes as a list of constants, so I added it to the `MVCFramework.Commons.pas` unit (these are all of the most frequently used codes, as defined in `http://www.w3.org/Protocols/rfc2616/rfc2616-sec10.html`). Therefore, you can write `HTTP_STATUS.OK` or `HTTP_STATUS.NOTFOUND` without remembering the status code. This is simple but very useful. You can also use it if you want to use other HTTP frameworks.

THTTPClient's methods which directly map HTTP verbs

This is a small recap of the `THTTPClient` method that maps to HTTP standard verbs. As you can see, all the methods return an instance of `IHTTPResponse`, which is a reference counted interface reference.

Send the `Delete` command to the URL:

```
function Delete(AURL: string; AResponseContent: TStream = nil;
   AHeaders: TNetHeaders): IHTTPResponse;
```

Send the `Options` command to the URL:

```
function Options(const AURL: string; const AResponseContent:
   TStream = nil; const AHeaders: TNetHeaders = nil):
   IHTTPResponse;
```

Send the `Get` command to the URL:

```
function Get(AURL: string; AResponseContent: TStream = nil;
   AHeaders: TNetHeaders = nil): IHTTPResponse;
```

`CheckDownloadResume` checks whether the server has the download resume feature. This is not a one-to-one mapping to the standard HTTP verb, but this method relies on the `HEAD` verb with a very small value for the `RANGE` header. If the server responds with the data, then it supports the download resume. So, this is a handy method to use in the case of a large download to check whether the download can be resumed (or split):

```
function CheckDownloadResume(AURL: string): Boolean;
```

Just like the `CheckDownloadResume`, this is not a standard HTTP verb, but it's a handy method that sends the `GET` command to the URL, adding the `RANGE` header, which is used to get a part of the remote resource. It is used to resume interrupted downloads or to split large downloads:

```
function GetRange(AURL: string; AStart: Int64; AnEnd: Int64 = -1;
  AResponseContent: TStream = nil;
  AHeaders: TNetHeaders = nil): IHTTPResponse;
```

Send the `TRACE` command to the URL. The HTTP `TRACE` method returns the contents of client HTTP requests in the entity-body of the `TRACE` response. Please note that attackers could leverage this behavior to access sensitive information, such as cookies or authentication data, contained in the HTTP headers of the request (there is more information about the `TRACE` vulnerability in the next section):

```
function Trace(AURL: string; AResponseContent: TStream = nil;
  AHeaders: TNetHeaders = nil): IHTTPResponse;
```

Send the `HEAD` command to the URL:

```
function Head(AURL: string; AHeaders: TNetHeaders = nil):
  IHTTPResponse;
```

Post a raw file without multipart information:

```
function Post(AURL: string; ASourceFile: string;
  AResponseContent: TStream = nil; AHeaders: TNetHeaders = nil):
  IHTTPResponse; overload;
```

Post `TStrings` values, adding multipart information:

```
function Post(AURL: string; ASource: TStrings;
  AResponseContent: TStream = nil; AEncoding: TEncoding = nil;
  AHeaders: TNetHeaders = nil): IHTTPResponse; overload;
```

Post a stream without multipart information:

```
function Post(AURL: string; ASource: TStream;
  AResponseContent: TStream = nil; AHeaders: TNetHeaders = nil):
  IHTTPResponse; overload;
```

Post a multipart form data object. This is used to mimic an HTML `FORM` submit:

```
function Post(AURL: string; ASource: TMultipartFormData;
  AResponseContent: TStream = nil; AHeaders: TNetHeaders = nil):
  IHTTPResponse; overload;
```

Send the PUT command to the URL:

```
function Put(AURL: string; ASource: TStream = nil;
  AResponseContent: TStream = nil; AHeaders: TNetHeaders = nil):
  IHTTPResponse;
```

The THTTPClient class also supports standard HTTP verbs—PATCH and MERGE.

How to verify that HTTP TRACE is disabled

The TRACE command is a standard HTTP verb that is defined in the HTTP specification and was considered a safe command up until some years ago. However, due to information that was disclosed, combined with other cross-domain exploits, TRACE is no longer considered safe. See http://www.kb.cert.org/vuls/id/867593 for more information.

So, how can I verify that the TRACE command is disabled? The easiest way to do this is to use telnet. The following is a step-by-step guide to testing whether TRACE is enabled on your webserver:

1. Launch telnet with telnet myserver myport, for example:

   telnet localhost 80

2. Now, we can issue the TRACE command for a given URL, for example:

   TRACE /index.html HTTP/1.0

3. If you don't see any character on the console while writing, don't worry; it is telnet LOCALECHO, that's disabled. It will work.

4. If TRACE is enabled, you will get an output that looks something like this:

   ```
   HTTP/1.1 200 OK
   Date: Thu, 27 Aug 2015 12:41:14 GMT
   Server: Apache/2.4.12 (Win32)
   Connection: close
   Content-Type: message/http

   TRACE /index.html HTTP/1.0

   Connection closed by foreign host.
   ```

5. If TRACE is disabled, the output will look like this:

```
HTTP/1.1 405 Method Not Allowed
Date: Thu, 27 Aug 2015 12:42:44 GMT
Server: Apache/2.4.12 (Win32)
<some others http header>

Connection closed by foreign host.
```

Coping with the encoded internet world using System.NetEncodings

The internet is the land of encodings! URIs are encoded, HTML provides specific encodings, emails work because of MIME Encoding, and the REST service works because, in some way, the client and server can talk to each other using some sort of encoding! There are many kinds of encoding for different purposes, but in this recipe we will talk about encodings that are handled by the classes contained in the System.NetEncodings.pas unit.

Getting ready

The System.NetEncodings unit contains the following classes:

- TNetEncoding: This class is a factory for the actual encoding classes; moreover, it serves as a base class for all other classes
- THTMLEncoding: This class provides methods to encode and decode data in HTML format
- TURLEncoding: This class provides methods to encode and decode data in URL encoding
- TBase64Encoding: This class provides methods to encode and decode data in the base64 format

The base64 format is a binary-to-text encoding schema that represents binary data in an ASCII string format by translating it into a radix-64 representation. This kind of encoding is commonly used when there is a need to encode binary data that needs to be stored and transferred over media designed to deal with textual data. This kind of encoding mechanism ensures that the data remains intact without modification during transport. base64 is commonly used in a number of applications, including REST services, email via MIME, in-line data in web pages, and to store complex data in XML or JSON.

Here's an example of base64 encoding:

```
Input text: this is a text
Encoded text: dGhpcyBpcyBhIHRleHQ=
```

HTML is the markup language of the web, and there are different ways to instruct the browser about the character encoding of the document that must be displayed. The web server and the document itself can contain information about native encodings such as ASCII, ISO8859-1 or the most popular, which is UTF-8. These kinds of encoding are called native encoding or charset (handled in Delphi by the `TEncoding` class). In addition to these native character encodings, characters can also be encoded as character references, which can be numeric character references (decimal or hexadecimal) or character entity references.

Escaping also allows for characters that are not easily typed, or that are not available in the document's character encoding, to be represented within the element and attribute content. For example, the acute-accented e (é), a character typically found only on Western European and South American keyboards, can be written in any HTML document as the entity reference `&eacute;` or as the numeric references `&#xE9;` or é, using characters that are available on all keyboards and are supported in all character encodings.

Here's an example of HTML encoding:

```
Input text: Italian word for "why" is "perché"
Encoded text: Italian word for &quot;why&quot; is &quot;perché&quot;
```

Not all the characters can be used in URLs. This is a big problem when you need to pass non-ASCII text as a parameter (for example, `http://www.myserver.com/page.php?q=t/h/i/s` doesn't work). URL encoding is used to encode the parameters passed over to the URL and make them suitable to be sent in such a way. Here are some examples of URL encoding:

- Parameters that are not encoded in the URL:
  `http://localhost/index?expression=3*4/5`
- URL with correctly encoded parameter:
  `http://localhost/index?expression=3%2A4%2F5.`

In this recipe, we'll download a PNG image encoded as base64 text, and then we will reconstruct the binary stream on the client; the binary's contents will be shown on the form in a `TImage` control. The base64 representation is read by our beloved sample REST server that was introduced in the previous recipe, so you can open the project called `PeopleManagerServer.dproj`, which is located in the folder `Chapter03\RECIPE02\Server`. Ensure that TCP port `8080` is available on your system and that your local InterBase instance is running. Run the server project. This server provides a resource that returns a base64 fake photo of the person. So, if the following URL returns the person's information encoded as JSON `GET http://localhost:8080/people/500`, then the following URL returns the person's fake photo (just a PNG file where the person's name has been drawn on a gradient background): `GET http://localhost:8080/people/500/photo`. The photo resource is base64 encoded text. Try this for yourself with different IDs to see what the response looks like.

How it works...

Open the project at `Chapter03\RECIPE03\NetEncoding.dproj` and launch it. In the edit section on the left, write `500` (or another valid ID in your database). Now, the form should look like the following:

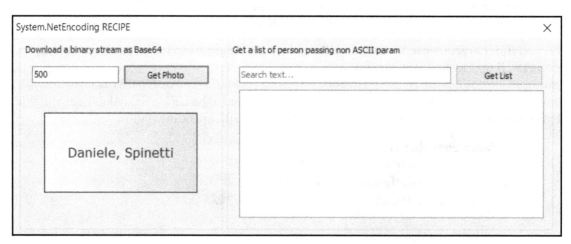

Figure 3.4: The image has been downloaded as base64 text from a REST service
and then decoded and loaded in the TImage control

Stop the program and see the code under the **Get Photo** button:

```
procedure TMainForm.btnGetPhotoClick(Sender: TObject);
var
  LHTTP: THTTPClient;
  LResponse: IHTTPResponse;
  LPNGStream: TMemoryStream;
  LPNGImage: TPngImage;
  LURLFormat: string;
begin
  LHTTP := THTTPClient.Create;
  try
    //create the correct URL
    LURLFormat := 'http://localhost:8080/people/%s/photo';
    LResponse := LHTTP.Get(
      Format(LURLFormat,
        [EditPersonID.Text]));
    //check for a valid response
    if LResponse.StatusCode <> 200 then
    begin
      ShowMessage(LResponse.StatusText);
      Exit;
    end;
    LPNGStream := TMemoryStream.Create;
    try
      //convert response string from base64 to a binary stream
      if TNetEncoding.Base64
          .Decode(LResponse.ContentStream, LPNGStream) = 0 then
        raise Exception.Create('Invalid Base64 stream');
      LPNGImage := TPngImage.Create;
      try
        LPNGStream.Position := 0;
        //load the binary stream into the PNGImage...
        LPNGImage.LoadFromStream(LPNGStream);
        //...and assign it to the TImage control
        Image1.Picture.Assign(LPNGImage);
      finally
        LPNGImage.Free;
      end;
    finally
      LPNGStream.Free;
    end;
  finally
    LHTTP.Free;
  end;
end;
```

There's more...

Coping with encodings is not simple, but in the Unicode and distributed world we live in now, we cannot avoid it, so it's worth correctly understanding how (at least) the most popular encoding works and where they can (or must) be used.

Another interesting application of `TNetEncoding.Base64` is related to the Data URI scheme (`https://en.wikipedia.org/wiki/Data_URI_scheme`). Using base64 and the Data URI scheme, you can embed images, scripts, and CSS directly into your HTML. This technique should be used carefully because, if the HTML changes but the image doesn't, using embedded images means that the browser cannot properly use the client cache and thus the pages download time increases.

Also, if you generate HTML content, you should encode the content using `TNetEncoding.HTML` to be sure that all browsers can correctly display complex characters or symbols. This also protects your site from XSS attacks (`https://en.wikipedia.org/wiki/Cross-site_scripting#Safely_validating_untrusted _HTML_input`).

Saving space using System.Zip

Historically, Delphi contains `TZCompressionStream` and `TZDecompressionStream` to respectively compress and decompress streams of bytes using the `zlib` format. These classes are quite useful but are quite low-level, since it is simply a stream compressor. In this recipe, we'll use a high-level class to compress and decompress folders and files. It is quite limited in terms of possibilities (you can compress and decompress files and folders, nothing more), but it's very simple to use. Just keep in mind that this class is very specialized, so if you need a compression library to work with network protocols or on the fly with compression/decompression, don't use this. But if you need a no-brain solution to compress something, this is the way to do it.

The ZIP file format doesn't need presentation. However, a recap could be useful:

ZIP is a free lossless data compression format, supports various compression algorithms, one of which is based on a variant of the LZW algorithm (DEFLATE is the most common). Designed in 1989 by Phil Katz for PKZIP, as an alternative to the ARC compression format by Thom Henderson. A `.zip` file may contain one or more files or folders; each file is compressed separately, which allows you to quickly extract individual files (sometimes even partially damaged files) to the detriment of the overall compression. A `.zip` file using the file extensions `.zip`, is recognized by the PK header (ASCII encoding) and the MIME media type `application/zip`.

In this recipe, we'll see a simple Zipper/UnZipper tool which is implemented using the `TZipFile` class, declared in the `System.Zip.pas` unit.

How it works...

Open the project at `Chapter03\RECIPE04\ZipUnZip\ZipUnZip.dproj` and run it. The GUI is quite simple; just click on the button and see the results:

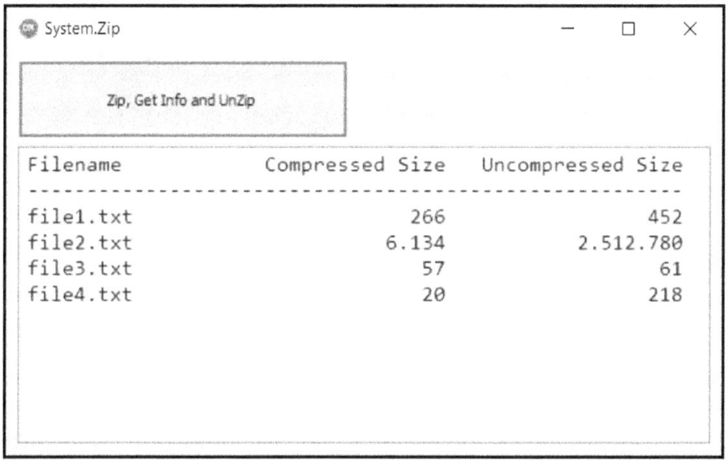

Figure 3.6: The Zip tool

This little program is a showcase for the `TZipFile` class. In the recipe folder, there is also a test folder called `FolderToZip` with four files inside it. These files are zipped, inspected, and unzipped by the code under the button. Let's see it:

```
procedure TMainForm.btnZipUnZipClick(Sender: TObject);
var
  ZF: TZipFile;
  I: Integer;
begin
  MemoSummary.Clear;
  // this single statement zip recursively a folder in a zip file
  TZipFile.ZipDirectoryContents('MyFolder.zip', 'FolderToZip');

  // let's inspect the content of the zip file...
  ZF := TZipFile.Create;
  try
    // open the zip file to read information
    ZF.Open('MyFolder.zip', TZipMode.zmRead);
```

```
    // create table headers...
    MemoSummary.Lines.Add('Filename'.PadRight(15) + 'Compressed
Size'.PadLeft
        (20) + 'Uncompressed Size'.PadLeft(20));
    MemoSummary.Lines.Add(''.PadRight(55, '-'));

    // loop through the compressed file and extract information
    // about name, original size and compressed size
    for I := 0 to ZF.FileCount - 1 do
    begin
MemoSummary.Lines.Add(TEncoding.Default.GetString(ZF.FileInfo[I].FileName)
        .PadRight(15) + FormatFloat('###,###,##0',
        ZF.FileInfo[I].CompressedSize).PadLeft(20) +
FormatFloat('###,###,##0',
        ZF.FileInfo[I].UncompressedSize).PadLeft(20));
    end;
  finally
    ZF.Free;
  end;
  // now actually uncompress the file into a folder
  TZipFile.ExtractZipFile('MyFolder.zip', 'UnzippedFolder');
end;
```

Yes, it is quite simple. Obviously, if you don't need to inspect the ZIP content, all we need is as follows:

```
procedure TMainForm.btnZipUnZipClick(Sender: TObject);
begin
  TZipFile.ZipDirectoryContents('MyFolder.zip', 'FolderToZip');
  TZipFile.ExtractZipFile('MyFolder.zip', 'UnzippedFolder');
end;
```

I told you; it is not very flexible, but it is really simple to use. So, if you need to decompress an update file for your program and you only need decompression, you can use the `TZipFile` class; thus, you don't need to use external components anymore.

There's more...

There are a lot of compression libraries available for Delphi, but before including a third-party dependency in your project, it's better to check whether the RTL has something usable for our job.

For example, sometimes we need to compress our data, either to reduce network traffic when transmitting it or just to save space in storage media.

As explained in the previous chapters, Delphi wraps the zlib library into two TStream descendant classes: TZCompressionStream and TZDecompressionStream. To use them, you need to include the System.Zlib.pas unit in your use list. In the recipe folder, there is a bonus project that shows you how to compress and decompress generic streams (which can be a file stream, a memory stream, or anything that can be wrapped as a stream).

Here's the relevant code:

```
procedure TMainForm.Compress(const ASrc, ADest: TStream);
var
  LCompressor: TZCompressionStream;
begin
  LCompressor := TZCompressionStream.Create(ADest);
  try
    LCompressor.CopyFrom(ASrc, 0);
  finally
    LCompressor.Free;
  end;
end;


procedure TMainForm.Decompress(const ASrc, ADest: TStream);
var
  LDecompressor: TZDecompressionStream;
begin
  LDecompressor := TZDecompressionStream.Create(ASrc);
  try
    ADest.CopyFrom(LDecompressor, 0);
  finally
    LDecompressor.Free;
  end;
end;
```

Decoupling your code using a cross-platform publish/subscribe mechanism

The publish/subscribe pattern, also known as the Observer pattern, is a very popular design pattern. It comes under a lot of different names, but the final scope is always the same alert when something interesting happens to it. In this recipe, we'll see a utilization of the `TMessageManager` class, which is the publish/subscribe mechanism implemented in the `System.Messaging.pas` unit.

Getting ready...

What exactly does the cross-platform `TMessageManager` class do? Put simply, it allows you to listen for events and assign actions to run when those events occur. Just like in VCL or FireMonkey, you know about mouse and keyboard events that occur on certain user interactions. These are very similar, except that we can emit events (or send messages) on our own, when we want to, and not necessarily based on user interaction or other mechanisms inside other components. `TMessageManager` is based on the publish/subscribe model because we can subscribe to a particular type of message and then publish it.

The other important question is this—*Why would you use the event model?* You already know that, in Delphi, the event model allows you to decouple built-in UI code (the **TButton** code, for example) from the business code (your own code), but another benefit to events is that they are a very loose way of coupling parts of your code together. An event can be emitted, but if no code is listening for it, that's okay; it will just be passed unnoticed. This means removing listeners (or event emissions) never results in compile or runtime errors.

`TMessageManager` is the class in charge of the application's message handling, and it manages message dispatching. Its `DefaultManager` property returns an object that acts as an application-wide notification center and is widely used in mobile development to listen for system-generated events (such as `OrientationChanged` and so on). You can call `TMessageManager.DefaultManager` to access the singleton instance of `TMessageManager`. However, in many cases, you will be happy to know that it is possible to create many instances of `TMessageManager`. Once you have an instance of `TMessageManager` (retrieved by the default instance as a singleton or a new instance that's created with `TMessageManager.Create`), you can call `TMessageManager.SubscribeToMessage` to subscribe message-handling methods to specific types of message. The events hooked to the subscription may be methods of an object or anonymous methods.

After you subscribe a method to a type of message, every time there is a call to
TMessageManager.SendMessage with a message of the target type, the subscribed
methods are called. A simple interaction with one publisher and two subscribers is
implemented in the following code:

```
program HelloMessaging;

{$APPTYPE CONSOLE}
{$R *.res}

uses
  System.SysUtils,
  System.Messaging;

begin

  // subscribe to a String message on the default message manager
  TMessageManager.DefaultManager.SubscribeToMessage(TMessage<String>,
    procedure(const Sender: TObject; const AMessage: TMessage)
    begin
      WriteLn('Called callback1 with value: ',
        TMessage<String>(AMessage).Value);
    end);

  // subscribe to a String message on the default message manager
  TMessageManager.DefaultManager.SubscribeToMessage(TMessage<String>,
    procedure(const Sender: TObject; const AMessage: TMessage)
    begin
      WriteLn('Called callback2 with value: ',
        TMessage<String>(AMessage).Value);
    end);

  // send a String message to the default message manager
  WriteLn('Let''s send a message to the subscribers...');
  TMessageManager.DefaultManager.SendMessage(nil,
    TMessage<String>.Create('Hello Messaging'));

  Readln; // wait for a return...

end.
```

With this in mind, let's say that our customers ask for an application with a floating tool window showing information about what's happening in the main form. For the sake of simplicity, let's say that the secondary forms must show the text entered in the main form memo. A first naïve approach may imply a deep connection between the first form and all the secondary form instances so that, when the memo changes, the main form knows all the secondary forms available and updates the content of the **TLabel**. This approach is really wrong—we are coupling the main form with all other forms, and it is also difficult to maintain. Using the `TMessageManager`, instead, all the forms are decoupled from the main form, and so the code is easily to maintain and new features can be simply added.

Let's see the logic's schema:

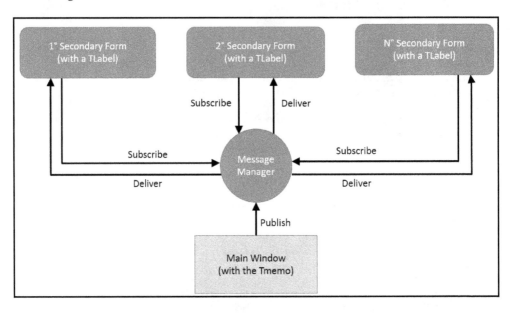

Figure 3.7: The message manager nicely decouples the components of our system

How it works...

Open the recipe project named `Chapter03\RECIPE05\Messaging.dproj`. The main form is quite simple. There is a memo where we'll write text and a button that will open a secondary form. The code under these two controls is as follows:

```
uses
    MyMessageManagerU, SecondaryFormU;

procedure TMainForm.btnOpenClick(Sender: TObject);
```

```
begin
  TSecondaryForm.Create(Application).Show;
end;

procedure TMainForm.memTextChange(Sender: TObject);
begin
  MessageManager.SendMessage(Self,
    TStringMessage.Create(memText.Lines.Text));
end;
```

The `btnOpenClick` event handler simply creates and shows a new `TSecondaryForm` instance. The `memTextChange` event handler sends a message of type `TStringMessage` (a type defined in the `MyMessageManager` unit) to the private `TMessageManager` instance. Where is this instance created? Open the `MyMessageManagerU.pas` unit:

```
unit MyMessageManagerU;

interface

uses
  System.Messaging;

type
  //Define a simple message derived from TMessage<String>.
  //Now it is useful only to make the code more readable, but in
  //the future this class can be used to create more complex
  //messages without change the methods interface
  TStringMessage = class(TMessage<String>)
  end;

//the factory to get the message manager
function MessageManager: TMessageManager;

implementation

var
  //private variable to hold the reference to the message manager
  LMessageManager: TMessageManager = nil;

function MessageManager: TMessageManager;
begin
  if not Assigned(LMessageManager) then
  begin
    LMessageManager := TMessageManager.Create;
  end;
  Result := LMessageManager;
end;
```

```
initialization

finalization

//free the message manager at the program termination
LMessageManager.Free;

end.
```

With this simple unit, we can simply write the following:

```
MessageManager.SubscribeToMessage(...
```

We can use this to subscribe to this specialized message manager to send messages:

```
MessageManager.SendMessage(...
```

It's simple and powerful.

Let's see the consumer of these messages in the following TSecondaryForm code:

```
implementation

uses
  MyMessageManagerU, System.Messaging;

procedure TSecondaryForm.btnOpenFormClick(Sender: TObject);
begin
  //just like in the main form, also here we can open other
  //secondary forms
  TSecondaryForm.Create(Application).Show;
end;

procedure TSecondaryForm.FormCreate(Sender: TObject);
begin
  //At the FormCreate we've to register on the MessageManager to
  //be sure that the TStringMessage will be
  //delivered to this instance too.
  FRegID := MessageManager.SubscribeToMessage(TStringMessage,
    procedure(const Sender: TObject; const AMessage: TMessage)
    begin
      lblText.Caption := TStringMessage(AMessage).Value;
    end);
end;

procedure TSecondaryForm.FormClose(Sender: TObject;
  var Action: TCloseAction);
begin
```

```
//at the form close, we have to UnSubscribe
MessageManager.Unsubscribe(TStringMessage, FRegID, False);
Action := TCloseAction.caFree;
end;
```

Now, run the project, click four times on the main form's button, and rearrange the windows so that you can see all of them. Write some text in the memo and you should see something similar to the following screenshot:

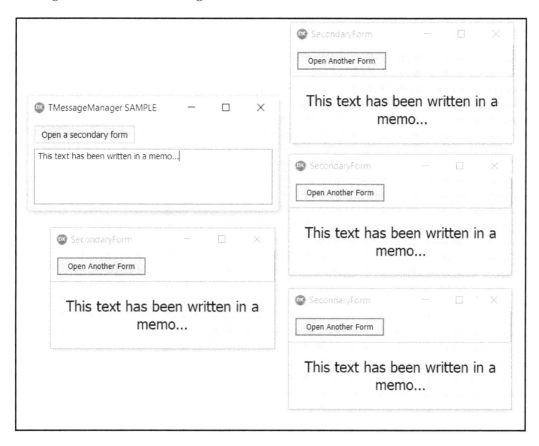

Figure 3.8: The message is delivered to all the subscribers

There's more...

Messaging is a really large topic. The same concepts we saw in this recipe can be applied also to distributed systems using the *Enterprise Message Bus* (see `https://en.wikipedia.org/wiki/Enterprise_service_bus`).

Also, within our program, messaging is a great way to decouple our classes and to make software easier to maintain and improve. Another real-world use case for object types is within the boundaries of a class. Remember that publishing a message is a synchronous process. Your call to `SendMessage` will not return until each subscriber's `MessageListener` code is run in turn. Therefore, your publisher will take longer for the `SendMessage` call to return. Regarding multithreading, consider that `TMessageManager` is not usable in a multithreading environment as-is. If you want to create a more complex multi-threaded messaging system, you have to manually handle the synchronization between threads.

Going Cross-Platform with FireMonkey

4

In this chapter, we will cover the following recipes:

- Giving a new appearance to the standard FireMonkey controls using styles
- Creating a styled TListBox
- Impressing your clients with animations
- Using master/details with LiveBindings
- Showing complex vector shapes using paths
- Using FireMonkey in a VCL application
- Reinventing your GUI, also known as mastering FireMonkey controls, shapes, and effects

Introduction

The FireMonkey framework is the app development and runtime platform behind Delphi and C++Builder. FireMonkey was introduced in these products with version XE2 (September 2011) and is the first native GPU-powered application platform. The IT world is becoming more multiplatform with each passing year. FireMonkey is a key technology for Embarcadero because it is designed to build multidevice, truly native apps for Windows, Mac, Android, and iOS.

This chapter explains some of the great features of FireMonkey. These recipes are applicable to the latest RAD Studio versions. FireMonkey is relatively young compared to VCL, so if you have an older version of RAD Studio, some things may not work as expected but the fundamental things are still valid. What is exposed in these recipes will be useful on every platform that's supported by the framework. Some OS-related features may not be available everywhere, but most of the concepts are usable on Microsoft Windows, Mac OS X, Android, and iOS. These are ready-to-use recipes that will be useful in everyday development.

Giving a new appearance to the standard FireMonkey controls using styles

Since Version XE2, RAD Studio has included FireMonkey. FireMonkey is an amazing library. It is a really ambitious target for Embarcadero, but it's important due to its mid and long-term strategy. VCL is and will remain a Windows-only library, while FireMonkey has been designed to be completely OS and device independent. You can develop one application and compile it anywhere (if anywhere is in Windows, OS X, Android, and iOS; let's say that it's a good part of anywhere).

One of the main features of FireMonkey is customization through styles. A styled component doesn't know how it will be rendered on the screen, because the style is in charge of it. By changing the style, you can change any aspect of the component without changing its code. The relation between the component code and style is similar to the relationship between HTML and CSS; one is the content and another is the display. In terms of FireMonkey, the component code contains the actual functionalities that the component has, but the aspect is completely handled by the associated style. All the `TStyledControl` child classes support styles.

Getting ready

Let's say you have to create an application to find a holiday house for a travel agency. Your customer wants a nice-looking application to search for a dream house for their customers. Your graphic design department (if present) decided to create a watermark effect, as shown in the following screenshot, and you have to create such an interface. How do we do that? Let's have a look:

Figure 4.1: This is the UI we want

How to do it...

In this case, we require some step-by-step instructions, so here they are:

1. Create a new FireMonkey desktop application (navigate to **File | New | Multi-Device Application**).

2. Drop a **TImage** component on the form. Set its **Align** property to **Client**, and use the **MultiResBitmap** property and its property editor to load a nice-looking picture.

3. Set the **WrapMode** property to **Fit** and resize the form to let the image cover the entire form.

4. Now drop a **TEdit** component and a **TListBox** component over the **TImage** component. Name the **TEdit** component `EditSearch` and the **TListBox** component `ListBoxHouses`.

5. Set the `Scale` property of the **TEdit** and **TListBox** components to the following values:
    - **Scale.X**: 2
    - **Scale.Y**: 2

6. Your form should now look like the following:

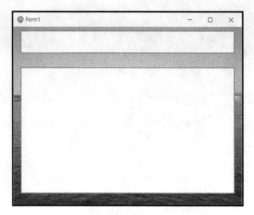

Figure 4.2: The form with standard components

The actions to be performed by the users are very simple. They should write some search criteria in the **Edit** field and click on **Return**. Then, the listbox will show all the houses available for that criteria (with a contains search). In a real application, you require a database or a web service to query, but this is a sample so you'll use fake search criteria on fake data.

7. Add the `RandomUtilsU.pas` file from the `Commons` folder of the project and add it to the uses clause in the main form.

8. Create an **OnKeyUp** event handler for the **TEdit** component and write the following code inside it:

```
procedure TForm1.EditSearchKeyUp(Sender: TObject;
    var Key: Word; var KeyChar: Char; Shift: TShiftState);
var
  I: Integer;
  House: string;
  SearchText: string;
begin
  if Key <> vkReturn then
    Exit;

  // this is a fake search...
  ListBoxHouses.Clear;
  SearchText := EditSearch.Text.ToUpper;
  //now, gets 50 random houses and match the criteria
  for I := 1 to 50 do
  begin
```

```
House := GetRndHouse;
if House.ToUpper.Contains(SearchText) then
   ListBoxHouses.Items.Add(House);
end;
if ListBoxHouses.Count > 0 then
   ListBoxHouses.ItemIndex := 0
else
   ListBoxHouses.Items.Add('<Sorry, no houses found>');
ListBoxHouses.SetFocus;
end;
```

9. Run the application and try to familiarize yourself with its behavior.

10. Now, you have a working application, but you still need to make it transparent. Let's start with the **FireMonkey Style Designer** (**FSD**).

 Up until XE8, the FSD was probably the least usable part of the RAD Studio IDE. However, after a little improvement in RAD Studio 10 Seattle, in RAD Studio 10.1 Berlin, the designer was completely redesigned and is now much better.

11. Right-click on the **TEdit** component. From the contextual menu, choose **Edit Custom Style** (general information about styles and the style editor can be found at http://docwiki.embarcadero.com/RADStudio/en/FireMonkey_Style_Designer and http://docwiki.embarcadero.com/RADStudio/en/Editing_a_FireMonkey_Style).

12. Delphi opens a new tab that contains the FSD. However, to work with it, you need the **Structure** pane to be visible as well (navigate to **View** | **Structure** or press *Shift + Alt + F11*).

13. In the **Structure** pane, all of the styles used by the **TEdit** control are available. You should see a **Structure** pane that's similar to the following screenshot:

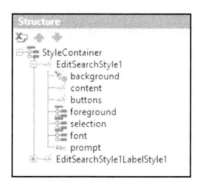

Figure 4.3: The Structure pane showing the default style for the TEdit control in RAD Studio 10.1 Berlin

14. In the **Structure** pane, open the `EditSearchStyle1` node, select the background subnode, and go to the **Object Inspector**.

15. In the **Object Inspector** window, remove the content of the **SourceLookup** property.

 The background part of the style is **TActiveStyleObject**. A **TActiveStyleObject** style is able to show a part of an image by default and another part of the same image when the component that uses it is active, checked, focused, mouse-hovered, pressed, or selected. The image to be used is in the **SourceLookup** property. Our **TEdit** component must be completely transparent in every state, so we removed the value of the **SourceLookup** property.

16. Now, the **TEdit** component is completely invisible. Click on **Apply and Close** and run the application. As you can confirm, the edit works, but it is completely transparent. Close the application.

17. When you opened the FSD for the first time, a **TStyleBook** component was automatically dropped on the form, and this contains all your custom styles. Double-click on it and the style designer will open again.

18. The edit, as you saw, is transparent, but it is not usable at all. You need to at least see least where to click and write. Let's add a small bottom line to the edit style, just like a small underline.

19. To perform the next step, you need the **Tool Palette** window and the **Structure** pane to be visible. Here is my preferred setup for this situation:

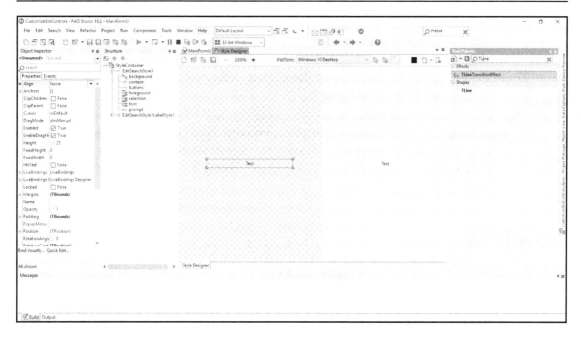

Figure 4.4: The Structure pane and the Tool Palette window are visible at the same time using the docking mechanism; you can also use floating windows if you wish

20. Now, search for a **TLine** component in the **Tool Palette** window. Drag-and-drop the **TLine** component onto the `EditSearchStyle1` node in the **Structure** pane. Yes, you have to drop a component from the **Tool Palette** window directly onto the **Structure** pane.

21. Now, select the **TLine** component in the **Structure** pane (do not use the FSD to select the components; you have to use **Structure** pane nodes). In **Object Inspector**, set the following properties:
    - **Align**: **Contents**
    - **HitTest**: **False**
    - **LineType**: **Bottom**
    - **Opacity**: 0.6

22. Close the FSD tab (or click on **Apply and Close** for versions prior to RAD Studio 10.1 Berlin).

23. Run the application. Now, the text is underlined with a small black line that makes it easy to identify that the application is transparent. Stop the application.

24. Now, you have to work on the listbox; it is still 100% opaque.

25. Right-click on the **ListBoxHouses** option and click on **Edit Custom Style**.

26. In the **Structure** pane, there are some new styles related to the **TListBox** class. Select the **listboxhousesstyle1** option, open it, and select its child style, **background**.

27. In the **Object Inspector**, change the **Opacity** property of the **background** style to `0.6`. Click on **Apply and Close**.

28. That's it! Run the application, write `Calif` in the **Edit** field, and press **Return**. You should see a nice-looking application with a semitransparent user interface showing your dream houses in California (just as shown in the screenshot in the *Getting ready* section of this recipe). Are you amazed by the power of FireMonkey styles?

How it works...

The trick used in this recipe is simple. If you require a transparent UI, just identify which part of the style of each component is responsible for drawing the background of the component. Then put the **Opacity** setting to a level less than 1 (0.6 or 0.7 could be enough for most cases). Why not simply change the `Opacity` property of the component? Because if you change the `Opacity` property of the component, the whole component will be drawn with that opacity. However, you only need the background to be transparent; the inner text must be completely opaque. This is the reason why you changed the style and not the component property.

In the case of the **TEdit** component, you completely removed the painting when you removed the **SourceLookup** property from **TActiveStyleObject**, which draws the background.

As a rule of thumb, if you have to change the appearance of a control, check its properties. If the required customization is not possible using only the properties, then change the style.

See also

- If you are new to FireMonkey styles, it is more than likely that most concepts in this recipe will have been difficult to grasp. If so, check the official documentation on the Embarcadero DocWiki at the following URL: `http://docwiki.embarcadero.com/RADStudio/en/Customizing_FireMonkey_Applications_with_Styles`.

Creating a styled TListBox

As you saw in the previous recipe, it is possible to style styled controls and completely change their appearance. While in the VCL, the **TListBox** control is a mere wrapper over the corresponding control in the MS Windows API; in FireMonkey, the **TListBox** component is a completely different beast. A **TListBox** component contains a list of **TListBoxItem**, and a **TListBox** item is a **TStyledControl** descendant. This means that every single item in a **TListBox** component can be styled! This feature opens a huge set of new possibilities regarding the use of this control.

Getting ready

In this recipe, you'll see a set of styled **TListBoxItem** components that, when added to **TListBox**, change their appearance completely. Let's say you have a listbox containing a log of events that happened in a monitored remote system. Some events are simply informative, while other events can denote a malfunction. Different kinds of event are shown with different graphics in the listbox. Here are the events:

Type	Appearance
Normal	This is the default option for **TListBoxItem**.
Hint	This has blue colored text on a white background. The text is left aligned but indented by 40 pixels.
Warning	This has black colored text on a white background. There is a small yellow flag on the left-hand side.
Error	This has red colored text over a white background. There is a small red flag on the left.

What you require is shown in the following screenshot:

Figure 4.5: The listbox with some types of event logged

To achieve this result using VCL, you usually rely on the owner drawing controls or some third-party ones. However, with FireMonkey, all of these customizations are a matter of style, so they are simpler, faster to implement, reusable, and more flexible.

How to do it...

Let's start creating our stunning FireMonkey GUI:

1. Create a new FireMonkey desktop application.
2. Drop four **TButton** components, a **TListBox** component, and a **TStyleBook** component onto the form.
3. Double-click on the **TStyleBook** component and open the style editor.
4. Show the **Structure** pane (navigate to **View** | **Structure** or press *Shift + Alt + F11*).
5. Drop three **TLayout** components to create three different styles.
6. Set the **StyleName** property of the three **TLayout** components as follows:
   - **errorlistboxitem**
   - **hintlistboxitem**
   - **warninglistboxitem**

   The **StyleName** property allows you to reference the style from your form, so we've created three new styles that are usable from the main form.

7. In every **TLayout**, drop a **TText** so that every **TLayout** contains a **TText**. Set the **TText**. Align the property to **Client**.
8. Set the **StyleName** property for each **TText** to **eventtext**. Pay attention; every **TText** in each **TLayout** has the same value in the **StyleName** property. This allows you to use the **StylesData** property independently of the applied style. Therefore, you can write `StylesData['eventtext'] := 'Hello World'`, and the control that had `StyleName` equal to `eventtext` will be correctly assigned.
9. Now, let's work on each style. Select the **hintlistboxitem** style from the **Structure** pane.
10. Now, select the inner **eventtext** component (a **TText** component) and set its **TextSettings.FontColor** property to **Blue** and its **Margins.Left** property to **40**.
11. Select the **warninglistboxitem** style from the **Structure** pane.
12. Now, drop a **TImage** component onto the style at the same level of **TText**.
13. Set **TImage.Align** to **MostLeft**.

14. Load a small 32 x 32 icon showing a small yellow flag in its **MultiResBitmap** property (some free icons are provided with the code in this book).

15. Set **TImage.Width** to **40**.

16. Select the **errorlistboxitem** style from the **Structure** pane.

17. Set the **TText.TextSettings.FontColor** property to **Red**.

18. Now, drop a **TImage** component into the style at the same level of **TText**.

19. Set **TImage.Align** to **MostLeft**.

20. Load a small 32 x 32 icon showing a small red flag in its **MultiResBitmap** property.

21. Set **TImage.Width** to **40**.

22. Now, your **Structure** pane should look like this:

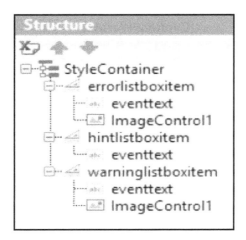

Figure 4.6: Structure pane for creating FireMonkey GUI

23. Click on **Apply and Close** on the style designer toolbar.

24. Now, the **TStyleBook** component contains all the custom styles. However, currently those styles are not used. Let's use them.

25. Select the form and set **TStyleBook** to StyleBook1.

26. Go to the form class declaration and add the following private method:

```
procedure TForm1.AddEvent(EventType, EventText: String);
var
  LBItem: TListBoxItem;
begin
  LBItem := TListBoxItem.Create(ListBox1);
```

```
  LBItem.Parent := ListBox1;
  if EventType.Equals('normal') then
  begin
    LBItem.Text := EventText;
  end
  else
  begin
    LBItem.StyleLookup := EventType + 'listboxitem';
    LBItem.StylesData['eventtext'] := EventText;
  end;
  ListBox1.AddObject(LBItem);
end;
```

27. Set the button names and captions to the following values:
    - btnNormal (**Caption: Normal**)
    - btnHint (**Caption: Hint**)
    - btnWarning (**Caption: Warning**)
        - btnError (**Caption: Error**)

28. Create four event handlers, one for each **TButton** component, as shown in the following code:

```
procedure TForm1.btnNormalClick(Sender: TObject);
begin
  AddEvent('normal', 'This is a normal event');
end;


procedure TForm1.btnHintClick(Sender: TObject);
begin
  AddEvent('hint', 'This is an HINT');
end;


procedure TForm1.btnWarningClick(Sender: TObject);
begin
  AddEvent('warning', 'WARNING! This is a WARNING!');
end;


procedure TForm1.btnErrorClick(Sender: TObject);
begin
  AddEvent('error', 'ERROR! This is an ERROR!');
end;
```

29. Hit *F9* and try to click on the buttons; you should see something like the **Structure** pane shown earlier.

How it works...

By clicking on each button, a new **TListBoxItem** is created (in the **AddEvent** method). Depending on the event type, the correct style is selected from the **TStyleBook** component. There is no need to directly refer to **TStyleBook**, since FireMonkey looks automatically to the form's **TStyleBook**. The **StyleLookup** property sets the style used for **TListBoxItem**, while the **StylesData** indexed property contains the values for every style component with **StyleName**. By setting `StylesData['eventtext']`, you are actually setting the **Text** property of the inner **TText** component.

See also

- FireMonkey styles are really powerful. The style designer makes working with styles quite simple and once you grasp the foundations of using FireMonkey, styles are addictive! Some links to go deeper with styles are as follows:
    - `http://docwiki.embarcadero.com/RADStudio/en/FireMonkey_Style_Designer`
    - `http://docwiki.embarcadero.com/RADStudio/en/Customizing_FireMonkey_Applications_with_Styles`
    - `http://docwiki.embarcadero.com/RADStudio/en/Working_with_Native_and_Custom_FireMonkey_Styles`

Impressing your clients with animations

Animations are a nice thing. A well done animation, not too intrusive and with good visual information, can explain what is happening on the UI better than a thousand words. In this recipe, you will implement a dual list with the `include<>exclude` paradigm so that what is removed from one list is included in the other list and vice versa. You will be using FireMonkey animations.

FireMonkey animations are really simple to use. Some kinds of property type can be animated. Some of these types are color, bitmap, gradient, and floating point number. The most used animation engine is the **TFloatAnimation**. This is used to animate floating point values such as **Opacity**, **Position.X**, **Position.Y**, **Width**, **Height**, and many more.

How to do it...

What you want to create is shown in the following screenshot:

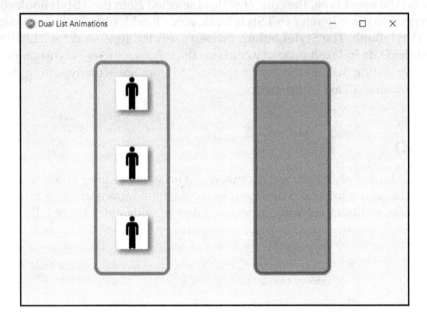

Figure 4.7: The dual list selection form

There are three images in the left-hand side gray list and zero images in the red list on the right-hand side. Click on an image; the clicked image will slide to the opposite list (gray to red or red to gray) using a nice animation. The steps to reproduce the images are as follows:

1. Create a new FireMonkey desktop application.
2. Drop two **TRectangle** components on the form. Align the first one on the left-hand side and call it **LeftRect**, and the second one to the right-hand side and call it **RightRect**, as shown in the screenshot in the *Getting ready* section of this recipe.
3. Set the properties of the left-hand side rectangle to this:
   - **Fill.Color**: #FFE0E0E0
   - **Fill.Kind**: Solid
   - **Stroke.Thickness**: 5
   - **Opacity**: 0.5
   - **XRadius**: 10
   - **YRadius**: 10

4. Set the properties of the right-hand side rectangle to this:
   - **Fill.Color**: **Red**
   - **Fill.Kind**: **Solid**
   - **Stroke.Thickness**: 5
   - **Opacity**: 0.5
   - **XRadius**: 10
   - **YRadius**: 10

5. Now, drop three **TImage** components on the form and align them in the left-hand side **TRectangle**.

6. Load some kind of picture or icon into **TImages**. You can use the same images for each **TImage** (as I did) or different images for each **TImage**. It depends on what kind of information you want to transfer to your user.

In the included source code, you can find an image called `blackman.png`, which is the one that I used.

7. Now, for each image, drop **TShadowEffect**. The effect must be owned by the **TImage** component so that in the **Structure** pane, the **TImage** component contains a subnode named **TShadowEffect**. Perform the same action for each **TImage**.

8. Now, set the **Distance** property to 5 for each **TShadowEffect**.

9. Our UI is created. Now, you have to write come code. In the `FormCreate` event, write the following code and declare `FLeftLimit` and `FRightLimit` as private class members of the type `Single`:

```
procedure TDualListForm.FormCreate(Sender: TObject);
begin
  FLeftLimit := LeftRect.ParentedRect.CenterPoint.X - Image1.Width
/ 2;
  FRightLimit := RightRect.ParentedRect.CenterPoint.X -
Image1.Width / 2;
end;
```

10. In the `FormCreate` event handler, `FLeftLimit` and `FRightLimit` are calculated. The objective is that, when the image is clicked, it should start from the left-hand side rectangle and should move to the right-hand side rectangle. If the image is clicked a second time, it should return to the left-hand side (with the same animation).

11. Now, let's use the same event handler for all three **TImage** components. Create the event handler by double-clicking on the **TImage1** component. Fill the event handler with this code:

```
procedure TDualListForm.Image1Click(Sender: TObject);
var
  LImage: TImage;
begin
  LImage := (Sender as TImage);

  if LImage.Tag = 0 then
  begin
    LImage.Tag := 1;
    //Slide the image to the right rectangle
    TAnimator.AnimateFloat(LImage, 'Position.X',
        FRightLimit, 0.8,
        TAnimationType.Out, TInterpolationType.Elastic)
  end
  else
  begin
    LImage.Tag := 0;
    //Slide the image to the left rectangle
    TAnimator.AnimateFloat(LImage, 'Position.X', FLeftLimit, 0.8,
      TAnimationType.Out, TInterpolationType.Elastic);
  end;

  //let's make the image a little bigger to mimic
  //a sort of 3D space using Scale property. Check that
  //this animation is delayed and will happen when the move
  //to the right (or to the left) is already started.
  TAnimator.AnimateFloatDelay(LImage, 'Scale.X', 1.2, 0.2, 0.2);
  TAnimator.AnimateFloatDelay(LImage, 'Scale.Y', 1.2, 0.2, 0.2);

  //Back to the original dimension using delay
  TAnimator.AnimateFloatDelay(LImage, 'Scale.X', 1, 0.2, 1);
  TAnimator.AnimateFloatDelay(LImage, 'Scale.Y', 1, 0.2, 1);
end;
```

12. As you can see, I've used the `TImage.Tag` property to keep track of the current position. It would be better to have an external data model to hold this kind of visual state, instead of putting this information in the graphical components, but for this demo, it's okay.

13. Now, connect the same `TImage1.OnClick` event handler to `TImage2.OnClick` and `TImage3.OnClick` as well. In this way, you can centralize the behavior in a single event handler.

14. Run (navigate to **Run** | **Run** or press *F9*) and start clicking on the images.

How it works...

This recipe is very simple and is a good example of how animations can be used to gain not only a visual wow effect (that may probably even disturb your user in some cases), but also some informative content.

The approach is simple: when the user clicks, the program will communicate using animations to mimic the real, physical world. An example of this is an eBook reader application on your smartphone. Is it strictly required to show a *page turning animation* when you change pages? No! However, it makes it clear to the user what is happening. Your animation should be used for the same purpose.

See also

Some useful basic information about animations can be read on the Embarcadero DocWiki:

- `http://docwiki.embarcadero.com/RADStudio/en/FireMonkey_Animation_Effects`
- `http://docwiki.embarcadero.com/RADStudio/en/Using_FireMonkey_Animation_Effects`

Using master/details with LiveBindings

When you have a customer with his/her orders or an invoice with his/her items, you have a **master/details** (**M/D**) relationship. In this recipe, you will learn how to use the new LiveBindings technology to show an M/D relationship.

As explained in the Embarcadero wiki:

> *"LiveBindings is a data-binding feature supported by both the VCL and FireMonkey frameworks in RAD Studio. LiveBindings is an expression-based framework, which means it uses bindings expressions to bind objects to other objects or to dataset fields."*

LiveBindings is a very nice technology and can be used in VCL applications as well, but its main targets are FireMonkey applications. Indeed, it is the only way to perform automatic data binding in the FireMonkey framework. If you don't know what LiveBindings is or what its strengths are, I suggest you stop here and read the article in the Embarcadero wiki at `http://docwiki.embarcadero.com/RADStudio/en/LiveBindings_in_RAD_Studio`.

What we want to do in this recipe is create a simple but complete FireMonkey application that handles a sort-of M/D relationship. Usually, this kind of thing involves the use of some databases. In this case, however, we are abandoning the SQL-based approach (which uses two or more datasets) in favor of a purely object-oriented approach. In other words, you will use a list of objects instead of a simple SQL query, and the relationships are child objects that are contained in the main object, not another query. Keep in mind that, although for this simple example the OOP approach may seem unnecessary, when you deal with a lot of logic, the classic dataset approach rapidly becomes unmanageable, while the OOP approach tends to be really stable, easily maintained, and understood by a third programmer.

Getting ready

The simple UML class diagram generated by Delphi is shown next:

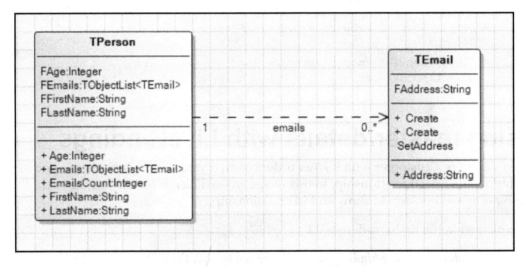

Figure. 4.8: The UML class diagram for the recipe

The main list of objects contains the **TPerson** instances. Each **TPerson** instance, as shown in the preceding diagram, contains a variable number of emails. So, the **TPerson** class has a property called **Emails** that is a list of **TEmail** instances. Instead of filtering all of the emails and showing only the emails related to the selected person (which usually happens in classic SQL programming), you'll show only the emails of that person; no filters are involved. The emails are already tied to the person. In this recipe, the difference between the TDataset approach and the object-oriented approach will be clear. The final application is shown in the following screenshot:

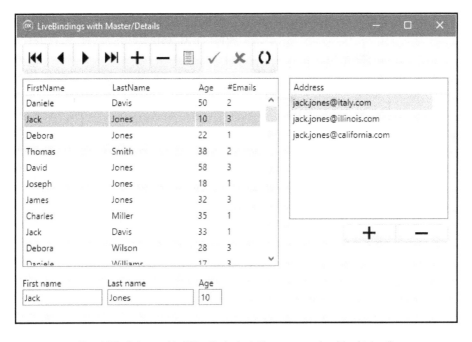

Figure 4.9 The final aspect of the M/D application that is able to manage people and the related emails

How to do it...

Let's start:

1. Create a new FireMonkey HD desktop application and name the main form
    `MainForm`.
2. Drop two **TGrid** components onto the form and name them `grdPeople` and
    `grdEmails`. Set the **Options.AlternatingRowBackground** property to **True** for
    both the components. Set the **Options.RowSelection** to **True** for`grdPeople`.
3. Drop two **TProtypeBindSource** components onto the form and name them
    `bsPeople` and `bsEmails`.

4. Double-clicking on **bsPeople** shows its field definitions. Using the **Add** button (the first button from the left-hand side), add four fields, as shown in the following screenshot:

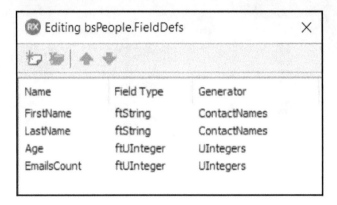

Figure 4.10: bsPeople field definitions

5. Close the field definition of **bsPeople**.
6. Double-clicking on **bsEmails** shows its field definitions. Using the **Add** button (the first button from the left-hand side), add the **Address** field, as shown in the following screenshot:

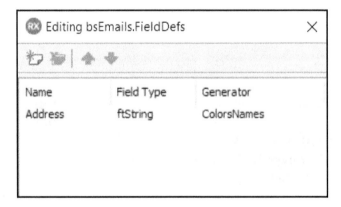

Figure 4.11: bsEmails field definitions

7. Close the field definition of **bsEmails**.
8. Drop a **TBindNavigator** component on the form and connect its **DataSource** property to **bsPeople**.

9. Drop another **TBindNavigator** component on the form and connect its **DataSource** property to **bsEmails**. Then, set all the elements inside its **VisibleButtons** property to **False** and set only **nbInsert** and **nbDelete** to **True** (this will allow you to insert or remove any email from a person).

10. Now, drop three **TEdit** components on the form and name them **EditFirstName**, **EditLastName**, and **EditAge**.

11. Our UI is almost ready. Add some labels and arrange the controls, as shown in the *Fig 4.2*.

12. Now, the interesting part begins.

13. Navigate to **View | LiveBindings Designer**.

14. The window shows the famous **LiveBindings Designer**. All the **LiveBindings Enabled** controls with their properties will be shown.

15. On the left-hand side toolbar, there is a set of buttons that are useful for changing the disposition and zoom of the diagram. Use the buttons; they will save your sanity!

16. Identify the **bsPeople** element and drag and drop all of its elements on `grdPeople`. You can also only drag the ***** column, but columns are not created at design time. So, if you want (and, usually, you will) to change the aspect of the grid columns, drag every field one by one.

17. Perform the same action (as done for **bsPeople**) with **bsEmails** and `grdEmails`.

18. Now, you have to connect the editable field of the **bsPeople** component to the **TEdits** component.

19. Connect `bsPeople.FirstName` to `EditFirstName.Text`.

20. Connect `bsPeople.LastName` to `EditLastName.Text`.

21. Connect `bsPeople.Age` to `EditAge.Text`.

22. Do not connect the `EmailsCount` field. This field is a read-only field, mapped to a read-only property, and is used to show the number of email addresses related to the current person. This technique can be quite useful when you have to show how many rows are contained in an invoice, how many orders are related to a customer, and so on.

23. If you run the application now, you will see that some fake data is generated. You will also notice that there is no M/D relationship between people and emails. We are about to fix this in a moment. Close the application and go back to Delphi.

24. Add a new unit, name it `BusinessObjectsU.pas`, and add the following code to it:

```
unit BusinessObjectsU;

interface

uses System.Generics.Collections;

type
  TEmail = class
  private
    FAddress: String;
    procedure SetAddress(const Value: String);
  public
    constructor Create; overload;
    constructor Create(AEmail: String); overload;
    property Address: String read FAddress write SetAddress;
  end;

  TPerson = class
  private
    FLastName: String;
    FAge: Integer;
    FFirstName: String;
    FEmails: TObjectList<TEmail>;
    procedure SetLastName(const Value: String);
    procedure SetAge(const Value: Integer);
    procedure SetFirstName(const Value: String);
    function GetEmailsCount: Integer;
  public
    constructor Create; overload;
    constructor Create(const FirstName, LastName: string; Age:
Integer); overload; virtual;
    destructor Destroy; override;
    property FirstName: String read FFirstName write SetFirstName;
    property LastName: String read FLastName write SetLastName;
    property Age: Integer read FAge write SetAge;
    property EmailsCount: Integer read GetEmailsCount;
    property Emails: TObjectList<TEmail> read FEmails;
  end;

implementation

uses System.SysUtils;

constructor TPerson.Create(const FirstName, LastName: string; Age:
Integer);
```

```pascal
begin
  Create;
  FFirstName := FirstName;
  FLastName := LastName;
  FAge := Age;
end;

// Called by LiveBindings to insert a new Person
constructor TPerson.Create;
begin
  inherited Create;
  FFirstName := '<name>';
  //initialize the emails list
  FEmails := TObjectList<TEmail>.Create(true);
end;

destructor TPerson.Destroy;
begin
  FEmails.Free;
  inherited;
end;

function TPerson.GetEmailsCount: Integer;
begin
  Result := FEmails.Count;
end;

procedure TPerson.SetLastName(const Value: String);
begin
  FLastName := Value;
end;

procedure TPerson.SetAge(const Value: Integer);
begin
  FAge := Value;
end;

procedure TPerson.SetFirstName(const Value: String);
begin
  FFirstName := Value;
end;

constructor TEmail.Create(AEmail: String);
begin
  inherited Create;
  FAddress := AEmail;
end;
```

```
  // Called by LiveBindings to insert a new Email
  constructor TEmail.Create;
  begin
    Create('<email>');
  end;

  procedure TEmail.SetAddress(const Value: String);
  begin
    FAddress := Value;
  end;

end.
```

25. Now, go to the `TMainForm` declaration and add the following code in the `private` section:

```
private
  FPeople: TObjectList<TPerson>;
  bsPeopleAdapter: TListBindSourceAdapter<TPerson>;
  bsEmailsAdapter: TListBindSourceAdapter<TEmail>;
  procedure PeopleAfterScroll(Adapter: TBindSourceAdapter);
  procedure LoadData;
```

26. Create the `PeopleAfterScroll` and `LoadData` methods in the implementation section (you can use *Ctrl* + *Shift* + *C* to generate the empty method body; check all the others keyboard shortcuts at `http://docwiki.embarcadero.com/RADStudio/en/Default_Keyboard_Shortcuts`):

```
procedure TMainForm.LoadData;
var
  I: Integer;
  P: TPerson;
  X: Integer;
begin
  for I := 1 to 100 do
  begin
    // create a random generated person
    P := TPerson.Create(GetRndFirstName, GetRndLastName, 10 +
Random(50));

    // add some email addresses (1..3) to the person
    for X := 1 to 1 + Random(3) do
    begin
      P.Emails.Add(TEmail.Create(P.FirstName.ToLower + '.' +
P.LastName.ToLower
        + '@' + GetRndCountry.Replace(' ', '').ToLower + '.com'));
    end;
```

```
      FPeople.Add(P);
   end;
end;

procedure TMainForm.PeopleAfterScroll(Adapter: TBindSourceAdapter);
begin
bsEmailsAdapter.SetList(bsPeopleAdapter.List[bsPeopleAdapter.Curren
tIndex].Emails, False);
   bsEmails.Active := True;
   bsEmails.First;
end;
```

27. On the main form, create the `FormCreate` and `FormDestroy` event handlers with this code:

```
procedure TMainForm.FormCreate(Sender: TObject);
begin
   Randomize;
   FPeople := TObjectList<TPerson>.Create(True);
   LoadData;
   bsPeopleAdapter.SetList(FPeople, False);
   bsPeople.Active := True;
end;

procedure TMainForm.FormDestroy(Sender: TObject);
begin
   FPeople.Free;
end;
```

The `Randomize` function is used to initialize the Random number generator. If you don't do this, every time you run the program, it will generate the same sequence. More information on this can be found at the following link: `http://docwiki.embarcadero.com/Libraries/en/System.Randomize`.

28. Now, show the main form, select **bsPeople**, and create the event handler for the `OnCreateAdapter` event. This event is called when the `TPrototypeBindSource` method has to decide whether to use fake randomly generated data or your real data. You have to handle this event and plug the code to provide your data. Write the following code in the event handler:

```
procedure TMainForm.bsPeopleCreateAdapter(Sender: TObject; var
ABindSourceAdapter: TBindSourceAdapter);
begin
```

```
bsPeopleAdapter := TListBindSourceAdapter<TPerson>.Create(self,
nil, False);
  ABindSourceAdapter := bsPeopleAdapter;
  bsPeopleAdapter.AfterScroll := PeopleAfterScroll;
end;
```

29. On the main form, select **bsEmails** and create the event handler for the
OnCreateAdapter event:

```
procedure TMainForm.bsEmailsCreateAdapter(Sender: TObject; var
ABindSourceAdapter: TBindSourceAdapter);
begin
  bsEmailsAdapter := TListBindSourceAdapter<TEmail>.Create(self,
nil, False);
  ABindSourceAdapter := bsEmailsAdapter;
end;
```

30. If you run the application, you should see a working form showing an M/D
relationship, or better a *has a* relationship, because a person has a list of emails.
Stop it and add a small trick.

31. If you try to add a new email, the new line will be added in the **TGrid**
component. I hate data entry directly into grids! In some cases, it is a great
feature, but in many cases it only shows up a badly designed UI (this is not the
case if you are developing a spreadsheet!). So, let's create
TBindSourceNavigator to show a dialog to add a new email.

32. Select the TBindSourceNavigator component named bnEmails, create an
event handler for the BeforeAction event, and then write the following code:

```
procedure TMainForm.bnEmailsBeforeAction(Sender: TObject; Button:
TNavigateButton);
var
  email: string;
begin
  if Button = TNavigateButton.nbInsert then
    if InputQuery('Email', 'New email address', email) then
    begin
      bsEmailsAdapter.List.Add(TEmail.Create(email));
      bsEmails.Refresh; // refresh the emails list
      bsPeople.Refresh; // refresh the email count
      Abort; //inhibit the normal behavior
    end;
end;
```

33. Now, run the application and try to add a new email; you'll see that a nice dialog comes up.
34. That's all folks!

How it works...

There are a few concepts involved in LiveBindings, but these concepts must be well-understood so that you can create a working application. Let's analyze this application.

At the beginning, the TPrototypeBindSource components are initialized with the TListBindSource<T> instances so that they show actual data instead of fake data. Then, in the FormCreate event handler, you create the actual list of objects that will contain your people and load some data into it using the LoadData method. This method loads some random data, but in a real application it should read data from some query or from some web service. This is one of the strengths of LiveBindings; you can visualize your data, whatever its origin. You are no longer tied to TDataSet!

After loading the data, you can set the **bsPeople** list of objects to your people and then activate it. This is okay for one single list of data, but how do you handle the M/D relationship?

In the bsPeople.OnCreateAdapter event, you must set an AfterScroll event handler for bsPeopleAdapter (the internal adapter used by TPrototypeBindSource). This event is called when the selected person changes. Therefore, you can handle data visualization on the email grid from this event. The code in this event handler is self-explanatory:

```
procedure TMainForm.PeopleAfterScroll(Adapter: TBindSourceAdapter);
begin
  //sets the email object list to the emails of the
  //selected person in the bsPeopleAdapter. The adapter
  //is automatically deactivated when the underlying list is
  //changed, so you don't need to set its Active property to False
bsEmailsAdapter.SetList(bsPeopleAdapter.List[bsPeopleAdapter.CurrentIndex].
Emails, False);
  //here the bsEmails is no more active, let's activate it
  bsEmails.Active := True;
  bsEmails.First;
end;
```

Usually, working with the internal adapter of `TPrototypeBindSource` is a bit messy because you have to write something like this:

```
//sets a new list of objects as data source
(bsPeople.InternalAdapter as TListBindSourceAdapter<TPerson>).
SetList(MyList);
```

Saving a reference when you are creating the actual adapter in the `OnCreateAdapter` method saves a lot of casting and makes code more readable. There are other solutions, but I really like this one.

There's more...

LiveBindings is a relatively new technology. It has changed a lot since its introduction in Delphi XE2, at least in terms of its high-level components. The good old Delphi programmer seems to not completely understand its power (probably because `TDataSet` along with VCL really does a good job for classic client/server applications), but there is still time to explore its capabilities. However, when you use FireMonkey, LiveBindings is mandatory, so I strongly suggest you try it because, sooner or later, you will have to use it for some mobile stuff or some general FireMonkey applications.

There are many things to say about LiveBindings—we've only scratched the surface. For example, if you are building a big project and you have to handle or show different kinds of recurrent entity, such as customers, orders, invoices, or users, you can create a `TListBindSourceAdapter<T>` descendant, compile it in a package, and install it in the tool palette so that, every time you require it, you can simply drag and drop it onto your data module or form.

See also

Here are some links where you can find more information about LiveBindings:

- *XE3 Visual LiveBindings: User defined objects* at `http://blogs.embarcadero.com/jimtierney/2012/12/11/31961`
- *LiveBindings GridColumns* at `http://www.youtube.com/watch?v=K6Xu90Rtbys`
- *TBindSourceDB* at `http://www.malcolmgroves.com/blog/?p=1072`
- *TAdapterBindSource and binding to Objects* at `http://www.malcolmgroves.com/blog/?p=1084`
- *Updating Objects via an Adapter* at `http://www.malcolmgroves.com/blog/?p=1186`
- *Formatting your Fields* at `http://www.malcolmgroves.com/blog/?p=1226`

- *XE3 Visual LiveBindings: Samples* at `http://blogs.embarcadero.com/jimtierney/2012/10/21/31944`
- If you are interested in the LiveBindings core, you can read an old article of Daniele Teti's that is still valid at `http://www.danieleteti.it/2011/08/30/in-the-core-of-livebindings-expressions-of-rad-studio-xe2/`

Showing complex vector shapes using paths

One of the biggest advantages of FireMonkey compared to VCL is its vector-based nature. Various visual parts can be created in FireMonkey using vector-based graphics (even if in some cases, using a bitmapped approach can be faster). In terms of vectorial graphics, there is a nice language called **Scalable Vector Graphics** (**SVG**) that allows you to define primitive shapes using a set of coordinates and not a raster image. This means that you can stretch the image without losing its resolution because the image is not actually stretched; it's completely redrawn using the new coordinates. That's it; the SVG file is made up of coordinates and mathematical formulae to join them.

Inside the SVG language, there is an element called SVG path. The path element is used to define a path. So, what's a path?

A path is a sequence of instructions to draw something using primitives. Think of an SVG path as a language into another language (let's say a sort of internal DSL).

The following commands are available for path data:

- M: This represents the `moveto` command (without drawing)
- L: This represents the `lineto` command (like M but with drawing)
- H: This represents the `horizontal lineto` command
- V: This represents the `vertical lineto` command
- C: This represents the `curveto` command
- S: This represents the `smooth curveto` command
- Q: This represents the `quadratic Bézier curve` command
- T: This represents the `smooth quadratic Bézier curveto` command
- A: This represents the `elliptical Arc` command
- Z: This represents the `closepath` command

All of these commands can also be expressed with lowercase letters. Uppercase letters mean absolutely positioned and lowercase letters mean relatively positioned.

So, the M50 0 L100 100 L0 100 Z path means:

- Move the pen to X50, Y0
- Draw a line from the current point to X100, Y100
- Draw a line from the current point to X0, Y100
- Close the path drawing a line to the origin point (X150, Y0)

It draws a triangle, which looks as follows:

Figure 4.12: The triangle drawn by the sample path data

Getting ready

In the FireMonkey framework, there is a component called **TPath** (it is defined in the FMX.Objects.pas unit; do not confuse it with the **TPath** component defined in the System.IOUtils.pas unit). The **TPath** component is able to interpret and show an SVG path. In this recipe, you'll see how to use it to draw complex vector shapes and fonts.

Let's say you want to monitor a continuous stream of data, maybe a value read from some kind of hardware or some value related to finance stock quotes. You want fresh data to be pushed from the right-hand side and the oldest data removed from the left-hand side. At any time, you can see the last 20 values, scrolling from right to left. This is shown in the following screenshot:

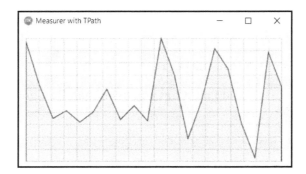

Figure 4.13: Scrolling data in a line graph: new data is pushed from the right-hand side and old data is removed from the left-hand side

Usually, in order to write something like this, you require some third-party components or you have to write a lot of code to write all the values and axes and deal with proportional issues. Using the **TPath** component, you don't have to do all of this! The **TPath** component with a proper SVG PATH is completely in charge of stretching and redrawing your graphic in order to fit the drawing area.

How to do it...

Let's create this application step by step:

1. Create a new FireMonkey desktop application.
2. Drop a **TPanel** component onto the form. In the **TPanel** component, put a **TPath** component and set its **Align** property to **alClient**. Now, the **TPath** component should fit into the **TPanel** component.
3. Drop another **TPath** component onto the first one and again set its **Align** property to **alClient**.
4. Now, you should have **TPanel** with two nested **TPath** components inside it.
5. Show the structure of the form (*Shift + Alt + F11*).
    1. Name the first **TPath** component PathValues and the second **TPath** component PathAxis.
6. Drop a **TTimer** component on the form and double-click on it. In the OnTimer event handler, write the following code:

```
procedure TMainForm.Timer1Timer(Sender: TObject);
begin
  FValuesQueue.Add(Trunc(Random * 100));
  RefreshGraph;
end;
```

7. Set the **Timer Interval** property to **50**.

8. Now, go to the code editor and declare a private form instance variable:

```
FValuesQueue: TList<Integer>;
```

9. Create `FormCreate` and `FormDestroy` event handlers and fill them with the following code:

```
procedure TMainForm.FormCreate(Sender: TObject);
var
  I: Integer;
  svggrid: string;
begin
  FValuesQueue := TList<Integer>.Create;
  for I := 0 to 19 do
    FValuesQueue.Add(0);

  svggrid := '';
  for I := 0 to FValuesQueue.Count - 1 do
    svggrid := svggrid + ' M' + I.ToString + ',0 V100';
  for I := 1 to 10 do
    svggrid := svggrid + ' M0,' + IntToStr(I * 10) + ' H20';
  PathAxis.Data.Data := svggrid;
end;

procedure TMainForm.FormDestroy(Sender: TObject);
begin
  FValuesQueue.Free;
end;
```

10. So far, you've declared and initialized your data container (the `TList<Integer>` item named `FValuesQueue`); now, let's do something with its data. Create a private procedure named `RefreshGraph` and fill it with the following code:

```
procedure TMainForm.RefreshGraph;
var
  I: Integer;
  svg: string;
begin
  svg := 'M0,100 ';
  if FValuesQueue.Count > 19 then
  begin
    svg := svg + 'L0,' + (100 - FValuesQueue.First).ToString;
    FValuesQueue.Delete(0); //remove the first
  end;

  for I := 0 to FValuesQueue.Count - 1 do
```

```
begin
    svg := svg + ' L' + I.ToString + ',' + (100 -
FValuesQueue[I]).ToString;
   end;

   svg := svg + ' L' +
     IntToStr(FValuesQueue.Count - 1) + ' 100 ';
   PathValues.Data.Data := svg;
  end;
```

11. Run the application.
12. Are you disappointed with the performance? In this case, the debugger load on the execution speed is heavy. So, to check the real drawing speed, run it without the debugger (*Shift + Ctrl + F9*).
13. You should now see the graph scrolling at a good speed.

How it works...

The architecture is simple—the timer is the (fake) data producer that fills the list. Then, the list is used to draw the graph. After drawing the graph, the first list element is removed, waiting for the next one.

In a real-world application, some fine-tuning may be necessary; in this case a classic producer/consumer pattern is more suited to doing this compared to a simple **TTimer**. However, in this sample, a normal **TTimer** component is enough.

A good thing to note is that you have a fixed coordinate system when drawing the values in the graph. You don't have to worry about form size, relative, or absolute coordinates, and so on. All the details are handled by the **TPath** component.

So, if you'd like to add another scrolling graph of a different size, you could use the same SVG PATH data to show the same graph on another area.

Let's add another `TPanel->TPath->TPath` triad on the form and make the **TPanel** component bigger than the previous one. With a little change in the code (the full code is available), you can have something like the following:

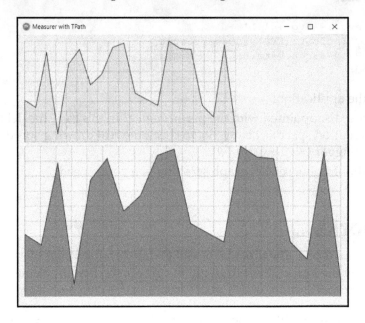

Figure 4.14: Another scrolling graph showing the same values added without changing the drawing code

There's more...

The `SVG PATH` data can be very useful. If you require complex `SVG PATH` data, I suggest that you use a proper editor to generate the path. There is a nice online editor that can generate this kind of information called **Method Draw**, and it's available at `http://editor.method.ac/`. The `SVG PATH` data can be also used to drive animations using the `TPathAnimation` component.

The producer/consumer cited in this recipe is a classic concurrency pattern. You can find more information on this at `http://javarevisited.blogspot.it/2012/02/producer-consumer-design-pattern-with.html`.

Using FireMonkey in a VCL application

As you probably know, VCL is incompatible with FireMonkey. What does this mean? Embarcadero explains this in the DocWiki:

"FireMonkey (FMX) and the Visual Component Library (VCL) are not compatible and should not be used together in the same module. That is, a module should be exclusively one or the other, either FireMonkey or VCL. The incompatibility is caused by framework differences between FireMonkey (FMX) and VCL."

However, there is still something that can be done to use FireMonkey functionalities in a VCL application.

It's very probable that a VCL application could benefit from using some components or functionalities that are only present in the FireMonkey framework. So, what could be the solution? One solution is to create a Windows DLL that contains all the FireMonkey code and exposes a set of raw functions to access them. Then, the VCL application can load the DLL and call the exposed functions. Let's see this in action.

This recipe requires familiarity with some advanced Delphi concepts, so there will not be a step-by-step section; I'll only talk about the project code.

How to do it...

Let's begin!

1. Open the recipe project group called `UsingFMXfromVCL.groupproj`. The group contains two projects:
    - A VCL application (`vclmainproject.exe`), which is your legacy application
    - A DLL project (`fmxproject.dll`), which contains all the FireMonkey stuff

2. To get an idea about projects, navigate to **Project** | **Build all Projects**, select the `vclmainproject.exe` file, and hit *F9* to run it. The `fmxproject.dll` file has been compiled in the same folder of `vclmainproject.exe`. You should see the following form:

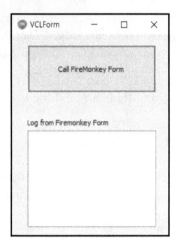

Figure 4.15: The VCL form that will use the DLL containing the FireMonkey code

3. By clicking on the **Call FireMonkey Form** button, you can call the FireMonkey DLL, which will show a FireMonkey form that can send the main form some information using a callback (we'll talk about this in a moment). The callback makes your project a little bit difficult, but being able to send something to the caller is a fundamental part of any integration.

4. If you click on the button and play with the FMX controls, you should get something like this:

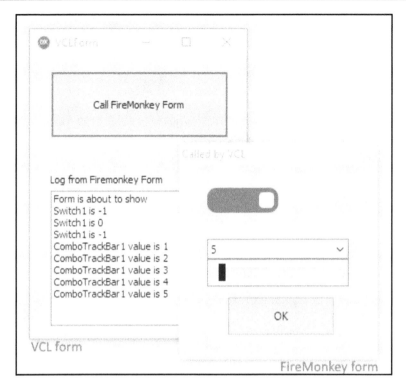

Figure 4.16: The FireMonkey form being used by the VCL application

How it works...

The CommonsU.pas unit is shared between the VCL and FMX projects and contains the declaration for the callback function, as shown here:

```
type
    TDLLCallback = procedure(const Value: String);
```

The DLLImportU.pas unit is only used by the VCL project (because it needs to import the DLL functions). It is really simple and refers to the TDLLCallBack declaration:

```
unit DLLImportU;

interface

uses
    CommonsU;
```

```
procedure Execute(const Caption: String; Callback: TDLLCallback); stdcall;
external 'fmxproject';

implementation

end.
```

These two files are the *bridge* between the VCL project and the FMX project. Now, let's see how the VCL project calls the FireMonkey DLL.

Using the **Project Manager**, select the VCL project main form. The `Button Click` event handler calls the `Execute` external function with the following code:

```
procedure MyCallBack(const Value: String);
begin
  VCLForm.ListBox1.Items.Add(Value);
  VCLForm.ListBox1.Update;
end;

procedure TVCLForm.btnCallFMXClick(Sender: TObject);
begin
  Execute('Called by VCL', MyCallBack);
end;
```

Notice that the `MyCallBack` procedure is not a form method, but a simple procedure. The reason for this is because inside it I used an instance name of the form, `VCLForm`, and it cannot use the implicit `Self` reference. Also, a normal string and a function pointer are passed to the `Execute` function. Notice that the function pointer is `MyCallBack` and not `MyCallBack()` (with parentheses, it means call the procedure; without parentheses, it means the address of).

The VCL project doesn't require further explanation. Let's switch to the FMX DLL. Using the **Project Manager**, select the `fmxproject.dll` file and navigate to **Project | View Source**.

The `library` project file contains the exported functions and the startup code to show the FMX form. Its code is shown here:

```
library fmxproject;

uses
  System.ShareMem, Winapi.Windows,
  System.SysUtils, System.Classes,
  FMXMainForm in 'FMXMainForm.pas' {Form1},
  CommonsU in 'CommonsU.pas';
```

```
{$R *.res}

procedure Execute(const Caption: String; Callback: TDLLCallback); stdcall;
var
  frm: TForm1;
begin
  frm := TForm1.Create(nil);
  try
    frm.Caption := Caption; //use the passed string as Caption
    frm.FCallback := Callback; //link callback as form property
    frm.ShowModal;
  finally
    frm.Free;
  end;
end;


{ This is exported function that will be used by the VCL form }
exports Execute;

begin

end.
```

As you can see, the `callback` pointer has been assigned to a `form` property to be accessible from it. How will the FMX form use the `callback` pointer? In this recipe, it uses the `callback` pointer to send some information about the components on it to the main VCL form.

This is the relevant code for the main VCL form:

```
type
  TForm1 = class(TForm)
    btnClose: TButton;
    Switch1: TSwitch;
    ComboTrackBar1: TComboTrackBar;
    procedure btnCloseClick(Sender: TObject);
    procedure Switch1Switch(Sender: TObject);
    procedure ComboTrackBar1Change(Sender: TObject);
    procedure FormShow(Sender: TObject);
    procedure FormClose(Sender: TObject; var Action: TCloseAction);
  private
  public
    {This is the function pointer to the main VCL form callback}
    FCallback: TDLLCallback;
  end;

implementation
```

```
{$R *.fmx}

procedure TForm1.ComboTrackBar1Change(Sender: TObject);
begin
  //send the value of TComboTrackBar
  FCallback('ComboTrackBar1 value is ' + ComboTrackBar1.Value.ToString);
end;

procedure TForm1.FormClose(Sender: TObject; var Action: TCloseAction);
begin
  //inform the main form about FMX form closing
  FCallback('Form is about to close');
end;

procedure TForm1.FormShow(Sender: TObject);
begin
  //inform the main form about FMX form showing
  FCallback('Form is about to show');
end;

procedure TForm1.Switch1Switch(Sender: TObject);
begin
  //inform the main form about the state of the Switch
  FCallback('Switch1 is ' + Switch1.IsChecked.ToString);
end;
```

The FMX-side code is not complex, and you can use whatever complex data structure you want to send information from the FMX form to the VCL form. A good and simple solution for this is to define a simple textual protocol to allow a single callback to bring multiple types of information. For this kind of thing, I used to use a JSON-serialized string. If the data that you have to transfer is a lot, or the data transfer rate is critical, you can use a specific record and use a pointer to it to share information between the DLL and the main program (in this way, you save the JSON generate/parse time). If the values sent by the callback are many, you can queue the values and process them as soon as possible; this is something like the producer/consumer design pattern.

There's more...

This recipe follows the *official* approach and uses two different projects (one from VCL and one from FireMonkey) to use the FireMonkey framework from a VCL application.

What if you have a legacy project where you'd like to use a FireMonkey DLL, but the legacy project is not in Delphi VCL (let's say it is in C#, Visual C++, Python, or any other language that can load a DLL)? You can still use the same approach, but you cannot use Delphi-specific data types. So, your strings should be `PChar` and so on. You can find more information on this at `http://delphi.about.com/od/objectpascalide/a/dlldelphi.htm`.

Just to be clear, keep in mind that mixing FireMonkey and VCL forms in the same application isn't officially supported. However, there are a number of libraries that aim to integrate VCL and FireMonkey forms in the same project.

Here's a short list, in no particular order:

- *TFireMonkeyContainer* at `https://parnassus.co/open-source/tfiremonkeycontainer/`
- *MonkeyMixer* by *LaKraven Studios Ltd* at `https://github.com/LaKraven/MonkeyMixer`
- *Delphisorcery* at `https://bitbucket.org/sglienke/dsharp` (using `DSharp.Windows.FMXAdapter.pas`)

In this recipe, you used a function pointer as a callback. If you want to know more about this type and other types of callback, check out the following link: `http://www.delphi-central.com/callback.aspx`.

Reinventing your GUI, also known as mastering Firemonkey controls, shapes, and effects

As you have surely understood by now, FireMonkey is a completely new graphics library that allows a completely new way of thinking about your GUI. During my FireMonkey training, one of the first exercises that I gave to the class was: *Please look for the strangest FireMonkey control in the Tool Palette*. It is quite a strange exercise, but the reason for this is really important: you must realize as soon as possible that FireMonkey is not a cross-platform VCL; it is a new beast with new possibilities and new things to know. So, you have to rethink your GUI architecture because many *patterns* used in the last 5, 10, 15, or more years of VCL development now may be simply obsolete or are no longer the best things to do. For instance, you have to display a pie chart with some user interaction and some nice visual effects. When the user moves the mouse over a pie slice, the slice gets highlighted and some information is shown to the user. How can you achieve this?

In VCL, there are two ways: write your own code (a lot of it) or use a third-party control. In FireMonkey, there is also a third way: use primitive shapes to create it using no third-party controls and a very small amount of code. This is a recurring pattern in FireMonkey. This is the reason because your FireMonkey application needs fewer third-party controls than the VCL one. To show you how this is possible, in this recipe we'll create a simple pie chart with such aspects and user behavior. Here's a screenshot of the application we'll build:

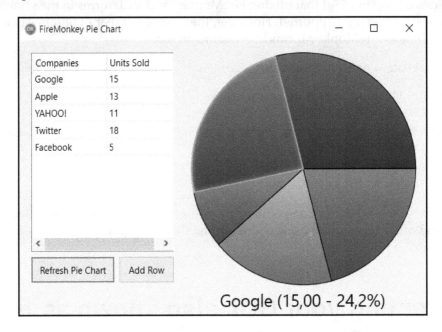

Figure 4.17: The pie chart generated using only FireMonkey components

Getting ready

The first steps in designing your GUI is to slow down and think about it. We need a pie chart, correct? So, is there a primitive shape packaged as a component in the FireMonkey library that's able to draw a slice of a pie? Yes, there is the `TPie` control. But we need a variable number of slices, and all the slices must be stacked one over the other to make a full pie. Where we could put these dynamically created slices? Obviously, in a `TLayout`! This is the basic idea behind the recipe. Let's talk about the recipe code in detail.

How it works...

Open `RECIPE07` in `Chapter 4`, named `PieChart.dproj`. The main form is quite simple; there is a `TStringGrid` to allow the user to write the data, a `TButton`, which actually starts the (re)generation of the chart, another button to add a new row in the grid, a `TLabel` to show some additional information when the user moves the mouse over the single slice, and the `TLayout` we already talked about. The following is the main form at design time:

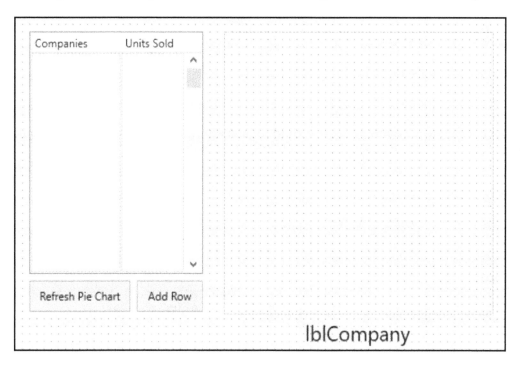

Figure 4.18: The main form at design time

In the `FormCreate` event handler, the initialization code creates the `TDictionary<String, Extended>`, which will contain the data represented in the chart. Moreover, considering this is a sample, the data is generated randomly and used to fill the grid. Here's the `FormCreate` code:

```
procedure TMainForm.FormCreate(Sender: TObject);
begin
  Randomize;
  FDataDict := TDictionary<String, Extended>.Create;
  sgData.RowCount := 5;
  sgData.BeginUpdate;
  try
```

```
    sgData.Cells[0, 0] := 'Google';
    sgData.Cells[1, 0] := RandomRange(2, 20).ToString;
    sgData.Cells[0, 1] := 'Apple';
    sgData.Cells[1, 1] := RandomRange(2, 20).ToString;
    sgData.Cells[0, 2] := 'YAHOO!';
    sgData.Cells[1, 2] := RandomRange(2, 20).ToString;
    sgData.Cells[0, 3] := 'Twitter';
    sgData.Cells[1, 3] := RandomRange(2, 20).ToString;
    sgData.Cells[0, 4] := 'Facebook';
    sgData.Cells[1, 4] := RandomRange(2, 20).ToString;
  finally
    sgData.EndUpdate;
  end;
  lblCompany.Text := '';
end;
```

As you can see, the generated data is about some companies, so each row in the grid will contain the **Company name** and **Units Sold** for something. Now, some interesting things will happen under the **Refresh Pie Chart** button. Here's the code:

```
procedure TMainForm.btnRefreshPieClick(Sender: TObject);
var
  LPie: TPie;
  LCurrAngle, LGrad: Single;
  LIdx: Integer;
  LRefColor: TAlphaColor;
  LPair: TPair<String, Extended>;
begin
  // Loads the data from the string grid and put them in the
  // dictionary using the company name for the key and the
  // units sold as the value.
  LoadData(FDataDict);

  // Get the total for all the companies using an
  // higher order function
  FTotalValue := HigherOrder.Reduce<Extended>(FDataDict.Values.ToArray,
    function(A, B: Extended): Extended
    begin
      Result := A + B;
    end, 0);

  // remove all the TPie already present into the TLayout
  // The first time there aren't child, but from the second time
  // yes, so let's remove all the TPie from the TLayout
  lytPie.DeleteChildren;

  LCurrAngle := 0;
  LIdx := 0;
```

```
    lytPie.BeginUpdate;
    try
      // looping through the dictionary and create each TPie
      for LPair in FDataDict do
      begin
        // some math to know how many degree each pie slide must be
        LGrad := 360 * LPair.Value / FTotalValue;

        // Build the pie slice, che the BuildPieSlice for details
        LPie := BuildPieSlice(LPair.Key);
        LPie.StartAngle := LCurrAngle;
        LPie.EndAngle := LCurrAngle + LGrad;
        LCurrAngle := LCurrAngle + LGrad;
        LRefColor := GetColor(LIdx);

        // Setup some nice gradients color to give
        // a sort of fake 3D effect to each slice
        SetupGradient(LPie.Fill, LRefColor);

        // Let's give some dynamicity to the chart
        // with some effects
        InjectEffects(LPie, GetColor(LIdx));
        Inc(LIdx);
      end;
    finally
      lytPie.EndUpdate;
    end;
  end;
```

There's quite a lot of stuff in this method! Let's analyze the interesting parts of it.

After having loaded the data into the dictionary, we're using a higher order function to summarize the total for the units sold by all of the companies. This function is a part of RECIPE01 in Chapter 2. It is a Reduce function, used here to perform a sum. This is done so that the result of the function is the total of all the units sold.

Having the total, we can start with the loop to create each TPie and add it into the TLayout. Each TPie starts its slice where the previous one ends, so that the nth slice has its StartAngle equal to the EndAngle of the previous slice.

Here's the code for the BuildPieSlice method and the other methods used by it:

```
function TMainForm.BuildPieSlice(AIdentifier: String): TPie;
begin
  //create the pie as child of TLayout
  Result := TPie.Create(lytPie);
  Result.Parent := lytPie;
```

```
    //setup some events to give interactivity
    Result.OnMouseEnter := OnPieEnter;
    Result.OnMouseLeave := OnPieLeave;
    //We need the company name to which each pie is referred to
    Result.TagString := AIdentifier;
    //Align to the parent so that all the TPie are aligned
    //and stacked one over the other
    Result.Align := TAlignLayout.Contents;
  end;

  procedure TMainForm.OnPieEnter(Sender: TObject);
  var
    LKey: String;
    LPie: TPie;
    LValue, LPercValue: Extended;
  begin
    LPie := Sender as TPie;
    //move the current TPie to the front so that the effects
    //applied effect are not covered by the other TPie
    LPie.BringToFront;
    LKey := LPie.TagString;
    //gets the value for the company to which the pie
    //refers to and calculate the percentage over the total
    LValue := FDataDict.Items[LKey];
    LPercValue := LValue / FTotalValue * 100;
    //display the data in the TLabel
    lblCompany.Text := Format('%s (%.2f - %2.1f%%)',
      [LKey, LValue, LPercValue]);
  end;

  procedure TMainForm.OnPieLeave(Sender: TObject);
  begin
    //clear the TLabel when the mouse leave the TPie
    lblCompany.Text := '';
  end;
```

The color of each TPie is provided by a function contained in the ColorsUtils.pas unit. This unit contains a static array of colors and some functions to manipulate them. Here's the code:

```
const
  Colors: TArray<TAlphaColor> = [$FF4285F4,$FFFBBC05,
          $FF34A853,$FFEA4335,$FFA90FF4,$FF9F3C00];

function GetColor(AIndex: Integer): TAlphaColor;
begin
   //Gets a color from the list
   Result := Colors[AIndex mod Length(Colors)];
```

```
end;

function GetDarkerColorByPercent(AColor: TAlphaColor; ADarkerPercent:
Integer): TAlphaColor;
begin
  //returns a color which is a
  //percentage darker than the input
  Result := InterpolateColor(AColor, TAlphaColorRec.Black, ADarkerPercent /
  100);
end;

function GetLighterColorByPercent(AColor: TAlphaColor; ADarkerPercent:
Integer): TAlphaColor;
begin
  //returns a color which is a
  //percentage lighter than the input
  Result := InterpolateColor(AColor, TAlphaColorRec.White, ADarkerPercent /
  100);
end;
```

Nothing complex here, but the RTL `InterpolateColor` function is quite interesting. This function is used by the `TColorAnimation` effect (and other classes) and has this prototype:

```
function InterpolateColor(const Start, Stop: TAlphaColor; T: Single):
TAlphaColor;
```

The function interpolates the color value between colors `Start` and `Stop` at time moment `T`. So, when `T = 0`, the result is equal to `Start` and when `T = 1`, the result is equal to `Stop`. Interesting, but we need a function that is able to create a lighter or a darker color, starting from one given color. How can we use the `InterpolateColor` to do that? Quite simple.

What is the darkest color in the world? Black (which, as you know, strictly speaking is not even a color). So, if I need to create a color which is 20 percent darker than standard red, I can use `InterpolateColor` to create a new color that *tends* to be dark, but it is only at 20% of the transition. The code looks like the following:

```
TwentyPercDarker := InterpolateColor(TAlphaColorRec.Red,
TAlphaColorRec.Black, 0.2);
```

We can use the same approach to create a lighter color. What's the lightest color in the world? White! So, the code here is:

```
TwentyPercLighter := InterpolateColor(TAlphaColorRec.Red,
TAlphaColorRec.White, 0.2);
```

Simple and effective!

There are two more interesting methods to explain in the main form. The first one is `SetupGradient`. Given a `TBrush` and a reference color, this method sets up the brush to use the color gradient, which starts from the reference color and ends at a color that's 50 percent darker:

```
procedure TMainForm.SetupGradient(ABrush: TBrush; ARefColor: TAlphaColor);
begin
  ABrush.Kind := TBrushKind.Gradient;
  ABrush.Gradient.Color := ARefColor;
  ABrush.Gradient.Color1 := GetDarkerColorByPercent(ARefColor, 50);
end;
```

The last method is `InjectEffects`, which adds effects and animations to the `TFmxObject` that's passed in. These effects are triggered when the mouse moves over the control and are:

- An inner glow
- A color animation of the **Stroke**, from black to a reference color
- A size animation of the **Stroke** thickness to make the border more evident

Here's the code that injects the effect to the `TFmxObject` that's passed in:

```
procedure TMainForm.InjectEffects(AComponent: TFmxObject; ARefColor:
TAlphaColor);
var
  LEffect: TInnerGlowEffect;
  LColorAnimation: TColorAnimation;
  LBoldAnimation: TFloatAnimation;
begin
  // Glow effect when MouseOver
  LEffect := TInnerGlowEffect.Create(AComponent);
  LEffect.Enabled := False;
  LEffect.Trigger := 'IsMouseOver=True';
  LEffect.Parent := AComponent;
  LEffect.GlowColor := TAlphaColorRec.White;
  LEffect.Opacity := 0.5;
  LEffect.Softness := 0.5;

  // Stroke.Color animation when MouseOver
  LColorAnimation := TColorAnimation.Create(AComponent);
  LColorAnimation.PropertyName := 'Stroke.Color';
  LColorAnimation.Enabled := False;
  LColorAnimation.Trigger := 'IsMouseOver=True';
  LColorAnimation.TriggerInverse := 'IsMouseOver=False';
  LColorAnimation.Parent := AComponent;
  LColorAnimation.StartValue := TAlphaColorRec.Black;
```

```
LColorAnimation.StopValue := GetLighterColorByPercent(ARefColor, 20);

// Stroke.Thickness animation when MouseOver
LBoldAnimation := TFloatAnimation.Create(AComponent);
LBoldAnimation.PropertyName := 'Stroke.Thickness';
LBoldAnimation.Enabled := False;
LBoldAnimation.Trigger := 'IsMouseOver=True';
LBoldAnimation.TriggerInverse := 'IsMouseOver=False';
LBoldAnimation.Parent := AComponent;
LBoldAnimation.StartValue := 1;
LBoldAnimation.StopValue := 2;
end;
```

With some code and a bit of cleverness, we've created a not-so-bad pie chart using only the basic FireMonkey shapes, effects, and animation. The same concepts can be used in other situations. For instance, now that we know how to create this pie chart, how difficult could it be to create a histogram? In actuality, it's very simple! You have to use TRectangle instead of TPie and the linear distance instead of degrees! Yes, FireMonkey is very flexible and really repays the time spent getting confident with it.

Now, launch the program, click on the **Refresh Pie Chart** button, and see the chart. Move the mouse over the slices and see how the behaviors that have been implemented make the GUI nice to look at. You can also add a new row in the grid, put in other data, and hit the **Refresh Pie Chart** button once more to see how it changes.

There's more...

It's quite important to know all the layouts that are available in FireMonkey. As the Embarcadero documentation states:

> *"FireMonkey layouts are containers for other graphical objects that can be used to build complex interfaces with visual appeal. The FireMonkey layouts extend the functionality of TControl to control the arrangement, sizing, and scaling of their child controls, and offer the possibility to manipulate a group of controls as a whole."*

In this recipe, we spoke about the basic TLayout; however, there are a lot of layouts. Here's a link so that you can understand the basics about them: http://docwiki.embarcadero.com/RADStudio/en/FireMonkey_Layouts_Strategies.

The way you arrange controls inside the layouts is also important. This link can help you get in touch with the basics of FireMonkey: http://docwiki.embarcadero.com/RADStudio/en/Arranging_FireMonkey_Controls.

5

The Thousand Faces of
Multithreading

In this chapter, we will cover the following topics:

- Synchronizing shared resources with TMonitor
- Talking with the main thread using a thread-safe queue
- Synchronizing multiple threads using TEvent
- Communication made easy with Delphi Event Bus
- Displaying a measure on a 2D graph like an oscilloscope
- Using the Parallel Programming Library in the real world: Tasks
- Using the Parallel Programming Library in the real world: Futures
- Using the Parallel Programming Library in the real world: Parallel For/Join

Introduction

Multithreading can be your biggest problem if you cannot handle it with care. One of the fathers of the Delphi compiler used to say:

> *"New programmers are drawn to multithreading like moths to a flame, with similar results."*

> *– Danny Thorpe*

In this chapter, we will discuss some of the main techniques to handle single or multiple background threads. We'll talk about shared resource synchronization and thread-safe queues and events. The last three recipes will talk about the **Parallel Programming Library** introduced in Delphi XE7, and I hope that you will love it as much as I love it. Multithreaded programming is a huge topic. So, after reading this chapter, although you will not become a master of it, you will surely be able to approach the concept of multithreaded programming with confidence, and will have the basics to move to more specific stuff when (and if) you need to do so.

Synchronizing shared resources with TMonitor

TMonitor is a record used to synchronize threads. Just to be clear, we are talking about System.TMonitor, not Vcl.Forms.TMonitor.

Since Delphi 2009, the TObject instance size has been doubled to make room for an additional 4 bytes. What are these 4 bytes for? They provide TMonitor support!

Now, every TObject descendant can be used as a lock. The type that allows this is the System.TMonitor record, which implements a generic monitor synchronization structure.

Getting ready

In this recipe, you'll face one of the classic multithreading problems—concurrent access to a shared file. Specifically, you'll have a lot of threads writing some information on a file—the same file—and all the threads have to be synchronized for this. Otherwise, the file will not be accessible due to locking, which will cause exceptions in your program code. This problem can be solved in a multitude of ways, but TMonitor offers the simplest solution. Let's start.

How to do it...

Follow these step-by-step instructions to synchronize shared resources with TMonitor:

1. Create a new **VCL Forms Application** (navigate to **File** | **New** | **VCL Forms Application**).

2. Drop a **TButton**, **TListBox**, and **TTimer** component on the form.
3. Name the **TButton** component as `btnStart`, and change the value of **Caption** to `Multiple writes on a shared file`.
4. Add a new unit to the project, call it `FileWriterThreadU.pas`, and add the following code to it:

```
unit FileWriterThreadU;

interface

uses
  System.Classes, System.SyncObjs, System.SysUtils, System.IOUtils;

type
  TThreadHelper = class helper for TThread
  public
    function WaitFor(ATimeout: Cardinal): LongWord; platform;
  end;

  TFileWriterThread = class(TThread)
  private
    FStreamWriter: TStreamWriter;
  protected
    procedure Execute; override;
  public
    constructor Create(AStreamWriter: TStreamWriter);
  end;

implementation

{$IF Defined(MSWINDOWS)}
uses
  Winapi.Windows;
{$IFEND}

constructor TFileWriterThread.Create(AStreamWriter: TStreamWriter);
begin
  FStreamWriter := AStreamWriter;
  inherited Create(False);
end;

procedure TFileWriterThread.Execute;
var
  I: Integer;
  NumLines: Integer;
begin
  inherited;
```

```
      NumLines := 11 + Random(50);
      for I := 1 to NumLines do
      begin
        TThread.Sleep(200);
        //here we are locking the shared resource
        TMonitor.Enter(FStreamWriter);
        try
          FStreamWriter.WriteLine(Format('THREAD %5d - ROW %2d',
      [TThread.CurrentThread.ThreadID, I]));
        finally
          //unlock the shared resource
          TMonitor.Exit(FStreamWriter);
        end;
        if Terminated then
          Break;
      end;
  end;

  function TThreadHelper.WaitFor(ATimeout: Cardinal): LongWord;
  begin
  {$IF Defined(MSWINDOWS)}
    Result := WaitForSingleObject(Handle, ATimeout);
  {$ELSE}
    raise Exception.Create('Available only on MS Windows');
  {$IFEND}
  end;

  initialization

  Randomize; // we'll use Random function in the thread

  end.
```

5. Go back to the form and add the following units to the `interface uses` section:
   - `System.Generics.Collections`
   - `FileWriterThreadU`

6. In the `private` section of the form, declare the following variables:

```
private
  FOutputFile: TStreamWriter;
  FRunningThreads: TObjectList<TFileWriterThread>;
```

7. In the `FormCreate` and `FormClose` event handlers, add the following code:

```
procedure TMainForm.FormCreate(Sender: TObject);
begin
  FRunningThreads := TObjectList<TFileWriterThread>.Create;
```

```
    FOutputFile :=
TStreamWriter.Create(TFileStream.Create('OutputFile.txt',
        fmCreate or fmShareDenyWrite));
end;

procedure TMainForm.FormClose(Sender: TObject; var Action:
TCloseAction);
var
   Th: TFileWriterThread;
begin
   for Th in FRunningThreads do
     Th.Terminate;
   FRunningThreads.Free; // Implicit WaitFor...
   FOutputFile.Free;
end;
```

With the preceding code, you created a data structure to hold the thread list and file access. The `FOutputFile` variable is your shared resource for all the threads.

8. Create the **OnClick** event handler for `btnStart` and add the following code to it:

```
procedure TMainForm.btnStartClick(Sender: TObject);
var
   I: Integer;
   Th: TFileWriterThread;
begin
   for I := 1 to 10 do
   begin
     Th := TFileWriterThread.Create(FOutputFile);
     FRunningThreads.Add(Th);
   end;
end;
```

The preceding code creates 10 threads that will contend for the shared resource `FOutputFile`.

9. Now, threads can run without problems, but the UI doesn't have any information about their jobs. We want to check whether a thread is still running, or is already terminated. So, let's create the event handler for the `Timer1.OnTimer` event using the following code:

```
procedure TMainForm.Timer1Timer(Sender: TObject);
var
   Th: TFileWriterThread;
begin
   ListBox1.Items.BeginUpdate;
   try
```

```
      ListBox1.Items.Clear;
      for Th in FRunningThreads do
      begin
        if Th.WaitFor(0) = WAIT_TIMEOUT then
          ListBox1.Items.Add(Format('%5d RUNNING', [Th.ThreadID]))
        else
          ListBox1.Items.Add(Format('%5d TERMINATED', [Th.ThreadID]))
      end;
    finally
      ListBox1.Items.EndUpdate;
    end;
  end;
```

The preceding code will iterate over the thread list and check the state of each of them. The resultant check will fill the `ListBox1` component.

10. Run the application, and click on the button (the only button present on the form). You should see something like the following:

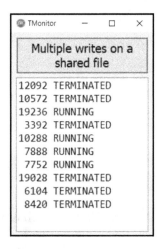

Figure 5.1: The main form showing thread statuses

11. `ListBox1` contains thread statuses. When all threads terminate, you can open the file and see that each of them wrote information without interference from the others; no crashes, no data loss. Your multithreading application is working alright.

12. If you want to see the file while the threads are writing it, you can use one of the Unix tail clone options for Windows suggested in the *Creating a Windows service* recipe of `Chapter 1`, *Delphi Basics*.

How it works...

The `btnStart` event creates 10 threads and puts each of them in a simple generic list declared as `TObjectList<TFileWriterThread>`. This list will be used to iterate over the threads when terminating or checking the status of threads. Threads are not configured with `FreeOnTerminate`, because we require a live reference to check their status.

The real work is done in the `Execute` method of `TFileWriterThread`. Let's check it out:

```
procedure TFileWriterThread.Execute;
var
  I: Integer;
  NumLines: Integer;
begin
  inherited;
  //decide how many numbers to write
  NumLines := 11 + Random(50);
  for I := 1 to NumLines do
  begin
    //wait a bit of time to simulate a higher workload
    TThread.Sleep(200);
    //acquire the lock on FStreamWriter.
    TMonitor.Enter(FStreamWriter);
    try
      //only one thread at time can execute this code
      FStreamWriter.WriteLine(Format('THREAD %5d - ROW %2d',
          [TThread.CurrentThread.ThreadID, I]));
    finally
      //Be sure to release the lock. Otherwise all threads
      //will hang waiting for acquire the lock
      TMonitor.Exit(FStreamWriter);
    end;
    //if thread is terminated exit from the loop
    if Terminated then
      Break;
  end;
end;
```

Another important piece of code is under the **TTimer** event handler:

```
procedure TMainForm.Timer1Timer(Sender: TObject);
var
  Th: TFileWriterThread;
begin
  ListBox1.Items.BeginUpdate;
  try
    ListBox1.Items.Clear;
    for Th in FRunningThreads do
    begin
      //check if the thread if still running. Method WaitFor has
      //been introduced by a class helper in the
      //FileWriterThreadU.pas file, it is not part of TThread
      if Th.WaitFor(0) = WAIT_TIMEOUT then
        ListBox1.Items.Add(Format('%5d RUNNING', [Th.ThreadID]))
      else
        ListBox1.Items.Add(Format('%5d TERMINATED', [Th.ThreadID]))
    end;
  finally
    ListBox1.Items.EndUpdate;
  end;
end;
```

The `WaitFor` method used in the **TTimer** event handler is not part of the standard `TThread` class, but has been introduced using a class helper. This is because the standard `WaitFor` method present on the `TThread` class doesn't provide a timeout for waiting, so it waits forever. If you want to check whether a thread is terminated, or if you simply want to have the GUI responsive while waiting for the thread termination, you cannot do it using the `WaitFor` method. So, we add a new `WaitFor` method that provides a timeout. When you are calling `WaitFor(0)`, you are only asking whether a thread is still running. This is another good utilization of class helpers.

There's more...

Monitors are not a Delphi-specific concept; Wikipedia mentions it as follows:

> *"Monitors were invented by C. A. R. Hoare and Per Brinch Hansen, and were first implemented in Brinch Hansen's Concurrent Pascal language."*

To have a clear understanding of what a monitor is and what's its main utilization, please read the Wikipedia article at `http://en.wikipedia.org/wiki/Monitor_` `%28synchronization%29`.

As a plus, a `TMonitor` class used in a smart way allows you to create a sort of new language construct. Consider the following code:

```
procedure ExecWithLock(const ALockObj: TObject; const AProc: TProc);
begin
  System.TMonitor.Enter(ALockObj);
  try
    AProc();
  finally
    System.TMonitor.Exit(ALockObj);
  end;
end;
```

Using the preceding code, it is possible to write something like the following:

```
ExecWithLock(Obj,
procedure
begin
  //Here you have thread safe access to Obj
end);
```

Cool, isn't it?

Talking with the main thread using a thread-safe queue

Using a background thread and working with its private data is not difficult, but safely bringing information retrieved or elaborated by the thread back to the main thread to show it to the user (as you know, only the main thread can handle the GUI in VCL, as well as in FireMonkey) can be a daunting task. An even more complex task would be establishing a generic communication between two or more background threads. In this recipe, you'll see how a background thread can talk to the main thread in a safe manner using the `TThreadedQueue<T>` class. The same concepts are valid for communication between two or more background threads.

Getting ready

Let's talk about a scenario. You have to show data generated from some sort of device or subsystem; let's say a serial, a USB device, a query polling on the database data, or a TCP socket. You cannot simply wait for data using **TTimer**, because this would freeze your GUI during the wait, and the wait can be long. You have tried it, but your interface became sluggish... you need another solution!

 Performing a long task on the main thread and being unresponsive for a long period can cause an **Application Not Responding (ANR)**. ANR in an application is annoying and frustrating to users. So, the OS shows a dialogue box to inform the user that the application is not responding, and will give the user the choice to either wait or to force close the app! This scenario must be avoided at all costs!

In the Delphi RTL, there is a very useful class called TThreadedQueue<T>, which is, as the name suggests, a particular parametric queue (a FIFO data structure) that can be safely used from different threads. How to use it? In the programming field, there is mostly no single solution that applies to all situations, but the following one is very popular. Feel free to change your approach if necessary; however, this is the approach used in the recipe code:

1. Create the queue within the main form
2. Create a thread and inject the form queue to it
3. In the thread Execute method, append all generated data to the queue
4. In the main form, use a timer or some other mechanism to periodically read from the queue and display data on the form

How to do it...

Open the recipe project called ThreadingQueueSample.dproj. This project contains the main form with all the GUI-related code, and another unit with the thread code.

The FormCreate event creates the shared queue with the following parameters that will influence the behavior of the queue:

- QueueDepth = 100: This is the maximum queue size. If the queue reaches this limit, all the push operations will be blocked for a maximum of PushTimeout, then the Push call will fail with a timeout.

- `PushTimeout = 1000`: This is the timeout in milliseconds that will affect the thread, which, in this recipe, is the producer of a producer/consumer pattern.
- `PopTimeout = 1`: This is the timeout in milliseconds that will affect the timer when the queue is empty. This timeout must be very short, because the pop call is blocking in nature, and you are in the main thread that should never be blocked for an extended period.

The button labeled **Start Thread** creates a `TReaderThread` instance, passing the already created queue to its constructor (this is a particular type of dependency injection called **constructor injection**).

The thread declaration is really simple, and is as follows:

```
type
  TReaderThread = class(TThread)
  private
    FQueue: TThreadedQueue<Byte>;
  protected
    procedure Execute; override;
  public
    constructor Create(AQueue: TThreadedQueue<Byte>);
  end;
```

While the `Execute` method simply appends randomly generated data to the queue, note that the `Terminated` property must be checked often so the application can terminate the thread and wait a reasonable time for its actual termination. In the following example, if the queue is not empty, check the termination at least every 700 milliseconds:

```
procedure TReaderThread.Execute;
begin
  while not Terminated do
begin
    TThread.Sleep(200 + Trunc(Random(500)));
    // e.g. reading from an actual device
    FQueue.PushItem(Random(256));
  end;
end;
```

So far, you've filled the queue. Now, you have to read from the queue and do something useful with the read data. This is the job of a timer. The following is the code of the timer event on the main form:

```
procedure TMainForm.Timer1Timer(Sender: TObject);
var
  Value: Byte;
begin
```

```
  while FQueue.PopItem(Value) = TWaitResult.wrSignaled do
  begin
    ListBox1.Items.Add(Format('[%3.3d]', [Value]));
  end;
  ListBox1.ItemIndex := ListBox1.Count - 1;
end;
```

That's it! Run the application and see how we are reading the data coming from the threads and showing this data on the main form, as can be seen in the following screenshot:

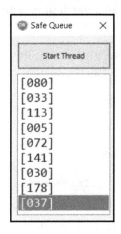

Figure 5.2: The main form showing data generated by the background thread

There's more...

The `TThreadedQueue<T>` is very powerful, and can be used to communicate between two or more background threads in a consumer/producer schema as well. You can use multiple producers, multiple consumers, or both. The following screenshot shows a popular schema used when the speed at which the data generated is faster than the speed at which the same data is handled. In this case, usually you can gain speed on the processing side using multiple consumers:

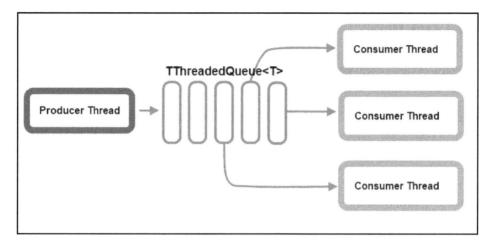

Figure 5.3: Single producer, multiple consumers

Synchronizing multiple threads using TEvent

The synchronization details we discussed so far were related to a data flow that is generated in the background thread context and has to be used in another thread. The other thread can be the main thread or another background thread. In this recipe, you'll use a simple synchronization mechanism called **event**, which can be useful when you have to notify a new state, not necessarily new data. Obviously, the new state could also mean there is new data to handle. In such cases, the state change alerts you about new data being available.

Getting ready

The recipe scenario is simple: you have a lot of running threads that are doing something for you. You want to know when all of them are terminated. In this case, you can use a TEvent object (this is a tiny wrapper around an OS Event object).

How to do it...

This recipe is a bit articulated, so we'll not discuss the steps needed to recreate it. Please open the recipe project code named `ThreadsTermination.dproj`; let's comment on it together.

The GUI is minimal; there is a button to run the threads, and a list box to show the current state of the threads. The `FormCreate` event initializes a list to hold the threads that will be used later. When you click on the button, the program launches five threads. Each thread waits for a random amount of time, and then generates a random number that should represent your output data. The main thread has to be notified about the thread termination. The thread code is as follows:

```
unit MyThreadU;

interface

uses
  System.Classes, System.SyncObjs;

type
  TMyThread = class(TThread)
  private
    FEvent: TEvent;
    FData: Integer;
  protected
    procedure Execute; override;
  public
    constructor Create(AEvent: TEvent);
    destructor Destroy; override;
    property Event: TEvent read FEvent;
    function GetData: Integer;
  end;

implementation

uses System.SysUtils;

constructor TMyThread.Create(AEvent: TEvent);
begin
  FEvent := AEvent;
  inherited Create(False);
end;

destructor TMyThread.Destroy;
begin
```

```
  FreeAndNil(FEvent);
  inherited;
end;

procedure TMyThread.Execute;
begin
  TThread.Sleep(2000 + Random(4000));
  FData := Random(1000);
  // This call sets the internal event state to signaled
  FEvent.SetEvent;
end;

function TMyThread.GetData: Integer;
begin
  Result := FData;
end;

end.
```

In the thread, the constructor is injected a `TEvent` instance. When the thread does its job, it calls the `SetEvent` method on the event instance. This call sets the internal event state to signaled. What's that for? It is required because the main thread is waiting for this change. To be more precise, it is waiting to know when all the threads have called their `SetEvent` methods. The following function is used to check whether there are any running threads:

```
function TMainForm.AreThereThreadsStillRunning: Boolean;
var
  H: THandleObject;
begin
  Result := TEvent.WaitForMultiple(Handles, 1, True, H) = wrTimeout;
end;
```

In the preceding code, the variable `Handles` is an array containing all the `Events` that have to be checked for termination.

The button event handler requires a bit of explanation. The code is as follows:

```
procedure TMainForm.btnStartClick(Sender: TObject);
var
  i: Integer;
  Evt: TEvent;
begin
  if (FThreads.Count > 0) and AreThereThreadsStillRunning then
  begin
    ShowMessage('Please wait, there are threads still running');
    Exit;
  end;
```

```
    FThreads.Clear;
    for i := 0 to High(Handles) do
    begin
      Evt := TEvent.Create;
      Handles[i] := Evt;
      FThreads.Add(TMyThread.Create(Evt));
    end;
    ListBox1.Items.Add('Threads running');
    Timer1.Enabled := True;
  end;
```

When the user clicks on the button, the application checks whether there are any running threads from previous clicks. If so, inform the user with a ShowMessage and exit. If there are no running threads, the code fills the thread list with five threads. Each thread has its own TEvent instance to talk to. The reference to the TEvent variable is passed to the threads, but the threads have a property of accessing it during its runtime.

What is the best way to read the thread status? In a **TTimer** class, the code under the **OnTimer** event is the following consider that this timer is normally disabled:

```
    procedure TMainForm.Timer1Timer(Sender: TObject);
    var
      th: TMyThread;
    begin
      if not AreThereThreadsStillRunning then
      begin
        Timer1.Enabled := False;
        ListBox1.Items.Add('All threads terminated');
        for th in FThreads do
        begin
          ListBox1.Items.Add(Format('Th %4.4d = %4d', [th.ThreadID,
    th.GetData]));
        end;
      end;
    end;
```

With this last procedure, you retrieved the thread status. When all threads finished running, you also retrieved the *calculated* value.

There's more...

The event object is used to send a signal to a thread indicating that a particular event has occurred inside another thread. The event does not carry information; it simply informs us that *something has happened*. It is simple, but can be useful in creating very complex synchronization mechanisms between two or more threads.

Events can be in a signaled state or not. If you want to have a deeper understanding of the event objects and their utilization, visit the following links:

- `http://msdn.microsoft.com/en-us/library/windows/desktop/ms682655(v=vs.85).aspx`
- `http://docwiki.embarcadero.com/RADStudio/en/Waiting_for_a_Task_to_Be_Completed`

Bonus recipe – AsyncTaskTests

Another useful scenario for applying events are asynchronous tests! Take the example where you need to make a REST request to an endpoint and verify the result. How would you develop the test? (In this case, we are talking about integration tests.)

In this recipe, I used DUnitX as the XUnit test framework. If you don't know what TDD is (I really hope you know it, because I'm a huge fan of TDD and it can really save your life!), or DUnitX or XUnit for that matter, I've provided you with the following links:

- `https://en.wikipedia.org/wiki/Test-driven_development`
- `https://en.wikipedia.org/wiki/XUnit`
- `http://docwiki.embarcadero.com/RADStudio/en/DUnitX_Overview`
- `https://en.wikipedia.org/wiki/Integration_testing`

If you use DUnitX, the context of execution of tests is the main thread. This means that you cannot put your tests in background threads, because the main context could end before the background thread has terminated, and therefore your tests will not be executed. In this case, the `TEvent` variable is very useful to signal to main context that the background task has terminated, and the effective test of the resulting data can be executed.

Here's the code:

```
procedure TAsyncTaskTests.TestAsyncTask;
var
  LEvent: TEvent;
  LResult: Boolean;
begin
  LResult := False;
  LEvent := TEvent.Create;
  TTask.Run(
    procedure
    var
      LHTTP: THTTPClient;
      LResp: IHTTPResponse;
```

```
    begin
      LHTTP := THTTPClient.Create;
      try
        LResp := LHTTP.Get
('http://api.timezonedb.com/v2/get-time-zone?format=json&by=zone&zone=Italy
/Rome');
        if LResp.StatusCode = 200 then
        begin
          LResult := not LResp.ContentAsString(TEncoding.UTF8).IsEmpty
        end
        else
        begin
          raise Exception.CreateFmt('Cannot get time. HTTP %d - %s',
            [LResp.StatusCode, LResp.StatusText]);
        end;
      finally
        LHTTP.Free;
        // SIGNAL THAT TASK HAS TERMINATED!!
        LEvent.SetEvent;
      end;
    end);
  // attend for max 5 seconds
  Assert.IsTrue(TWaitResult.wrSignaled = LEvent.WaitFor(5000),
    'Timeout request');
  Assert.IsTrue(LResult);
end;
```

You can find this bonus recipe opening under `Chapter 5\CODE\RECIPE03\ AsyncTaskTests.dproj`, or `Chapter 5\CODE\RECIPE03\ Recipe03.groupproj`.

Communication made easy with Delphi Event Bus

A typical Delphi application tends to be composed of many layers, modules, or structures, such as Forms, Views, Presenters, Data Modules, and Services. Effective communication between these components can become difficult if they are tightly coupled together.

In the lower level of your app architecture, such as the Data Module, when an action happens, you might want to send data to a higher level, such as the view (Form). To do this, you might want to create a listener interface, async tasks, or callbacks. All of these will work, but they have some major drawbacks:

- Direct or tight coupling of components
- Inflexible, changes are expensive
- Boilerplate code
- Propagation through all layers
- Repetition of code
- Difficulty in testing
- Increased risk of bugs

All of these things are brilliantly solved by modern **model-view-whatever** (**MVW**) frameworks; here's a nice article on this subject: `https://www.beyondjava.net/blog/ model-view-whatever/`. In this recipe, I'll show you how to solve these drawbacks by using an event bus framework—Delphi Event Bus.

Getting ready

In modern applications, we have to put user experience problem at first! What matters is how the application responds to user experience:

- Responsive design
- Non-blocking UI
- Offline-first (where possible)

The purpose is to create a login form, where, after entering the data, the authentication request is made through a REST call. The UI must show a waiting UI for the user while the authentication process is in progress, and finally inform the user of authentication, both in case of success and error.

This recipe uses a Delphi Event Bus open source project:

- An event bus framework based on a publish/subscribe mechanism
- Project website: `https://github.com/spinettaro/delphi-event-bus` written by Daniele Spinetti

Delphi Event Bus (**DEB**) is a publish/subscribe event bus framework for the Delphi platform. DEB is designed to decouple different parts/layers of your application while still allowing them to communicate efficiently. A really exciting feature of this framework is that you can choose the delivery mode of the events, in the main thread or a background thread. We'll use this amazing feature to communicate with UI.

To download DEB, go to the project website and clone the repository using a Git client. You can use the command-line version with the following command line (suppose you're in `D:\DEV\Delphi`):

```
git clone https://github.com/spinettaro/delphi-event-bus.git delphi-event-bus
```

Then, you have to add on your search paths:

- `D:\DEV\Delphi\delphi-event-bus\source`
- `D:\DEV\Delphi\delphi-event-bus\source\external`

For this recipe, we'll also use a new entry of RAD Studio 10.2.3—**FireMonkey UI Templates**. FireMonkey UI templates, designed to highlight FMX's multi-device capabilities, showcase best practices, and help new users to get started more quickly. They showcase UI paradigms that mobile application developers need in today's applications.

There is one that is right for us: **Login Screens**. So go on **GetIt (Tools)** | **GetIt Package Manager**, write `Login Screens` on search edit, and install it:

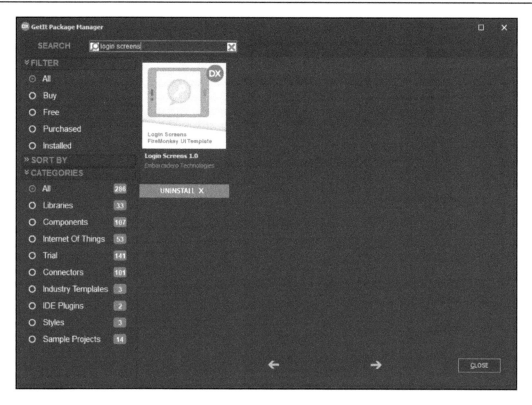

Figure 5.4: FireMonkey UI Template

How to do it...

1. Let's start by copying the `LoginProject1` project you have just downloaded (this is a fantastic feature of the FireMonkey UI Template! We're already up and running aren't we?).

2. Create `ServicesU` unit to the definition and creation of the service responsible for making the authentication call:

```
unit ServicesU;

interface

type

  IRESTAPIService = interface
    ['{53017C06-D7F7-4F2D-9FEA-43ED543E9BC6}']
```

```
    procedure DoLogin(const AUsername: String; const APassword:
String);
  end;

  TRESTAPIService = class(TInterfacedObject, IRESTAPIService)
  public
    procedure DoLogin(const AUsername: String; const APassword:
String);
  end;

function GetRESTAPIService: IRESTAPIService;

implementation

uses
  System.Threading, System.Classes, EventsU, EventBus,
System.SysUtils;

function GetRESTAPIService: IRESTAPIService;
begin
  Result := TRESTAPIService.Create;
end;

{ TRESTAPIService }

procedure TRESTAPIService.DoLogin(const AUsername, APassword:
String);
begin
  TTask.Run(
    procedure
    var
      LResult: boolean;
      LAuthEvent: TAuthenticationEvent;
    begin
      // simulate a long operation like HTTP request
      TThread.Sleep(3000);
      // Authentication is ok if username and password are equal
      LResult := AUsername = APassword;
      // create the event and post it with DEB default instance
      LAuthEvent := TAuthenticationEvent.Create(LResult);
      TEventBus.GetDefault.Post(LAuthEvent);
    end);
end;

end.
```

3. Create `EventsU` for the definition of authentication event needed to DEB publish/subscribe mechanism. `TAuthenticationEvent` is the type of event that will be triggered and will be submitted to all subscribers who have registered for this specific event:

```
unit EventsU;

interface

type

  TAuthenticationEvent = class(TObject)
  private
    FResult: Boolean;
    procedure SetResult(const Value: Boolean);
  public
    constructor Create(AResult: Boolean);
    property Result: Boolean read FResult write SetResult;
  end;

implementation

{ TAuthenticationEvent }

constructor TAuthenticationEvent.Create(AResult: Boolean);
begin
  inherited Create;
  Result := AResult;
end;

procedure TAuthenticationEvent.SetResult(const Value: Boolean);
begin
  FResult := Value;
end;

end.
```

4. Go to `LoginFrame` and add to `LoginFrame1` a `TRectangle` and a `TAniIndicator` within the rectangle:
    1. Align **Rectangle** to **Client**.
    2. Set **Opacity** to **0,7**.
    3. Set **Visible** to **False**.
    4. Align **AniIndicator** to **Client**.

5. Create a `WaitMode` method to change rectangle visibility, and enable **AniIndicator**. The following code is designed to simulate a wait user experience:

```
procedure TLoginFrame1.WaitMode(AValue: Boolean);
begin
  Rectangle1.Visible := AValue;
  AniIndicator1.Enabled := AValue;
end;
```

6. Create a public method `OnUserAuthentication` on **LoginForm**, and decorate it with **Subscribe Attribute**. In this way, we inform DEB instance, by using the `RegisterSubscriber` function, that there is a method that must be invoked when an event of type `TAuthenticationEvent` is triggered:

```
public
  { Public declarations }
  // ensure that UI event will be execute in Main Thread
  [Subscribe(TThreadMode.Main)]
  procedure OnUserAuthentication(AEvent: TAuthenticationEvent);
end;

procedure TForm1Login.OnUserAuthentication(AEvent:
TAuthenticationEvent);
begin
  try
    // put the wait mode to false, so rectangle invisible and
aniindicator disabled
    LoginFrame11.WaitMode(false);
    if AEvent.Result then
      ShowMessage('Login Ok. You will be redirected to the
dashboard...')
    else
      ShowMessage('Login KO. It seems that you have entered invalid
data,
      please check yuour data...')
  finally
    AEvent.Free;
  end;
end;
```

7. Finally, register `LoginForm` as the **Subscriber**, and define `AuthenticationRectBTNClick` (double-click on **Authentication Rect** in **LoginForm**). In this button click, we call the `DoLogin` method of `RESTAPIService`:

```
procedure TForm1Login.FormCreate(Sender: TObject);
begin
```

```
    TEventBus.GetDefault.RegisterSubscriber(Self);
  end;

  procedure TForm1Login.LoginFrame11AuthenticateRectBTNClick(Sender:
  TObject);
  begin
    LoginFrame11.WaitMode(True);
    GetRESTAPIService.DoLogin(LoginFrame11.EmailEdit.Text,
      LoginFrame11.PasswordEdit.Text);
  end;
```

How it works...

In the `FormCreate` event, the `LoginForm` is registered as the **Subscriber** in the publish/subscribe mechanism of DEB. This means that, thanks to the `OnUserAuthentication(AEvent: TAuthenticationEvent);` and `Subscribe(TThreadMode.Main)` attribute, when an event of type `TAuthenticationEvent` is fired in DEB, this method will be executed. Furthermore, specifying `TThreadMode` as a main, it is ensured that the thread in which that method will be executed is the main thread (UI manipulation must be executed in the main thread).

In the `DoLogin` method of `TRESTAPIService`, the authentication call has been implemented: `TThread.Sleep` simulates a lengthy operation like an HTTP call, and then authentication is successful, if the username corresponds to the password. Here's the magic; by invoking the `Post` method of the default instance of `TEventBus`, all those who signed up for the `TAuthenticationEvent` are informed; in this specific case, `LoginForm` in `OnUserAuthentication(AEvent: TAuthenticationEvent);`.

No boilerplate code, no direct or tight coupling of components, no repetition of code, `RESTAPIService` could be tested in isolation, no sync with the main thread... isn't all of this fantastic?

Run the application, fill the authentication data, click on `AuthenticateBTNRect` to make the WaitUI visible, and make the call through `RESTAPIService`:

Figure 5.5: Login Form

Following is the image of login form in wait mode:

Figure 5.6: Login Form in WaitMode

There's more...

Using publish/subscribe or message bus architecture prevents all the potential problems highlighted above from becoming drawbacks. It is a very good way of implementing effective communication between components in an application without any of them needing to be aware of the others immediately, with all the benefits this entails. In the end, Event Bus saves loads of time and makes your code more readable, maintainable, extensible, and flexible.

"If the only tool you have is a hammer, you tend to see every problem as a nail."

- Abraham Maslow

When developers discover the benefits of it, they tend to replace everything with Event Bus, every communication call! Abusing it makes your code hard to understand, and hard to debug. Nested events and long event chains are a common cause of unexpected behavior, and making it difficulty to debug.

See also

There are also other scenarios that DEB solves in an elegant way. Take a look at the `samples` section in the repository `https://github.com/spinettaro/delphi-event-bus`.

For this recipe, the new FireMonkey UI templates were used, and there are a lot of UI templates ready for use. You can find more information here:

- `http://docwiki.embarcadero.com/RADStudio/en/10.2_Tokyo_-_Release_3`
- `https://community.embarcadero.com/blogs/entry/new-in-10-2-3-firemonkey-ui-templates-login-screens`
- `https://community.embarcadero.com/blogs/entry/new-in-10-2-3-firemonkey-ui-templates-profile-screens`

Displaying a measure on a 2D graph like an oscilloscope

An **oscilloscope** is a type of electronic test instrument that allows the observation of constantly varying signal voltages. Usually, information is shown as a 2-dimensional plot graph of one or more signals as a function of time. In this recipe, you'll implement a type of oscilloscope to display data generated by a background thread. Obviously, in this recipe, you'll not create an accurate oscilloscope; rather, a nice real-world utilization of retrieving data and using it continuously in the GUI.

Getting ready

You'll need to use the `TThreadedQueue<Extended>` class to bring out data from the background thread to the main thread. The approach is similar to that shown in the recipe *Talking with the main thread using a thread-safe queue*, but, in this case, we have to show data in a complex manner-on a 2D graph, showing only the last data retrieved.

How to do it...

This recipe has a background thread acting like an *analog signal generator* that is able to generate a sine-style stream of data and a graph that plots this data. The resulting application is as follows:

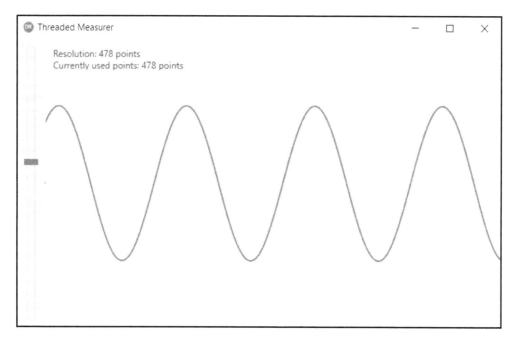

Figure 5.7: The main form showing a sine function generated by a background thread

You can adjust the resolution of the plot (number of points used to draw the sine) using the track bar on the left-hand side. Let's see the most important parts.

The thread used as a signal generator is very simple. As shown in the following code, it uses the `System.Math.Sin` function to generate a sine wave form. Roughly every 10 milliseconds, a new value is appended to the queue; this value is the sample you get from the measured system. The code is as follows:

```
procedure TSignalGeneratorThread.Execute;
var
  Value: Extended;
begin
  inherited;
  Value := 0;
  while not Terminated do
  begin
    TThread.Sleep(10);
    FQueue.PushItem(Sin(Value) * 100);
    Value := Value + 0.05;
    if Value >= 360 then
      Value := 0;
  end;
end;
```

Being a classic producer/consumer, this architecture has to deal with the classic problem of a queue being full and not accepting any data until someone starts to dequeue from it. At regular intervals, a **TTimer** dequeues all the values from the queue and appends them to a different queue living in the main thread.

This queue must have a fixed size, so if there are more values than what is defined by the resolution, the oldest values are dequeued until the queue size is equal to the maximum length permitted. This adjustment is done in the timer event handler with the following code:

```
procedure TMainForm.Timer1Timer(Sender: TObject);
var
  Value: Extended;
  QueueSize: Integer;
begin
  // put readed values in the display list...
  // max FMaxValuesCount values
  while FValuesQueue.PopItem(QueueSize, Value) = TWaitResult.wrSignaled do
  begin
    FDisplayList.Add(Value);
  end;
  // remove values from the head of the list...
  while FDisplayList.Count > FMaxValuesCount do
  begin
    FDisplayList.Delete(0);
  end;
```

```
  // RefreshGraph;
  pb.Repaint;
end;
```

The actual values are plotted on a simple 2D graph using TPaintBox as a canvas. Remember that only the main thread should repaint and call paint procedures. The following is the code in the OnPaint event used to draw the plot:

```
procedure TMainForm.pbPaint(Sender: TObject; Canvas: TCanvas);
var
  Values: TPolygon;
  I: Integer;
  XStep: Extended;
  YCenter: Integer;
begin
  // prepare scene
  Canvas.BeginScene;
  Canvas.Stroke.Kind := TBrushKind.Solid;
  Canvas.Stroke.Thickness := 1;

  // setup the canvas with a white background
  Canvas.Fill.Color := TAlphaColorRec.White;
  Canvas.FillRect(RectF(0, 0, Canvas.Width, Canvas.Height), 0, 0, [], 1);

  // write the blue top-left labels
  Canvas.Fill.Color := TAlphaColorRec.Blue;
  Canvas.FillText(RectF(10, 10, Canvas.Width, 40),
    'Resolution: ' + MaxValuesCount.ToString + ' points', False, 1, [],
    TTextAlign.Leading, TTextAlign.Leading);
  Canvas.FillText(RectF(10, 25, Canvas.Width, 40),
    'Currently used points: ' +
    FDisplayList.Count.ToString + ' points', False, 1, [],
    TTextAlign.Leading, TTextAlign.Leading);

  // preparing points to draw
  SetLength(Values, FDisplayList.Count);
  XStep := Canvas.Width / FDisplayList.Count;
  YCenter := Canvas.Height div 2;
  for I := 0 to FDisplayList.Count - 1 do
  begin
    Values[I].X := XStep * I;
    Values[I].Y := YCenter - FDisplayList[I];
  end;

  // setup the points aspect
  Canvas.Stroke.Thickness := 2;
  Canvas.Stroke.Color := TAlphaColorRec.Red;
  // draw the points
```

```
    DrawOpenPolygon(Canvas, Values, 1);

    // actually update the canvas
    Canvas.EndScene;
  end;
```

FireMonkey Canvas does not allow you to draw an *open* polygon. An open polygon is a shape composed of two or more line segments, where the first and the last points are not connected directly. Here, we need an open polygon; so how do we do it? Here's the code for the `DrawOpenPolygon` method:

```
procedure TMainForm.DrawOpenPolygon(const Canvas: TCanvas; const Points:
TPolygon; const AOpacity: Single);
var
  I: Integer;
  LPath: TPathData;
begin
  if Length(Points) = 0 then
    Exit;
  LPath := TPathData.Create;
  try
    LPath.MoveTo(Points[0]);
    for I := 1 to High(Points) do
      LPath.LineTo(Points[I]);
    Canvas.DrawPath(LPath, AOpacity);
  finally
    LPath.Free;
  end;
end;
```

There's more...

Showing dynamically changing data is always a challenge, and is a typical synchronization problem if you have to read from a blocking and very fast data source; however, using queues in an efficient way can help to reach the correct architecture. If you have very high concurrency (many consumers or many producers), or a very high producer speed compared to the consumer's speed, you may have some performance improvements using lock-free data structures.

Unluckily, in Delphi, there are no ready-to-use lock-free data structures; however, there are very good libraries, even open source, that implement it in the context of multithreaded programming. One of the most popular libraries, although, at the time of writing, it is only designed for the Windows platform, is the open source `OmniThreadLibrary` from Primož Gabrijelčič (`http://www.omnithreadlibrary.com/`).

Using the Parallel Programming Library in the real world: Tasks

Since RAD Studio XE7 Delphi and C++ Builder, developers have been able to use the **Parallel Programming Library** (**PPL**). What is PPL? PPL is a part of the Delphi RTL that provides facilities for multithreading (or parallel) programming.

PPL is available for all the platforms supported by Delphi, and provides a number of advanced features for running tasks, joining tasks, and waiting on groups of tasks to process. PPL is not only a different way to create threads, but is a different way to manage threads as well. Why? Because to manage all of these features (tasks, futures, parallel for, and joining), there is a thread pool that self-tunes automatically (based on the load on the CPUs), so you do not have to care about creating or managing threads for this purpose.

The good news is that PPL is quite simple to use and doesn't require big changes to your application. You can use this library by including `System.Threading` in your application or app. This unit is made up of several features that can be included in new and existing projects.

So the question is, *How and when can I use the PPL?* Well, in all cases where you usually need to use a thread, you should consider a task. This doesn't mean that you will not create threads any more, but, in many cases, you will end up using some sort of task instead of a normal thread.

In this recipe, we'll develop a reusable asynchronous library to accomplish a very recurrent need: start a background operation and be informed, in the main thread, when the background process ends successfully or not.

Using plain PPL, a very recurrent code is similar to the following:

```
procedure TMainForm.btnITaskClick(Sender: TObject);
var
  LTask: ITask;
begin
  LTask := TTask.Run(
    procedure
    var
      LResult: Integer;
    begin
      Sleep(1000); //do something useful here...
      LResult := Random(100); //some kind of "result"
      //Queue the execution in the main thread
      TThread.Queue(nil,
        procedure
```

```
      begin
        TaskEnd(LResult); //TaskEnd is called in the UI thread
      end);
    end);
  end;
```

This is quite simple, but things get a bit more complicated when you have many tasks running and you have to handle exceptions. So, a bit layer to increase usability can be useful. Here's the `Async.Run<T>`.

A complete call to the `Async.Run<T>` method is made up of three anonymous methods:

- **A background task**: This is a function returning some kind of data. It runs in a background thread using a PPL task.
- **A success callback**: This is a procedure that gets the result of the background task. It runs in the main UI thread.
- **An error callback**: This is a procedure that gets the exception raised by the background task, if any. It runs in the main UI thread.

This small library can be used in the following way:

```
Async.Run<String>(
  function: String
  begin
    //This is the "background" anonymous method. Runs in the
    //background thread, and its result is passed
    //to the "success" callback.
    //In this case the result is a String.
  end,
  procedure(const Value: String)
  begin
    //This is the "success" callback. Runs in the UI thread and
    //gets the result of the "background" anonymous method.
  end,
  procedure(const Ex: Exception)
  begin
    //This is the "error" callback.
    //Runs in the UI thread and is called only if the
    //"background" anonymous method raises an exception.
  end);
```

In this case, the data returned by the background function is a string, but, `Async.Run<T>` being a generic method, you can change the type to whatever you want.

Getting ready

In this recipe, we'll create the `Async` library and a testbed program for it. Our objective is to exercise the library with some use cases to see how it works.

How it works...

Open the project `AsyncTaskSample.dproj`, and then let's talk about the `AsyncTask.pas` unit.

Here's the unit with some comments:

```
unit AsyncTask;

interface

uses
  System.SysUtils,
  System.Threading; //The PPL unit

type
  //the "background" task
  TAsyncBackgroundTask<T> = reference to function: T;

  //the "success" callback
  TAsyncSuccessCallback<T> = reference to procedure(const TaskResult: T);

  //the "error" callback
  TAsyncErrorCallback = reference to procedure(const E: Exception);

  //the default "error" callback if the user does not provide it
  TAsyncDefaultErrorCallback = reference to procedure(const E: Exception;
    const ExptAddress: Pointer);

  //the main class
  Async = class sealed
  public
    class function Run<T>(Task: TAsyncBackgroundTask<T>; Success:
TAsyncSuccessCallback<T>;
      Error: TAsyncErrorCallback = nil): ITask;
  end;

var
   //default "error" callback. It is a public var so that the
   //programmer can override the default behaviour
   DefaultTaskErrorHandler: TAsyncDefaultErrorCallback = nil;
```

```
implementation

uses
  System.Classes;

class function Async.Run<T>(Task: TAsyncBackgroundTask<T>; Success:
TAsyncSuccessCallback<T>;
  Error: TAsyncErrorCallback): ITask;
var
  LRes: T;
begin
  //the background task starts here
  Result := TTask.Run(
  procedure
  var
    Ex: Pointer;
    ExceptionAddress: Pointer;
  begin
    Ex := nil;
    try
      LRes := Task(); //run the actual task
      if Assigned(Success) then
      begin
      //call the success callback passing the result
        TThread.Queue(nil,
          procedure
          begin
            Success(LRes);
          end);
      end;
    except
      //let's extend the life of the exception object
      Ex := AcquireExceptionObject;
      ExceptionAddress := ExceptAddr;

      //queue on the main thread to call the error callback
      TThread.Queue(nil,
        procedure
        var
          LCurrException: Exception;
        begin
          LCurrException := Exception(Ex);
          try
            if Assigned(Error) then
              Error(LCurrException) //call the "error" callback
            else
              DefaultTaskErrorHandler(LCurrException, ExceptionAddress);
          finally
```

```
      //free the exception object. It is necessary
      //because we "extended" the natural life
      //of the exception object beyond the except block
      FreeAndNil(LCurrException);
    end;
  end);
  end; //except
 end); //task.run
end;


initialization

//this is the default error callback
DefaultTaskErrorHandler :=
  procedure(const E: Exception; const ExceptionAddress: Pointer)
  begin
    ShowException(E, ExceptionAddress);
  end;

end.
```

Now that we know how the `Async.Run<T>` is implemented, let's see how to use it. Open the main form and check the code under each button. Let's start with the `btnSimple` button:

```
procedure TMainForm.btnSimpleClick(Sender: TObject);
begin
  Async.Run<Integer>(
    function: Integer //long operation in the background
    begin
      Sleep(2000);
      Result := Random(100);
    end,
    procedure(const Value: Integer) //show the result in the UI
    begin
      //write the result in a memo in the form
      Log('RESULT: ' + Value.ToString);
    end);
end;
```

The long and, in this case, fake operation is executed in the anonymous method as a function that returns an integer. When the function ends, its return value is passed to the other anonymous method, which is a procedure, and runs in the UI thread so that it can interact with the user. If you run the program and click on this button, you can verify that the UI is not frozen while the long operation (actually, a `Sleep(2000)` call) is running.

The second button is named btnWithException and shows how to handle exceptions that may be raised inside the background thread:

```delphi
procedure TMainForm.btnWithExceptionClick(Sender: TObject);
begin
  Async.Run<String>(
    function: String
    begin
      raise Exception.Create('This is an error message');
    end,
    procedure(const Value: String)
    begin
     // never called
    end,
    procedure(const Ex: Exception)
    begin
        Log('Exception: ' + sLineBreak + Ex.Message);
    end);
end;
```

Quite simple, isn't it? If something goes wrong in the background, the related exception object is passed to the error callback. Pay attention that the error block is not a standard Delphi exception block, it is just an anonymous method that gets an Exception object. So, for instance, a raise—call to reraise the current exception is not allowed.

The next button called btnExceptionDef shows the library's ability to handle the exception raised in the background even if the programmer doesn't handle it directly or forgot to do it:

```delphi
procedure TMainForm.btnExceptionDefClick(Sender: TObject);
begin
  Async.Run<String>(
    function: String
    begin
      raise Exception.Create('Handled by the default Exception handler');
    end,
    procedure(const Value: String)
    begin
     // never called
    end);
end;
```

Clicking on this button, you will see the Delphi standard exception message. In some cases, this can be sufficient; but if you need some custom handling, you can simply pass the specific callback or override the default behavior, assigning another default handler to the global variable DefaultTaskErrorHandler.

The last button actually does something useful: it gets the current time from a rest service. The button is called btnRESTRequest, and this is the code behind it:

```
procedure TMainForm.btnRESTRequestClick(Sender: TObject);
begin
  Async.Run<String>(
    function: String
    var
      LHTTP: THTTPClient;
      LResp: IHTTPResponse;
    begin
      LHTTP := THTTPClient.Create;
      try
        LResp := LHTTP.Get('http://worldclockapi.com/api/json/utc/now');
        if LResp.StatusCode = 200 then
        begin
          Result := LResp.ContentAsString(TEncoding.UTF8)
        end
        else
        begin
          raise Exception.CreateFmt('Cannot get time. HTTP %d - %s',
            [LResp.StatusCode, LResp.StatusText]);
        end;
      finally
        LHTTP.Free;
      end;
    end,
    procedure(const AJSONStringResp: String)
    var
      LJSONObj: TJSONObject;
    begin
      LJSONObj := TJSONObject.ParseJSONValue(AJSONStringResp) as
TJSONObject;
      try
        Log('Current Date Time: ' +
LJSONObj.GetValue('currentDateTime').Value);
      finally
        LJSONObj.Free;
      end;
    end,
    procedure(const Ex: Exception)
    begin
      Log('Exception: ' + sLineBreak + Ex.Message);
    end);
end;
```

At this point, the code should be clear. In the background task, the actual rest call is executed. If the server replies with a 200 OK HTTP status, then the response body is passed to the success callback, otherwise an exception is raised explaining that the error occurred.

There's more...

The PPL greatly simplifies multithreading programming; however, the biggest advantage is the `ThreadPool`, which does the dirty job of creating, destroying, and reusing the background threads. So please, don't see the PPL as a different way to create threads; it is a powerful mechanism to correctly handle multiple threads without saturating the CPUs. Remember that when you ask the PPL to start a task, the task may not start immediately. This is because the `ThreadPool` may decide to put your task in the waiting queue and actually start it ASAP, but not now. The great thing is that this is actually not a problem but a feature, because, otherwise, you will easily saturate the CPUs' resources. In my own experience, in complex situations, you cannot simply start a thread when you need it, but you have to inform the thread manager that you need a thread and then it can start it ASAP. This thread manager was absent in Delphi RTL before the PPL, so this is the reason why I greatly appreciate the PPL `ThreadPool`.

Here are some useful links to get started with the PPL concepts and classes:

- `http://docwiki.embarcadero.com/RADStudio/en/Using_the_Parallel_Programming_Library`
- `http://docwiki.embarcadero.com/RADStudio/en/Using_TTask_from_the_Parallel_Programming_Library`
- `http://www.danieleteti.it/using-dynamic-arrays-and-parallel-programming-library-part-1/`

Using the Parallel Programming Library in the real world: Futures

Futures are a great tool in the tool chest of every programmer. But, wait! What's a future?

Well, while a task can be seen as a sort of asynchronous procedure, a future can be seen as an asynchronous function; however, while using a task, the process is quite clear (it runs in the background and uses some sort of messaging to talk to the other thread), the future is a bit more complex. When should I get the return value of the future? Let's talk about futures with an example. You can use futures to run tasks on a separate thread and then forget about them, but often, you'll want to use the result of the task. The future function returns an `IFuture<T>` reference that you can use to request the result of type `T`. The reference is like the ticket that a dry cleaner gives you; at any time, you can use it to request your clean dress, but if your dress isn't clean yet, you'll have to wait. Similarly, you can use the reference value to request a future's result, but if the future isn't done computing the result, you'll have to wait.

In this recipe, we'll develop a simple application able to convert money between any currency, in this case, the Euros. So, you will set the source currency type, set the amount of money, and then, with the press of a button, you can convert the value in Euros. Quite simple; however, this application gets the currency rates from a web service, so there is some delay in every conversion. Let's talk about the process. Take a look at the following table:

You	Program
Select the source currency type	Do nothing
Write the amount of money to be converted	Do nothing
Press the button	Call the web service using the selected currency symbol Wait for the response (let's say some seconds) Parse the response Calculate the result Show the result in a `TEdit`

As you can see, the program waits for the input for a long time while you are writing the data. Should we do it better? Sure!

Getting ready

How could we improve the program flow to optimize the user wait periods, thereby making the program faster?

When the user selects the currency type, we could already have the conversion rates for that currency, but we cannot block the main thread while the user is using the UI because it is bad practice and upsets the user. So, what we want to do is to start the request for the conversion rates in the background and, in this case, a future is really the perfect solution.

We'll implement this process in our recipe:

You	Program
Select the source currency type	Run a future in parallel, which gets the conversion rates for the selected currency. The UI remains responsive.
Write the amount of money to be converted	Do nothing, but the future is running and likely will terminate before you push the button to get the conversion.
Press the button	Calculate the result using the future result. If the future is still running, the user has to wait, but usually this is not the case. Show the result in a `TEdit`.

How it works...

Let's open the project `CurrencyRatesCalculator.dproj`. The GUI is quite simple and is shown here:

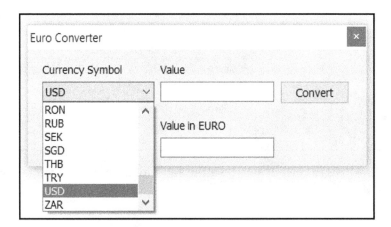

Figure 5.8: CurrencyRatesCalculator GUI

The first problem we have to cope with is: how many, and which currency types, will be available in the combo? Well, we'll use a nice, free RESTful service to get the conversion rates available at `https://exchangeratesapi.io/`, and this service has a set of APIs to interact with. For instance, to get the latest foreign exchange reference rates in JSON format, we can call the following request:

GET `https://exchangeratesapi.io/api/latest`

This obtains the following response (rates are quoted against the Euro by default):

```
{
  "base": "EUR",
  "date": "2018-07-23",
  "rates": {
    "AUD": 1.5821,
    "BGN": 1.9558,
    "BRL": 4.4322,
    "CAD": 1.5387,
    "CHF": 1.1614,
    "CNY": 7.9508,
    "CZK": 25.848,
    "DKK": 7.4509,
    "GBP": 0.8917,
    "HKD": 9.1948,
    "HRK": 7.3978,
    "HUF": 326.02,
    "IDR": 16976.63,
    "ILS": 4.2618,
    "INR": 80.673,
    "ISK": 124.6,
    "JPY": 130.18,
    "KRW": 1328.95,
    "MXN": 22.3736,
    "MYR": 4.7602,
    "NOK": 9.5533,
    "NZD": 1.7204,
    "PHP": 62.593,
    "PLN": 4.32,
    "RON": 4.6464,
    "RUB": 73.9197,
    "SEK": 10.3488,
    "SGD": 1.5979,
    "THB": 39.167,
    "TRY": 5.5624,
    "USD": 1.1716,
    "ZAR": 15.8205
  }
}
```

Therefore, in the rates object, we have all the currencies the service can handle. We've to issue a request in `FormCreate`, elaborate the response, and then set the `Combo.Items` property to the list of currencies available. Obviously, we do not want to block the user in the `FormCreate` event (it is really bad practice), so the program uses our nice `AsyncTask.pas` developed in the recipe to get the list without blocking the GUI. Here's the code in the `FormCreate` event handler:

```delphi
procedure TMainForm.FormCreate(Sender: TObject);
begin
  Async.Run<TStringList>(
    function: TStringList
    var
      LHTTP: THTTPClient;
      LResp: IHTTPResponse;
      LJObj: TJSONObject;
      LJRates: TJSONObject;
      I: Integer;
    begin
      LHTTP := THTTPClient.Create;
      try
        //send the request and parse the json response
        LResp := LHTTP.Get('https://exchangeratesapi.io/api/latest');
        LJObj :=
TJSONObject.ParseJSONValue(LResp.ContentAsString(TEncoding.UTF8)) as
TJSONObject;
        try
          //gets the json object 'rates' and
          //loop through the property names
          LJRates := LJObj.GetValue<TJSONObject>('rates');
          Result := TStringList.Create;
          for I := 0 to LJRates.Count - 1 do
          begin
            //add each names in the resulting TStringList
            Result.Add(LJRates.Pairs[I].JsonString.Value);
          end;
          //sort the list
          Result.Sort;
        finally
          LJObj.Free;
        end;
      finally
        LHTTP.Free;
      end;
    end,
    procedure(const Strings: TStringList)
    begin
      //set the list as the items property of the combobox
```

```
    cbSymbol.Items.Assign(Strings);
  end);
end;
```

Now, when the program starts, there are no delays and, as soon as the request terminates, the cbSymbol combo box fills up with all the available currency symbols. Now, we have to start a future when the user selects a currency from the list. Here's the relevant code:

```
procedure TMainForm.cbSymbolClick(Sender: TObject);
begin
  StartFuture;
end;

procedure TMainForm.StartFuture;
var
  LBaseSymbol: String;
begin
  EditResultInEuro.Clear;
  if cbSymbol.ItemIndex < 0 then
    Exit;

  LBaseSymbol := cbSymbol.Text;
  FConversionRate := TTask.Future<Currency>(
    function: Currency
    var
      LHTTP: THTTPClient;
      LResp: IHTTPResponse;
      LJObj: TJSONObject;
    begin
      LHTTP := THTTPClient.Create;
      try
        //send the request using the
        //selected currency symbol
        LResp :=
LHTTP.Get(Format('http://exchangeratesapi.io/api/latest?&base=%s&symbols=EU
R',
[LBaseSymbol]));
        //parse the response and get the rate
        LJObj :=
TJSONObject.ParseJSONValue(LResp.ContentAsString(TEncoding.UTF8)) as
TJSONObject;
        try
          Result := LJObj.GetValue<TJSONNumber>('rates.EUR').AsDouble;
        finally
          LJObj.Free;
        end;
      finally
        LHTTP.Free;
```

```
        end;
      end);
  end;
```

When the user selects a currency in the combo, the future starts. Then, the user writes a value in the `EditValue`, and the future is running in parallel. Then, the user clicks on the button to get the conversion. At this point, the future may be terminated or not, but our code is not affected by this; we are just reading the future return value, that's it. If the future is terminated, we simply read the value; if the future is not terminated, the read is blocked until it terminates. Here's the code under `btnConvert`:

```
procedure TMainForm.btnConvertClick(Sender: TObject);
begin
  if not Assigned(FConversionRate) then
  begin
    ShowMessage('Please, select a currency symbol');
    Exit;
  end;
  EditResultInEuro.Text :=
    //simply "read" from the future
    //No synchronization or checks are needed
    FormatCurr('€ #,###,##0.00', FConversionRate.Value *
    StrToFloat(EditValue.Text));
end;
```

Run the program, and check how the combo gets filled with the available currency types without slowing down the program startup. Now, select a currency symbol, write a value in the edit, and click the button. As you can see, there are no slowdowns. If your network is fast enough when you click on the button to get the conversion, the future is already finished, so that you don't advise the delay. Here's a screenshot of the program running:

Figure 5.9: The program running

There's more...

Futures are not as ubiquitous as tasks are, but in certain types of situation, you really can simplify your code using them. Remember that an `IFuture` is a descendant of `ITask`, so you can use all the methods already used in the previous recipe to check whether the future is finished; however, in some cases, such as that exposed in this recipe, this is simply not needed. Here are some links to documentation regarding futures:

- Using `TTask.IFuture` from the Parallel Programming Library (`http://docwiki.embarcadero.com/RADStudio/en/Using_TTask.IFuture_from_the_Parallel_Programming_Library`).
- A tutorial on *Using Futures from the Parallel Programming Library* (`http://docwiki.embarcadero.com/RADStudio/en/Tutorial:_Using_Futures_from_the_Parallel_Programming_Library`).

Using the Parallel Programming Library in the real world: Parallel For/Join

One of the first loops that any programmer becomes familiar with is the `for` loop. In this recipe, we'll see a particular type of `for` loop—the **parallel** one. To be clear, this parallel `for` loop is not a new language feature, but is kind of implemented as a static class method.

The parallel `for` loop is part of the **Parallel Programming Library**, and is implemented by the `TParallel` class. Here's one of its (overloaded) versions and a utilization example:

```
//declaration
class method TParallel.&For(ALowInclusive, AHighInclusive:
  Integer; const AIteratorEvent: TProc<Integer>): TLoopResult;

//used as follows
TParallel.&For(1,10,
  procedure(Index: Integer)
  begin
    //executed 10 times with index 1..10
  end);
```

What is different about the classic `for`? The difference is that the anonymous method passed to the `for` method is executed on different threads concurrently; this is the reason it's a parallel `for`, as the `for` block is executed in parallel. This means that you cannot be sure that execution with `index` 5 runs before execution with `index` 6; they are just parallel. Another important consideration is that all the code inside the `for` block must be synchronized with the shared resources eventually used, because actually, each of them runs in a different thread context.

A good question is—*What happens if I start a parallel for from 1 to 10,000? Will 10,000 threads be created?*. No, absolutely not. This will cause a huge performance degradation at first, and then will eventually crash your program (depending on your OS and the stack size defined for each thread). Here, the `TThreadPool` does its job queuing the execution and maintaining the thread's number at a right level. Therefore, the 100,000 executions will be serialized and executed in a finite number of threads, depending on available CPU cores and the current load on them.

Similar to the parallel `for`, there is the parallel `join`, which is really a variation of the parallel for. We'll use `TParallel.Join` in this recipe. What does `TParallel.Join` do? It gets an array of `TProc` to execute and returns an `ITask` interface to check when execution ends. Then, in the background, it executes them in parallel using the `TThreadPool`.

Here's a utilization sample:

```
procedure TMainForm.Button1Click(Sender: TObject);
var
  LProcs: ITask;
begin
  LProcs := TParallel.Join([
    procedure
    begin
      // do something
    end,
    procedure
    begin
      // do something
    end,
    procedure
    begin
        // do something
    end]);
  LProcs.Wait(INFINITE);
end;
```

In this example, three anonymous methods are executed in parallel. Then, the last call to wait waits indefinitely for their termination.

Now, let's say that we have to generate some sort of summary data and generate a file for each customer in the database. Doing it serially is quite simple, but we will not use all the CPU cores to elaborate; just one after the other. Considering that the power and the cores provided by the current devices (computer, smartphone, or tablets) is not an optimized approach, let's use the `TParallel.Join` to optimize it.

Getting ready

We've to generate one text file for each customer in the database containing a list of its sales. The filename should contain the customer code, which is something like `customer_00123.txt`. Consider that, in many cases, the connection and login phases are the slower operations when dealing with a short connection to the database. So, we have to handle this problem too. Luckily, **FireDAC** provides pooled connections, and they are a must when using concurrent database accesses in multithread scenarios. Let's start by checking the recipe's code.

How it works...

As already mentioned, we have to access the database using a connection pool. FireDAC has the concept of connection definition. A connection definition describes how to store and use FireDAC *connection parameters*, and what a connection definition means. To specify connection parameters, an application must use a connection definition. The connection definition is a set of parameters. Depending on the kind of connection definition, a connection may also be pooled.

FireDAC supports three definition types:

Connection	Definition
Temporary	This has no name, is not stored in a connection definition file, and is not managed by the `FDManager`. It is defined directly on `TFDConnection` component, and cannot be used as pooled.

Private	This has a unique name, is managed by the FDManager, but is *not* stored in a connection definition file. It can be used as a pooled connection inside the sample application.
Persistent	This has a unique name, is managed by the FDManager, and is stored in a connection definition file. It can also be used as pooled in different applications. In this case, each application has its own pool.

In this case, we can use a persistent or private connection definition. We'll go for a private connection definition. Here's the code (reusable at 100%) used to define a connection definition that can be used in a pool:

```
procedure TMainForm.DefinePrivateConnDef;
var
   LParams: TStringList;
begin
   LParams := TStringList.Create;
   try
    LParams.Add('Database=employee');
    LParams.Add('Protocol=TCPIP');
    LParams.Add('Server=localhost');
    LParams.Add('User_Name=sysdba');
    LParams.Add('Password=masterkey');
    LParams.Add('Pooled=true'); //can be pooled!
    FDManager.AddConnectionDef(CONNECTION_DEF_NAME, 'IB', LParams);
   finally
     LParams.Free;
   end;
end;
```

Now, we can use this connection definition in each TFDConnection by simply assigning its ConnectionDefName property to the name of the connection definition:

```
var
   LConn: TFDConnection;
begin
   LConn := TFDConnection.Create(nil);
   LConn.ConnectionDefName := CONNECTION_DEF_NAME;
   LConn.Open;
   //use the connection
```

In our case, we will call the DefinePrivateConnDef procedure in the FormCreate event.

Our GUI is quite minimal. We've only a button that starts the parallel processing. Let's see the code under this button:

```
procedure TMainForm.btnStartClick(Sender: TObject);
var
  LConn: TFDConnection;
  LQry: TFDQuery;
  LTasks: TArray<TProc>;
  LProcs: ITask;
  i: Integer;
begin
  LConn := TFDConnection.Create(nil);
  try
    LConn.ConnectionDefName := CONNECTION_DEF_NAME;
    LConn.Open;
    LQry := TFDQuery.Create(LConn);
    LQry.Connection := LConn;
    LQry.Open('SELECT * FROM CUSTOMER ORDER BY CUST_NO');
    LQry.FetchAll;
    //prepare the array to contain all the TProc
    SetLength(LTasks, LQry.RecordCount);
    i := 0;
    while not LQry.Eof do
    begin
      //define each TProc passing the CustNo to work on
      LTasks[i] := MakeProc(LQry.FieldByName('CUST_NO').AsInteger);
      LQry.Next;
      Inc(i);
    end;
  finally
    LConn.Free;
  end;
  //create the output folder
  TDirectory.CreateDirectory(OUTPUT_FOLDER);
  //TParallel.Join is blocking but we don't want block the GUI, so
  //we call it inside a task
  TTask.Run(
    procedure
    begin
      //start the parallel processing
      LProcs := TParallel.Join(LTasks);
      //wait for finish
      LProcs.Wait(INFINITE);
      //inform the user that we've finished
      TThread.Queue(nil,
        procedure
        begin
          ShowMessage('Summary files generated successfully');
```

```
              btnStart.Enabled := True;
          end);
      end);
    btnStart.Enabled := False;
  end;
```

Following the comments, the code should be quite simple to understand. Particularly interesting is the `MakeProc` function and what it really does.

`TProc` doesn't have parameters, but we've to pass a different value (the `CUST_NO` field) to each of them, so here we're using a second order function to configure an anonymous method using the *scope capture* feature. Moreover, this method contains the actual code that creates the file reading from the `SALES` table.

Here's the `MakeProc`:

```
function TMainForm.MakeProc(const CustNo: Integer): TProc;
begin
  Result :=
    procedure
    var
      LConn: TFDConnection;
      LQry: TFDQuery;
      LOutputFile: TStreamWriter;
      LFName: string;
    begin
      LConn := TFDConnection.Create(nil);
      try
        LConn.ConnectionDefName := CONNECTION_DEF_NAME;
        LFName := TPath.Combine(OUTPUT_FOLDER,
        Format('customer_%.8d.txt', [CustNo]));
        LOutputFile := TFile.CreateText(LFName);
        try
          //Some fake delay to mimic a heavy computation...
          Sleep(1000);
          LQry := TFDQuery.Create(LConn);
          LQry.Connection := LConn;
          //gets the data
          LQry.Open('SELECT * FROM SALES WHERE CUST_NO = ? ' +
            'ORDER BY ORDER_DATE, PO_NUMBER', [CustNo]);
          //write the output file one record by line
          while not LQry.Eof do
          begin
            //GetRow().DumpRow() is an handy FireDAC method
            //to dump a row from a table.
            //In a real word application you probably need
            //a more complex formatting code
```

```
            LOutputFile.WriteLine(LQry.GetRow().DumpRow(True));
            LQry.Next;
          end;
      finally
        LOutputFile.Free;
      end;
    finally
      LConn.Free;
    end;
  end;
end;
```

`MakeProc` returns a `TProc`, which captures the value of the `CustNo` parameter. So, when we run them using the `TParallel.Join`, each `TTask` has a different value for the `CustNo` variable. Nice!

Run the program and hit the button. Wait a couple of seconds, and then you will get all the files created in the output folder and message dialog, which informs you that all the files have been created, all in parallel, and using a pooled database connection. This is quite a number of concepts and power in so few lines of code, isn't it?

There's more...

There are a lot of topics here. Here are some interesting DocWiki pages about FireDAC connection definitions and multithreading:

- If you have not already done so, read the recipes about FireDAC in `Chapter 1`, *Delphi Basics—The amazing TFDTable – indices, aggregations, views and SQL*; and *ETL made easy – TFDBatchMode, and Data Integration made easy – TFDLocalSQL*

- Different kinds of connection definitions (`http://docwiki.embarcadero.com/RADStudio/en/Defining_Connection_(FireDAC)`)

- Information about the `TFDManager` class that is responsible for connection definitions and connection management (`http://docwiki.embarcadero.com/Libraries/en/FireDAC.Comp.Client.TFDManager`)

- Using FireDAC in multithreading (`http://docwiki.embarcadero.com/RADStudio/en/Multithreading_(FireDAC)`)

- Using `TParallel.For` (`http://docwiki.embarcadero.com/RADStudio/en/Using_TParallel.For_from_the_Parallel_Programming_Library`)

6
Putting Delphi on the Server

In this chapter, we will cover the following recipes:

- Developing web client JavaScript applications with WebBroker on the server
- Converting a console service application into a Windows service
- Serializing a dataset to JSON and back
- Serializing objects to JSON and back using RTTI
- Sending a POST HTTP request for encoding parameters
- Implementing a RESTful interface using WebBroker
- Controlling the remote application using UDP
- Using app tethering to create a companion app
- Creating DataSnap Apache modules
- Creating a WebBroker Apache module and publishing it under HTTPS
- Using a cross-platform HTTPS client
- Logging like a pro using LoggerPro

Introduction

In this chapter, we'll see how well Delphi can behave when it runs on the server. Most server-side technology today is scripted or managed, and in many cases, it's a good choice. However, Delphi can be used to create very powerful enterprise servers with no external dependencies and great performance, and to do all these things, you need much less hardware power and memory to run it compared to, let's say, a J2EE server. Moreover, we'll see how to handle some of the most common problems when facing web servers, such as serialization, mime types, HTML encoding, and so on.

Developing web client JavaScript applications with WebBroker on the server

Many Delphi developers think that if you need to develop a web solution, you have to look for something other than Delphi. So, they give up years of Delphi knowledge and start to create a web solution with another technology. Although there are cases where Delphi is not the best choice (for instance, if you are developing a classic website with server-side dynamic generated pages), in most scenarios, Delphi can behave even better than many of the web-only technologies available today. What you need is a good framework to work with. In this recipe, we'll look at the WebBroker technology, available since Delphi 4, and consume it from a JavaScript application. Let's start!

Getting ready

This recipe uses two external open source projects:

- **DelphiMVCFramework**:
    - A powerful Delphi framework to develop RESTful web services
    - Project website: `https://github.com/danieleteti/delphimvcframework`
    - Written by Daniele Teti and a lot of good contributors from all over the world (also me)
- **jTable**:
    - A JQuery plugin to create AJAX-based **Create**, **Retrieve**, **Update**, **and Delete** (**CRUD**) tables
    - Project website: `http://jtable.org/`
    - Written by Halil İbrahim Kalkan

We'll start by downloading these libraries and putting each ZIP file into a folder, let's say, `C:\DelphiBook\Libs`.

Version 3.0.0 of the **DelphiMVCFramework** (**DMVCFramework**) has brought several innovations. A very useful one is that it is not necessary to download the Git repository to use it. Just download the latest version (`https://github.com/danieleteti/delphimvcframework/releases`) as a ZIP file and you will be OK.

Once you have downloaded the ZIP file and put it into a folder in your filesystem, you have to configure your IDE to find the DMVC units:

1. Navigate to **Tools | Options | Environment Options | Delphi Options | Library.**
2. Then, click the **...** on the **Library Path** edit, and add the following paths one by one (change `C:\DEV\DMVCFramework` to the appropriate path on your machine):
   - `C:\DEV\DMVCFramework\sources`
   - `C:\DEV\DMVCFramework\lib\loggerpro`

To download jTable, go to the project's website and clone the repository using a Git client. There are a lot of Git clients. A good general-purpose solution is **TortoiseGit**, which is a well-integrated Windows shell extension that's able to access remote and local Git repositories directly from Windows Explorer (TortoiseGit is downloadable from `https://tortoisegit.org/`). You can also use the command-line version and then use the following command:

```
git clone https://github.com/hikalkan/jtable.git jtable
```

Or, you can use Delphi to directly download the repository. Navigate to **File | Open from version control | Git** to do this. Then, in the window that appears, write the following information and click **OK**:

Figure 6.1: The dialog used to download the JTable project from its GitHub repository

Now, the integrated Git client will clone the repository, downloading all the necessary files. At the end of the process, the wizard will ask about which project we want to open. Click **Cancel** and close the dialog box. The jTable files have been downloaded in `C:\DelphiBook\Libs\jtable`; configure the Delphi library path to point there.

Being hosted in the GitHub jTable, the latest version of the code is also available as a ZIP file. Look for a button labeled **Download ZIP** in the project page. However, the preferred way to get the source code is by cloning the Git repository, and I strongly suggest you get confident using Git.

 In the recipe project, there is a downloaded copy of the sources; however, you can use this procedure if you want to download a fresher version of the sources.

Now that you know how and where to retrieve the external projects used in this recipe, let's start with the explanation.

How it works...

Open the recipe project `PhoneBookServer.dproj` from this chapter's recipe folder. This is a WebBroker project. WebBroker is a technology that's been available since Delphi 4 to help create web server applications by exposing an HTTP/HTTPS interface. More information about WebBroker can be found at the following URLs:

- `http://docwiki.embarcadero.com/RADStudio/en/Creating_Internet_server_applications_Index`
- `http://docwiki.embarcadero.com/RADStudio/en/Using_Web_Broker_Index`

In this recipe, we'll see a simple CRUD for an InterBase database table. Here's the final application running in a browser:

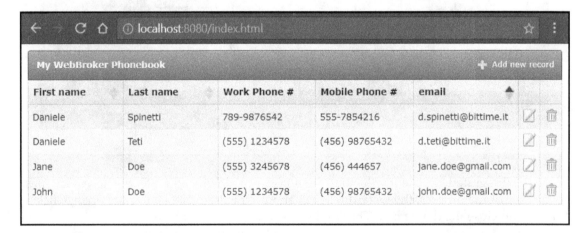

Figure 6.2: The final web application running in a browser

Take a look at the project folder. When you write WebBroker applications, the relative position of the static files used by the web application is important, and we have to deliver some static files to our clients:

Figure 6.3: The project folder layout

The DCU folder contains all the generated DCUs, while the www folder will be our document root for the static files. In the www folder, you have an index.html file and a lib folder. In the lib folder, there is a folder containing the jTable library. Our application is a web client app; that means that what the user sees in their browser is not completely generated by the server and then sent to the client, but the client has an initial HTML and then it will use JavaScript code to request data to the server using AJAX. When the server data is on the client (usually transferred as a JSON), the JavaScript code assembles data and HTML to generate the final DOM. In this recipe, we'll use jTable to avoid all the boring HTML writing to create a simple CRUD interface.

Let's start from the initial HTML file that's retrieved by the client. This is the file that starts our application, and the JavaScript inside it will download the actual data to show. If you open it using a normal text editor (this is better if done with syntax highlighting), you will see that the following files are loaded:

- jQuery library from the Google CDN
- jQuery-UI library from code.jquery.com
- jTable from a local copy
- jQuery-UI CSS for a specific theme from code.jquery.com
- jTable CSS theme from our local copy

These files are required by our web client app.

The jTable library allows you to generate a complete grid with embedded editing functionalities, only providing specific URLs to invoke. We'll provide the following URLs in the WebBroker server:

- /index.html: Delivers the main file.
- /getpeople: Returns a JSON array of JSON objects with the database data.
- /saveperson: Can be invoked to create or update a person on the database. If there is an ID field, then the person will be updated; otherwise, they will be created and a new ID will be provided by the database.
- /deleteperson: Deletes a person with a specified ID.

Note that this server is not a RESTful server. All the HTTP resources are invoked using a POST method. We are using plain WebBroker here, and DelphiMVCFramework is only used to easily serialize data retrieved from the database.

Back to Delphi and the recipe project. Open the WebModule and the **Show its Actions** property; you should see something similar to the following:

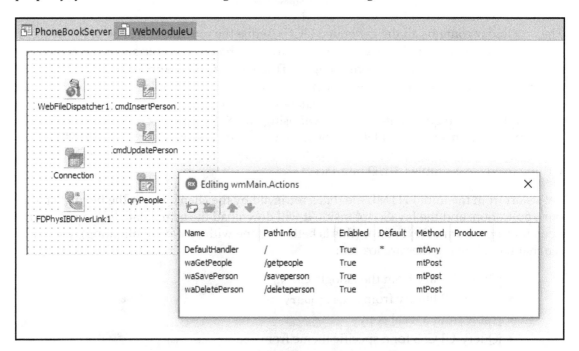

Figure 6.4: The WebModule and its actions

The `WebFileDispatcher` is configured to point to the `www` folder as its main root folder. In this way, all the files in that folder (that have permitted extensions) will be visible to the client.

FireDAC components are used to access the database. There's a **TFDConnection** pointing to a local InterBase database placed in the `DATA` folder (to run this project, you have to start the InterBase Service from the Service Control Panel). For each SQL statement, there is a component dedicated to it, apart from `DELETE`, which is executed directly on the connection.

At startup, we have to activate the database connection. Here's the `TFDConnectionBeforeConnect` event handler:

```
procedure TwmMain.ConnectionBeforeConnect(Sender: TObject);
begin
  Connection.Params.Values['Database'] :=
    TPath.GetDirectoryName(WebApplicationFileName) +
      '\..\DATA\SAMPLES.IB';
end;
```

We have to inform the internal web server where the static files are located and what the web root folder is. In our case, the web root folder is located at the same level of the executable and is called `www`, so in the `WebModuleCreate` event handler, we have to write the following code:

```
procedure TwmMain.WebModuleCreate(Sender: TObject);
begin
  FSerializer := TMVCJSONSerializer.Create;
  WebFileDispatcher1.RootDirectory := TPath.GetDirectoryName
    (WebApplicationFileName) + '\www';
end;
```

As you can see, in the preceding code in the `WebModuleCreate` event handler, the serializer is also instantiated, which will allow us to serialize the datasets in JSON objects.

Retrieving the people list

The client will issue a request to `/getpeople` and the server has to respond with a JSON array of JSON objects. This request is handled by the action `waGetPeopleAction`. The event handler contains the following code:

```
procedure TwmMain.wmMainwaGetPeopleAction(Sender: TObject; Request:
TWebRequest;
  Response: TWebResponse; var Handled: Boolean);
```

```
  var
    JPeople: TJSONArray;
    SQL: string;
    OrderBy: string;
  begin
    SQL := 'SELECT * FROM PEOPLE ';
    OrderBy := Request.QueryFields.Values['jtSorting'].Trim.ToUpper;
    if OrderBy.IsEmpty then
    begin
      SQL := SQL + 'ORDER BY FIRST_NAME ASC';
    end
    else
    begin
      if TRegEx.IsMatch(OrderBy, '^[A-Z,_]+[ ]+(ASC|DESC)$') then
      begin
        SQL := SQL + 'ORDER BY ' + OrderBy;
      end
      else
        raise Exception.Create('Invalid order clause syntax');
    end;
    // execute query and prepare response
    qryPeople.Open(SQL);
    try
      JPeople := TJSONArray.Create;
      while not qryPeople.Eof do
      begin
        JObj := TJSONObject.Create;
        // from MVCFramework.Serializer.JSON
        FSerializer.DataSetToJSONObject(qryPeople, JObj,
  TMVCNameCase.ncLowerCase, []);
        JPeople.Add(JObj);
        qryPeople.Next;
      end;
    finally
      qryPeople.Close;
    end;
    PrepareResponse(JPeople, Response);
  end;
```

This method executes a query on the PEOPLE table and then serializes the dataset returned using a serializer introduced by the MVCFramework.Serializer.JSON.pas unit (part of the DMVCFramework).

jTable can also handle sorting on the grid columns. To do this, send another request to the server with a parameter named `jSorting`, containing the field and the direction of the order in the form—`first_name asc` or `last_name desc`. This is a nice feature; however, we cannot simply concatenate this string to the SQL. We have to sanitize it to avoid an SQL injection attack. So, there is a regular expression to check whether the `jSorting` parameter contains only permitted characters and is composed of two words. We do not control if the field on which ordinate is a valid field because select will issue an error in that case.

The `PrepareResponse` method is needed to correctly prepare the response to communicate with jTable. If you want to understand the details, check out jTable's *Getting Started* section here: `http://jtable.org/GettingStarted`.

Creating or updating a person

jTable allows the user to create a new record or modify a record that has already been created. Here's the GUI used in the case of a modify request:

Figure 6.5: The edit dialog generated by the web client app

When the data has been filled in, the user can click **Save**, and then all the data is sent to the server in a POST request. This request is handled by the `waSavePersonAction` action, which is invoked with the `/saveperson` path. Here's the code used to create or update a record:

```
procedure TwmMain.wmMainwaSavePersonAction(Sender: TObject;
  Request: TWebRequest; Response: TWebResponse; var Handled: Boolean);
var
  InsertMode: Boolean;
  JObj: TJSONObject;
  LastID: Integer;
  HTTPFields: TStrings;
  procedure MapStringsToParams(AStrings: TStrings; AFDParams: TFDParams);
  var
    i: Integer;
  begin
    for i := 0 to HTTPFields.Count - 1 do
    begin
      if AStrings.ValueFromIndex[i].IsEmpty then
        AFDParams.ParamByName(AStrings.Names[i].ToUpper).Clear()
      else
        AFDParams.ParamByName(AStrings.Names[i].ToUpper).Value :=
AStrings.ValueFromIndex[i];
    end;
  end;
begin
  HTTPFields := Request.ContentFields;
  InsertMode := HTTPFields.IndexOfName('id') = -1;
  if InsertMode then
  begin
    MapStringsToParams(HTTPFields, cmdInsertPerson.Params);
    cmdInsertPerson.Execute();
    LastID := Connection.GetLastAutoGenValue('GEN_PEOPLE_ID');
  end
  else
  begin
    MapStringsToParams(HTTPFields, cmdUpdatePerson.Params);
    cmdUpdatePerson.Execute();
    LastID := HTTPFields.Values['id'].ToInteger;
  end;
  // execute query and prepare response
  qryPeople.Open('SELECT * FROM PEOPLE WHERE ID = ?', [LastID]);
  try
    JPeople := TJSONObject.Create;
    FSerializer.DataSetToJSONObject(qryPeople, JPeople,
TMVCNameCase.ncLowerCase, []);
    PrepareResponse(JPeople, Response);
```

```
    finally
       qryPeople.Close;
    end;
end;
```

The simple trick used in this code to determine if an insert or an update is requested is to check whether a field named ID is present in the POSTed fields. If an ID field is present, then we have to generate an update; otherwise, we have to generate an insert.

Deleting a person record is the simplest method. The code of the waDeletePerson action is invoked with the /deleteperson path. Here's the code:

```
procedure TwmMain.wmMainwaDeletePersonAction(Sender: TObject;
    Request: TWebRequest; Response: TWebResponse;
    var Handled: Boolean);
begin
    Connection.ExecSQL('DELETE FROM PEOPLE WHERE ID = ?',
[Request.ContentFields.Values['id']]);
    PrepareResponse(nil, Response);
end;
```

There is only one thing to note: we didn't use a specific command to issue the SQL statement, but we did so with the connection directly.

Running the application

Hitting *F9*, you should see a console window informing you that a server has been started. Open the browser and point it to http://localhost:8080. You should see what is shown in figure 6.1.

If not, try checking the following:

- Is port 8080 free?
- Is the InterBase database running correctly?
- Is the URL written correctly?

There's more...

This is only a small introduction to what you can do with WebBroker and a bounce of good JavaScript libraries. There are a lot of articles about WebBroker; some of them are a bit old but most are still applicable with the latest version of Delphi Seattle.

WebBroker can also create an ISAPI DLL for the Microsoft Internet Information Server and an Apache module DLL for the Apache HTTP web server. If you plan on deploying your web application on a production public server, you should consider putting your application behind a fully-fledged web server such as Apache or IIS.

Another solution is to use the simple web server created by Delphi and put a reverse proxy (`http://en.wikipedia.org/wiki/Reverse_proxy`) in front of it.

However, if you're using the application in your intranet, it is safe enough to publish it as a console application, or (better) a Windows service, directly on a server in your LAN.

Another bit of good news is that WebBroker and WebModules are really independent from the final program type where they will be linked. So, you can develop a console application, debug it, and then convert it into a Windows service, an Apache module, or an ISAPI DLL with a few clicks.

If you have trouble retrieving the DelphiMVCFramework project code, follow its *Getting Started* guide, which is contained in the *Developer Guide*, which you can find at `https://danieleteti.gitbooks.io/delphimvcframework/content/`.

Converting a console application into a Windows service

Writing and debugging a Windows service can be difficult and slow. In the *Creating a Windows service* recipe in `Chapter 1`, *Delphi Basics*, you learned how to do it from scratch, but in some cases you already have a console or VCL application that already does its job, but it would be much better if it could be recreated as a Windows service.

Getting ready

In this recipe, we'll take the WebBroker application created in the previous recipe as a console application, and convert it to a fully flagged Windows service. The same approach can be used for any type of service-like application that is not currently built as a service.

As a bonus, we'll learn that, if correctly architected, a project can be compiled as a console or VCL application and, without many changes, also as a Windows service. WebBroker is particularly well-architected to do so, so our application will benefit from that.

How to do it...

Perform the following steps:

1. Create a new **Service Application** by navigating to **File | New | Other... then Delphi Projects | Service Application**.
2. As soon as Delphi creates the project template, save all the files with the following names:
    - Save the project as `PhoneBookService.dproj`
    - Save the service module as `ServiceU.pas`

3. Show the object inspector for the service module and set the following properties:
    - `AllowPause = False`
    - `DisplayName = 'PhoneBookService'`

4. Now, add the WebModule from the developing (the file is named `WebModuleU.pas` and should be in the `Chapter06\CODE\RECIPE01` folder) to the project. This step allows us to reuse the code written for the console application for the service application.

5. Now, your **Project Manager** should look like the following screenshot:

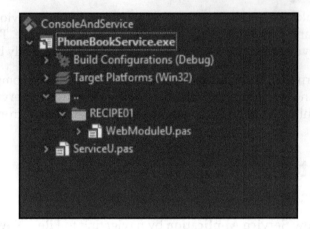

Figure 6.6: The Project Manager after adding WebModuleU.pas from the previous recipe

6. Now, we have to wire some things: open `ServiceU.pas` and add the `IdHTTPWebBrokerBridge` unit into the uses clause. This allows us to create an internal HTTP service in our Windows service.

7. Now, in the private part of the `TPhoneBook` declaration, add the following line:

```
private
    LServer: TIdHTTPWebBrokerBridge;
```

8. In the implementation section of `ServiceU.pas`, add the following uses clause:

```
uses
    Web.WebReq, WebModuleU;
```

9. Now, we have to handle the TCP server and the class registration for WebBroker. Let's create some `TPhoneService` event handlers.

10. Create the **OnCreate**, **OnStart**, and **OnStop** event handlers and fill them in with the following code:

```
procedure TPhoneBook.ServiceCreate(Sender: TObject);
begin
  if WebRequestHandler <> nil then
    WebRequestHandler.WebModuleClass := WebModuleClass;
end;
procedure TPhoneBook.ServiceStart(Sender: TService;
  var Started: Boolean);
begin
  LServer := TIdHTTPWebBrokerBridge.Create(nil);
```

```
      LServer.DefaultPort := 8080;
      LServer.Active := True;
  end;
  procedure TPhoneBook.ServiceStop(Sender: TService;
      var Stopped: Boolean);
  begin
      LServer.Free;
  end;
```

12. Build the project.

13. Copy the www folder from the previous recipe and put it at the same level as the compiled service.

14. OK, our service should be OK. Start a command prompt as the administrator, go to the folder where the service executable is, and write the following into the command line: PhoneBookService.exe/install.

15. A message dialog should inform you that the service has been installed correctly.

16. Now, go to the **Services** management console and you should see the new service named PhoneBookService listed among the others. Start it and navigate with your browser to the following URL: http://localhost:8080.

17. Now, you should see the WebBroker phone book page with some people listed in it.

18. If the phone page is not loading, the service probably isn't reaching the database. Check whether the database is running and whether the code under **OnBeforeConnect** of the database connection is set, to set the correct connection string.

How it works...

This recipe is really simple. All the dirty work is done by the WebBroker framework and by the TIdHTTPWebBrokerBridge class. As a general rule, when you have a TCP service that should listen while the service is running, simply start the TCP service in the **OnStart** event handler and stop it in the **OnStop** event handler. If your logic is more complex, you should be able to separate all the things that make the service available (start) and put them in the **OnStart** event handler, while all the things that make the service unavailable and free the resources (stop) should be put in the **OnStop** event handler.

If you need to also support a paused state, you have to find out what a paused state means for your service. For this recipe, a paused state is equal to a stopped state, so I simply removed the ability to pause the service.

There's more...

Every application may have a different way of being converted into a Windows service; however, you should be aware that your service runs in a different environment with respect to your normal application. Two notable differences are the following:

- Services can run out of any user context, and usually do. They usually run as a **Local System Account** (as with the service in this recipe), but can be configured to run as a particular user.
- The current folder for a service is not the folder where the executable is, but the `C:\Windows\System32` folder for 64-bit services, the same for 32-bit services when run on 32-bit machines, and `C:\Windows\SysWOW64` for 32-bit services that run on 64-bit machines.

Serializing a dataset to JSON and back

In the 90s, most of the Delphi program was always connected to the database server, in a fully-connected scenario. In this situation, dataset serialization was a nice topic. Today, the software world is heterogeneous—different operating systems, programs, and languages must still find a way to communicate and exchange data. We are in the IoT and big data era now!

Now a days, making your data available to other programs or getting data from other software running somewhere in the world is bread and butter, so you can understand that using a proprietary or exotic format is no longer enough. To better understand, the dear and old `Dataset.SaveToFile` may not be enough.

Let's say we have a JavaScript frontend for our Delphi application server. Your data should be *deDelphized* (I've just coined this word) and should be independent of the backend programming language or framework used. Delphi has a lot of serialization facilities, but there isn't a well-known way to serialize a dataset in JSON standard format and deserialize a standard JSON in a dataset (there are some units containing JSON serialization stuff, but the resultant JSON is very Delphi-oriented and not well-suited to being used to communicate with other non-Delphi programs).

In the DataSnap framework, there are classes devoted to doing this kind of thing and they are all contained in the unit `Data.DBXJSONCommon.pas`, but at the time of writing this book, they are not designed to be flexible enough to be used in heterogeneous scenarios. Don't be afraid; in this recipe, we'll solve all of these problems!

Getting ready

We'll use a DataSet helper provided by the unit `MVCFramework.DataSet.Utils.pas` of the previously mentioned **DelphiMVCFramework** (more information can be found here: `https://github.com/danieleteti/delphimvcframework`).

First, get **DelphiMVCFramework** using the Git repository (a simple guide is available here: `https://danieleteti.gitbooks.io/delphimvcframework/content/chapter_getting_started.html`). Then, create a new VCL project to do some experiments. This recipe is not a complete project but a set of demos showing what you can do with your datasets using the following helper.

The demo project is a simple list of buttons, a **TDBGrid**, and a **TMemo** to show the last JSON serialization that happened:

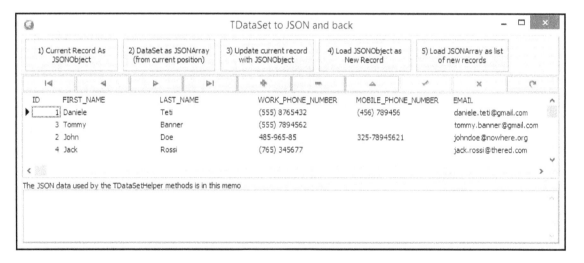

Figure 6.7: Demo for DataSet JSON serialization

How it works...

Under each button is a particular feature provided by the unit `MVCFramework.DataSet.Utils`. The helper serializes data in JSON format using a simple object or array of objects. To be clear, a single record will be serialized as a JSON object while a full dataset (or a set of records) is serialized as a JSON array containing JSON objects, one for each serialized record. In the uses clause of the form, there is a reference to the unit `MVCFramework.DataSet.Utils.pas`. This unit adds some method to each `TDataSet` descendant using a class helper, and in this project, we'll use some of them.

The first button converts the current dataset record (the dataset is called `qryPeople` and is owned by a data module called `dm`) into a JSON object; the code used is as follows:

```
Log := dm.qryPeople.AsJSONObject;
```

`Log` is a property used as a variable, but in its setter, it writes its new value to the memo. So, clicking button 1, you will have this situation in the form:

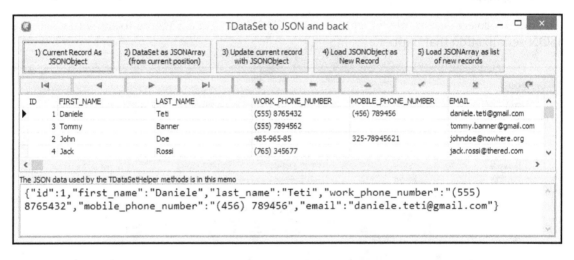

Figure 6.8: The memo showing the serialized version of the dataset's current record

It is really simple! You don't even have to know how to access the serialization engine, just include the `MVCFramework.DataSet.Utils` unit and all your datasets are able to serialize and deserialize themselves.

Button 2 serializes the dataset as a JSON array of JSON objects, starting from the current position:

```
Log := dm.qryPeople.AsJSONArrayString;
```

Go to the first record and click on button 2. The memo will show a JSON array like the following:

```
[{"id":508,"first_name":"Daniele","last_name":"Teti","work_phone_number":"(
555) 1234578","mobile_phone_number":"(456)
98765432","email":"d.teti@bittime.it"},{"id":509,"first_name":"Jane","last_
name":"Doe","work_phone_number":"(555)
3245678","mobile_phone_number":"(456)
444657","email":"jane.doe@gmail.com"},{"id":506,"first_name":"Daniele","las
t_name":"Spinetti","work_phone_number":"789-9876542","mobile_phone_number":
"555-7854216","email":"d.spinetti@bittime.it"},{"id":507,"first_name":"John
","last_name":"Doe","work_phone_number":"(555)
1234578","mobile_phone_number":"(456)
98765432","email":"john.doe@gmail.com"}]
```

As you can see, the helper takes care of null fields and serializes them as JSON null. Button 3 does an update on the record using a JSON object:

```
dm.qryPeople.LoadFromJSONObjectString(Log, TArray<String>.Create('id'));
```

It uses the previously serialized data (contained in the `Log` property) to update the current record. To use it, do the following:

1. Go to the first record.
2. Click button 1 (the memo fills with the serialized data as a JSON object).
3. Go to the record that you want to update.
4. Click button 3.
5. The record is updated!

You want to update all fields except the primary key, so as the second parameter of the `LoadFromJSONObjectString`, we have to pass an array of strings representing the name of the fields that we don't want to update in the dataset. In this case, we don't want to update the `id` field. So, when we call `qryPeople.Post`, the dataset sends an update to the database.

The fourth button is similar, but is used to create a new record starting from a JSON object. This is the code:

```
dm.qryPeople.LoadFromJSONObjectString(Log, TArray<String>.Create('id'));
```

To use it, do the following:

1. Go to the first record (or another record that you want to clone).
2. Click button 1 (the memo fills with the serialized data as a JSON object).
3. Click button 4.
4. A new record is created!

Obviously, you can use any JSON object to create a new record. To prove this, follow these steps:

1. Go to the first record.
2. Click button 1 (the memo fills with the serialized data as a JSON object).
3. Now, in the memo, change a JSON property, let's say, `last_name`. Look for `"last_name":"some string"` and change the value (`some string`) to something else.
4. Click the button.
5. A new record is created with the new value.

The JSON object can arrive from anywhere and can be put directly into your database using this simple `json->dataset` mapping. In the past year, I used a lot of these techniques in real-world web and mobile applications (the next recipe will focus on a more **object-oriented (OO)** approach compared to this one based on `TDataSet`).

The button 5 allows you to append a JSON array of JSON objects directly to the dataset:

```
dm.qryPeople.AppendFromJSONArrayString(Log, TArray<String>.Create('id'));
```

There's more...

Serialization and deserialization are huge topics. All the internet services, finally, depend on some kind of serialization. The average Delphi user is very skilled with some kinds of `TDataset` descendants (**TFDMemTable, TClientDataSet**, or other similar datasets) and normally tends to rely on some particular functionality that's present in the data access components suite that's been chosen. However, when the deserializer is not a Delphi program, some problems can arise. The unit analyzed resolves this kind of problem in a simple and elegant way (in my opinion). As a real example, the JSON format doesn't provide a specific type for dates and times.

If you try to blindly serialize the **TDate**, **TDateTime**, and **TTime** Delphi data types in JSON (using the underline double data type), you will get numbers that are perfectly valid for other Delphi programs but completely useless for JavaScript, Java, .NET, Python, and so on. So, the DataSet helper takes care of this and other problems using standard representation where JSON doesn't provide specific data types. In this case, all datetime data is serialized and deserialized using the ISO format that can be understood by all libraries and programming languages. Moreover, the helper is not dependent on the regional settings of the machine, so you can generate a JSON on an English-speaking PC and deserialize it on an Italian-speaking machine without problems with decimal separators, date format, currency formatting, and so on.

Serializing objects to JSON and back using RTTI

When you are using a domain model pattern (and you should do most of the time for non-trivial applications), the entities managed by your program are contained in objects. An object has a state and methods to change its state, just like any actual object in the real world.

Getting ready

Just like the datasets in the previous recipe, the need to serialize an object in a JSON object, send the object somewhere, and then recreate that object as it was before is very common. In this recipe, we'll use the `TJson` class and extend it with new functionalities.

How to do it...

Perform the following steps:

1. Create a new **VCL Forms Application**.
2. Drop four **TButton** and a **TMemo** on the form. Organize the **TButton** in a single row as a sort of toolbar and align the **TMemo** to cover the remaining part of the form.

3. Name the **TButton** components as follows:
   - btnObjToJSON
   - btnJSONtoObject
   - btnListToJSONArray
   - btnJSONArrayToList

4. Add a new unit to the project, name it JSON.Serializer.pas, and fill it in with the following code:

```
unit JSON.Serialization;

interface

uses
  REST.JSON, System.Generics.Collections, Data.DBXJSON,
System.JSON;

type
  TJSONUtils = class(TJSON)
  public
    class function ObjectsToJSONArray<T: class, constructor>
      (AList: TObjectList<T>): TJSONArray;
    class function JSONArrayToObjects<T: class, constructor>
      (AJSONArray: TJSONArray): TObjectList<T>;
  end;

implementation

uses
  System.SysUtils;

{ TJSONHelper }

class function TJSONUtils.JSONArrayToObjects<T>(AJSONArray:
TJSONArray)
  : TObjectList<T>;
var
  I: Integer;
begin
  Result := TObjectList<T>.Create(True);
  try
    for I := 0 to AJSONArray.Count - 1 do
      Result.Add(TJSON.JsonToObject<T>(AJSONArray.Items[I] as
TJSONObject));
  except
    FreeAndNil(Result);
```

```
      raise;
    end;
  end;

  class function TJSONUtils.ObjectsToJSONArray<T>(AList:
TObjectList<T>)
    : TJSONArray;
var
  Item: T;
begin
  Result := TJSONArray.Create;
  try
    for Item in AList do
      Result.AddElement(TJSON.ObjectToJsonObject(Item));
  except
    FreeAndNil(Result);
    raise;
  end;
end;

end.
```

5. Add another unit to the project and name it `PersonU.pas`.

6. The interface section of `PersonU.pas` must use the `REST.Json.Types` unit.

7. Now, declare a class as follows and let Delphi autocreate the property setters using *Ctrl + Shift + C*:

```
type
  TPerson = class
  public
        property ID: Integer;
        property FirstName: String;
        property LastName: String;
        property WorkPhone: String;
        property MobilePhone: String;
        property EMail: String;
  end;
```

8. After using *Ctrl + Shift + C*, go to the class's private section and add the attribute `JsonName` to the `FID` field, which is shown as follows:

```
private
  [JsonName('id')]
  FID: Integer;
```

9. Save the file and go back to the main form.

10. While on the `MainForm` code, hit *Alt + F11* and add
    the `JSON.Serializer.pas` unit to the interface uses clause; repeat this
    procedure and add the `PersonU.pas` unit.

11. Now, create a read/write property named `Log` in the main form; this property
    does not have an internal field but reads and writes its value from the
    `Memo1.Lines.Text` property, acting like a proxy for it.

12. To have some objects to work with, we need some fake data. So, create a method
    in the private section of the form called `GetPeople` with the following code:

    ```
    private
        function GetPeople: TObjectList<TPerson>;
    ```

13. Hit *Ctrl + Shift + C* and create the method body with the following code:

    ```
    function TMainForm.GetPeople: TObjectList<TPerson>;
    var
      P: TPerson;
    begin
      Result := TObjectList<TPerson>.Create(True);
      P := TPerson.Create;
      P.ID := 1;
      P.FirstName := 'Daniele';
      P.LastName := 'Spinetti';
      P.WorkPhone := '555-123456';
      P.MobilePhone := '(328) 7776543';
      P.EMail := 'me@spinettaro.it';
      Result.Add(P);
      P := TPerson.Create;
      P.ID := 2;
      P.FirstName := 'John';
      P.LastName := 'Doe';
      P.WorkPhone := '457-6549875';
      P.EMail := 'john@nowhere.com';
      Result.Add(P);
      P := TPerson.Create;
      P.ID := 3;
      P.FirstName := 'Jane';
      P.LastName := 'Doe';
      P.MobilePhone := '(339) 5487542';
      P.EMail := 'jane@nowhere.com';
      Result.Add(P);
      P := TPerson.Create;
      P.ID := 4;
      P.FirstName := 'Daniele';
      P.LastName := 'Teti';
      P.WorkPhone := '555-4353432';
    ```

```
    P.MobilePhone := '(328) 7894562';
    P.EMail := 'me@danieleteti.it';
    Result.Add(P);
  end;
```

14. Now, create the event handlers for the four buttons using the following code:

```
procedure TMainForm.btnJSONtoObjectClick(Sender: TObject);
var
  JObj: TJSONObject;
  Person: TPerson;
begin
  JObj := TJSONObject.ParseJSONValue(Log) as TJSONObject;
  try
    Person := TJSONUtils.JsonToObject<TPerson>(JObj);
    try
      ShowMessage(Person.FirstName + ' ' + Person.LastName);
    finally
      Person.Free;
    end;
  finally
    JObj.Free;
  end;
end;
procedure TMainForm.btnListToJSONArrayClick(Sender: TObject);
var
  People: TObjectList<TPerson>;
  JArr: TJSONArray;
begin
  People := GetPeople;
  try
    JArr := TJSONUtils.ObjectsToJSONArray<TPerson>(People);
    try
      Log := JArr.ToJSON;
    finally
      JArr.Free;
    end;
  finally
    People.Free;
  end;
end;
procedure TMainForm.btnObjToJSONClick(Sender: TObject);
var
  People: TObjectList<TPerson>;
  JObj: TJSONObject;
begin
  People := GetPeople;
  try
```

```
      JObj := TJSONUtils.ObjectToJsonObject(People[0]);
      try
        Log := JObj.ToJSON;
      finally
        JObj.Free;
      end;
    finally
      People.Free
    end;
end;
procedure TMainForm.btnJSONArrayToListClick(Sender: TObject);
var
  JArr: TJSONArray;
  People: TObjectList<TPerson>;
  Person: TPerson;
  S: String;
begin
  JArr := TJSONObject.ParseJSONValue(Log) as TJSONArray;
  try
    People := TJSONUtils.JSONArrayToObjects<TPerson>(JArr);
    try
      S := '';
      for Person in People do
       S := S + sLineBreak + Person.FirstName + ' ' +
Person.LastName;
    finally
      People.Free;
    end;
  finally
    JArr.Free;
  end;
  ShowMessage(S);
end;
```

15. Hit *F9* and see the application running.

How it works...

The Delphi RTL class TJSON contains two interesting methods:

```
class function ObjectToJsonObject(AObject: TObject): TJSOnObject;
```

This converts an object into its JSON representation, while this:

```
class function JsonToObject<T: class, constructor>(AJsonObject:
TJSOnObject): T;
```

Takes a JSON object and recreates the related object.

However, usually, we're dealing with a list of objects and an array of JSON objects. This is because the `JSON.Serialization.pas` unit extends the `TJSON` class, because we need to serialize and deserialize lists of objects, too.

Here's the public interface of TJSONUtils:

```
type
  TJSONUtils = class(TJSON)
  public
    class function ObjectsToJSONArray<T: class, constructor>(AList:
TObjectList<T>): TJSONArray;
    class function JSONArrayToObjects<T: class, constructor>(AJSONArray:
TJSONArray): ObjectList<T>;
  end;
```

With these four methods, we have been able to perform the following serializations:

- TObject: TJSONObject
- TJSONObject: TObject
- TObjectList<T>: TJSONArray of TJSONObject
- TJSONArray of TJSONObject: TObjectList<T>

What about the `JsonName` attribute on the `FID` field of `TPerson`? That attribute allows you to define a custom name for the serialized field. If you remove that attribute, the ID field is serialized as ID, which is ugly and not intuitive. The attribute allows you to serialize a nicer and standard ID as a JSON property name.

There's more...

The `TJSON` class allows you to define specific serialization and deserialization strategies based on data types and field names. If you want to serialize a field in a specific way, you can define a `JSONReflect` attribute on that field using the name of the class descendent from `TJSONInterceptor`. In the recipe folder, there is a bonus project called `JSONInterceptorSample`, which shows how even a stream can be serialized using an interceptor and the `JSONReflect` attribute.

Sending a POST HTTP request for encoding parameters

HTTP protocol supports some types of verbs. A verb is a way to ask a remote server something. Some of these verbs are GET, POST, PUT, DELETE, HEAD, PATCH, TRACE, and OPTIONS. For a detailed description of HTTP protocol, you can read the related RFCs at the following URLs:

- RFC7230, about HTTP/1.1 protocol: `https://tools.ietf.org/html/rfc7230`
- A specific section about the available verbs in the HTTP/1.1 protocol: `https://tools.ietf.org/html/rfc7231#section-4`

When you write a URL in the browser address bar and hit **Return**, you are issuing a GET request to the remote HTTP server. However, when you have to send form data to the server, usually, the HTML form uses the POST method. POST is designed to allow a uniform method of sending a block of data, such as the result of submitting a form to a data handling process or to post a message to a bulletin board, newsgroup, mailing list, or similar group of articles. In other words, while GET is intended to retrieve a resource from the server, POST is intended to transfer data from the client to the server. When sending data to the server, the client should inform it about the type of content (in the case of body data). This information is transferred in a specific request header called content-type. If you are sending a JSON, the content type should be `application/json`, while if a browser is sending data that a user wrote in an HTML form, the default content type is `application/x-www-form-urlencoded`. The content-type is sent by the client to inform the server about the type of content it is sending, and by the server to inform the client about the type of the content it is returning. For an overview about the different content types, check out the following link: `http://en.wikipedia.org/wiki/Internet_media_type`.

In this recipe, we'll show you how to send data to a remote web server using a POST method.

Getting ready

In this recipe, we'll use the web server created in the *Developing web client JavaScript applications with WebBroker* section on the server recipe, but this time, we're going to create a Delphi client to post data to that server. The data sent will be stored in the database and will be available through the already present web interface.

How to do it...

This recipe is really simple. So, start the WebBroker project created in the *Developing web client JavaScript applications with WebBroker* section on the server recipe (run the executable without debugging) and follow these instructions:

1. Create a new **VCL Forms Application**.
2. On the main form, drop five **TEdits**, one **TButton**, one **TRESTClient**, and one **TRESTRequest**. Organize the controls as in the following screenshot:

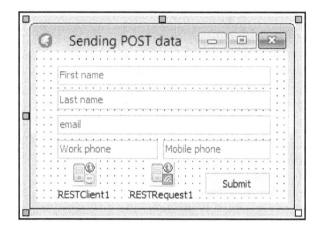

Figure 6.9: The client form used to send POST data to the web server

3. Give the **TEdits** meaningful names in order to avoid confusion in the next phase.
4. Set `RESTClient1.BaseURL` to `http://localhost:8080`.
5. Set the following properties on `RESTRequest1`:
    * `RESTRequest1.Client = RESTClient1`
    * `RESTRequest1.Method = rmPOST`
    * `RESTRequest1.Resource = 'saveperson'`
6. Double-click on the **TButton** and add the following code in its **OnClick** event:

```
procedure TMainForm.btnSubmitClick(Sender: TObject);
begin
  RESTRequest1.AddParameter('FIRST_NAME',
                             edtFirstName.Text);
  RESTRequest1.AddParameter('LAST_NAME',
                             edtLastName.Text);
  RESTRequest1.AddParameter('WORK_PHONE_NUMBER',
                             edtWorkPhone.Text);
```

```
RESTRequest1.AddParameter('MOBILE_PHONE_NUMBER',
                             edtMobilePhone.Text);
RESTRequest1.AddParameter('EMAIL', edtEmail.Text);
RESTRequest1.Execute;
end;
```

7. Run the program, write some data in the edits, and click the button. That's it! Your data has been saved on the database by the already created WebBroker application. Simple, isn't it?

The TREST* components were introduced in XE5 and are a fundamental part of a bigger strategic technology from Embarcadero. Moreover, while this recipe could also be realized easily with a simple TidHTTP, it's better to start by using TREST* or TNetHTTP* components. Why? Because TidHTTP relies on **OpenSSL** for its security layer while TNetHTTP* and TREST* (which internally use TNetHTTP*) use the native security layer offered by the operating system. So, when you need HTTPS support in your application, and believe me, it will happen, you can rely on optimal support and simpler deployment compared to the INDY SSL strategy. In the recipe folder, there is also a project that uses the TidHTTP component; you can choose which client library to use in your projects. Moreover, in the following chapters, we'll talk about the native HTTP/S components too.

How it works...

The URL where we have to send the data is http://localhost:8080/saveperson. The HTTP request is automatically created and sent to the server by the TRESTRequest and TRESTClient components. TRESTClient defines the endpoint for all the requests, while TRESTRequest defines the details for each different request. In this case, the BaseURL property contains the server name with the port (http://localhost:8080), while the request only has the Resource property set in the second part of the URL to saveperson.

What we're doing is adding a set of POST parameters with their values. Do you remember? RESTRequest1.Method is rmPOST, so it will be created and sent as a POST request. The parameter names depend on what the server expects, and we have to know the parameter names to correctly build a request.

As the name says, the TREST* components are mainly to be used with REST services, but can also be used with a normal HTTP service, as this recipe shows.

There's more...

The REST Client Library is very powerful. To learn more information about it and to find out how to use it when dealing with RESTful web services, read the following entry in the DocWiki: `http://docwiki.embarcadero.com/RADStudio/en/REST_Client_Library`.

If you want to see the REST Client Library in action with different kinds of services, check the `RESTDemo` sample here: `http://docwiki.embarcadero.com/CodeExamples/en/RESTDemo_Sample`.

Implementing a RESTful interface using WebBroker

What's REST? **Representational State Transfer** (**REST**) is an architectural style consisting of a coordinated set of architectural constraints applied to components, connectors, and data elements, within a distributed hypermedia system.

The term representational state transfer was introduced and defined in 2000 by Roy Fielding in his doctoral dissertation at UC Irvine. If you want to know more about REST, I strongly suggest you read Fielding's dissertation here: `https://www.ics.uci.edu/~fielding/pubs/dissertation/rest_arch_style.htm`.

So, how do you build a RESTful system in Delphi? There are a lot of solutions but, according to the mentioned definitions, RESTful is not a set of libraries or algorithms, it is an architectural style, and as such, it can be respected 100%, 60%, 30%, and so on. There is a sort of scale used to measure how much a system is RESTful or not. This scale was first introduced by Leonard Richardson at the QCon conference, so it is called the **Richardson Maturity Model** (**RMM**). To get all the benefits that a RESTful approach brings, you should aim for RMM level 3. Be happy; the system we'll develop in this recipe is compliant with RMM level 3.

Getting ready

Our REST service handles a resource that is stored in a database table called PEOPLE. It provides CRUD methods, plus some specific features to paginate the data. Remember that RESTful doesn't mean expose the method to do CRUD on a table, but expose the method to handle a resource. A resource can, or cannot, have a representation on a database table. Moreover, a resource can be also very complex with multiple nested objects, so while a table can be represented as a simple resource, a resource is generally not a mere table but an object graph stored on one, two, or more tables, or not stored at all. This is the HTTP REST interface that we'll implement:

HTTP VERB	URL	DESCRIPTION
GET	/people	Returns a JSON array containing one JSON object for each record present in the PEOPLE table. In each object, the property name is the name of the field, while the property values are the values of the fields.
GET	/people/($id) $id is a URL parameter	Returns a JSON object representing the specific person who has the ID = $id.
POST	/people	Creates a new person in the table people. Requires a request body containing the new person to create as a JSON object. The request content-type must be application/json.
PUT	/people/($id)	Updates the person with ID = $id with the data passed in the request body. Requires a request body containing the new person to update as a JSON object. The request content-type must be application/json.
DELETE	/people/($id)	Deletes the person with ID = $id
POST	/people/searches /people/searches?page=[x]	Returns a JSON array containing JSON objects. Executes a search over the PEOPLE table, returning only the records that match the filter passed as a JSON object in the request body. Requires a JSON object as request body. The parameter is passed as property "TEXT" in the request body, for example, {"TEXT":"ele"}.

This recipe uses DMVC, a Delphi open source framework based on WebBroker that allows you to create powerful RESTful web services. You can find the project code here: https://github.com/danieleteti/delphimvcframework.

Check out the project using the instructions on the website (check how to download a release as ZIP. You can also check out the recipe *Implementing a RESTful interface using WebBroker*, which describe how to download and use **DMVC**) and put it into a folder on your file system. There are no components or controls, only units. Now, you have to configure your IDE to find the DMVC units:

1. Navigate to **Tools** | **Options** | **Environment Options** | **Delphi Options** | **Library**.
2. Then, click the **...** on the **Library Path edit**, and add the following paths one by one (change `C:\DEV\DMVCFramework` to the appropriate path on your machine):

   - `C:\DEV\DMVCFramework\sources`
   - `C:\DEV\DMVCFramework\lib\loggerpro`

This recipe uses many DMVC features and could be a little confusing if you don't know the basics of REST and DMVC. If so, please read the following documentation before going any further:

- *Building Web Services the REST Way*: `http://www.xfront.com/REST-Web-Services.html`.
- *RESTful Web services*: The basics: `https://www.ibm.com/developerworks/webservices/library/ws-restful/`.
- The latest information about DelphiMVCFramework is available in the *Developer Guide*: `https://danieleteti.gitbooks.io/delphimvcframework/content/`.
- The *Developer Guide* is also available as a PDF here: `https://www.gitbook.com/download/pdf/book/danieleteti/delphimvcframework`.

A valuable resource for DelphiMVCFramework is its samples, so please check the `\Samples` folder in the project root folder. From this point onward, I won't repeat concepts and information already explained in the previously mentioned articles, so read them with care.

How to do it...

Perform the following steps:

1. Navigate to **Delphi Project** | **Web Broker** | **Web Server Application**.

2. Now, the wizard will ask you what type of web server application you want to create. This demo will be built as a console application. However, you can take advantage of the flexibility of WebBroker and add another type of application, for instance, an ISAPI DLL or a Windows service. At this point, select the standalone console application and click **Next**.

3. The wizard proposes a TCP port where the service will listen. Click on **Test** port; if the test port succeeds, use it, otherwise change the port until the test passes. In this recipe, port `8080` is used.

4. Click **Finish**.

5. Save all. Name the project `PeopleManager.dproj` and the WebModule `WebModuleU.pas`.

6. We will start from the business object classes. This web service will manage people, so let's create a new unit and declare the following class:

```
TPerson = class
public
  property ID: Integer;
  property FIRST_NAME: String;
  property LAST_NAME: String;
  property WORK_PHONE_NUMBER: String;
  property MOBILE_PHONE_NUMBER: String;
  property EMAIL: String;
end;
```

7. Hit *Ctrl + Shift + C* to autocomplete the declaration; save the file as `PersonBO.pas`. Note that in projects where you have a lot of different types of classes (`businessobjects`, `controllers`, `datamodules`, and so on), it can be useful to organize the units in different folders. So, I saved `PersonBO.pas` in a folder named `BusinessObjects`. Feel free to do this as well.

8. Now, it is time to create a DelphiMVCFramework controller. This is the class that contains all the code to handle the HTTP requests and responses. Here, there shouldn't be any business logic code.

9. Create a new unit, name it `PeopleControllerU.pas`, and save it into the `Controllers` folder.

10. Fill `PeopleControllerU.pas` with the following code:

```
unit PeopleControllerU;

interface

uses
  MVCFramework, PeopleModuleU, MVCFramework.Commons;
```

```pascal
type

  [MVCPath('/people')]
  TPeopleController = class(TMVCController)
  private
    FPeopleModule: TPeopleModule;
  protected
    procedure OnAfterAction(Context: TWebContext;
      const AActionNAme: string); override;
    procedure OnBeforeAction(Context: TWebContext; const
AActionNAme: string;
      var Handled: Boolean); override;
  public
    [MVCPath]
    [MVCHTTPMethod([httpGET])]
    procedure GetPeople;

    [MVCPath('/($id)')]
    [MVCHTTPMethod([httpGET])]
    procedure GetPersonByID(id: Integer);

    [MVCPath]
    [MVCHTTPMethod([httpPOST])]
    [MVCConsumes('application/json')]
    procedure CreatePerson;

    [MVCPath('/($id)')]
    [MVCHTTPMethod([httpPUT])]
    [MVCConsumes('application/json')]
    procedure UpdatePerson(id: Integer);

    [MVCPath('/($id)')]
    [MVCHTTPMethod([httpDELETE])]
    procedure DeletePerson(id: Integer);

    [MVCPath('/searches')]
    [MVCHTTPMethod([httpPOST])]
    [MVCConsumes('application/json')]
    procedure SearchPeople;
  end;

implementation

uses
  PersonBO, SysUtils, Data.DBXJSON, System.Math, System.JSON,
  MVCFramework.SystemJSONUtils;

{ TPeopleController }
```

```delphi
procedure TPeopleController.CreatePerson;
var
  Person: TPerson;
begin
  Person := Context.Request.BodyAs<TPerson>;
  try
    FPeopleModule.CreatePerson(Person);
    Context.Response.Location := '/people/' + Person.ID.ToString;
    Render(201, 'Person created');
  finally
    Person.Free;
  end;
end;

procedure TPeopleController.UpdatePerson(id: Integer);
var
  Person: TPerson;
begin
  Person := Context.Request.BodyAs<TPerson>;
  try
    Person.ID := id;
    FPeopleModule.UpdatePerson(Person);
    Render(200, 'Person updated');
  finally
    Person.Free;
  end;
end;

procedure TPeopleController.DeletePerson(id: Integer);
begin
  FPeopleModule.DeletePerson(id);
  Render(204, 'Person deleted');
end;

procedure TPeopleController.GetPersonByID(id: Integer);
var
  Person: TPerson;
begin
  Person := FPeopleModule.GetPersonByID(id);
  if Assigned(Person) then
    Render(Person)
  else
    Render(404, 'Person not found');
end;

procedure TPeopleController.GetPeople;
begin
  Render<TPerson>(FPeopleModule.GetPeople);
```

```
end;

procedure TPeopleController.OnAfterAction(Context: TWebContext;
  const AActionNAme: string);
begin
  inherited;
  FPeopleModule.Free;
end;

procedure TPeopleController.OnBeforeAction(Context: TWebContext;
  const AActionNAme: string; var Handled: Boolean);
begin
  inherited;
  FPeopleModule := TPeopleModule.Create(nil);
end;

procedure TPeopleController.SearchPeople;
var
  Filters: TJSONObject;
  SearchText: string;
  CurrPage: Integer;
begin
  Filters := TSystemJSON.StringAsJSONObject(Context.Request.Body);
  if not Assigned(Filters) then
    raise Exception.Create('Invalid search parameters');
  SearchText := TSystemJSON.GetStringDef(Filters, 'TEXT');
  if (not TryStrToInt(Context.Request.Params['page'], CurrPage)) or
    (CurrPage < 1) then
    CurrPage := 1;
  Render<TPerson>(FPeopleModule.FindPeople(SearchText, CurrPage));
  Context.Response.CustomHeaders.Values['dmvc-next-people-page'] :=
    Format('/people/searches?page=%d', [CurrPage + 1]);
  if CurrPage > 1 then
    Context.Response.CustomHeaders.Values['dmvc-prev-people-page']
:=
      Format('/people/searches?page=%d', [CurrPage - 1]);
end;

end.
```

11. This is quite long, but all of our RESTful interface is implemented in this unit. Now, we have to write the part that actually accesses the database. In this recipe, we'll use a simple design pattern called Table Data Gateway. **Table Data Gateway (TDG)** was defined for the first time by Martin Fowler in his fundamental, and highly recommended, book *Patterns of Enterprise Application Architecture* (`http://www.amazon.com/gp/product/0321127420`). TDG is defined as follows: An object that acts as a gateway to a database table. One instance handles all the rows in the table (`http://martinfowler.com/eaaCatalog/tableDataGateway.html`).

12. Let's create our TDG using a data module. Add a new data module, name it `PeopleModule`, and save it into the `Modules` folder as `PeopleModuleU.pas`.

13. Now, your **Project Manager** should look like the following:

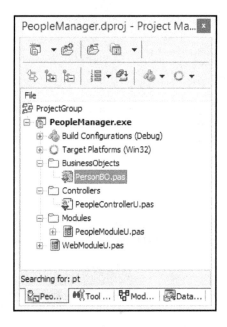

Figure 6.10: The Project Manager

14. Now, drop on the data module and link each other's components as follows (this is an extract of the `dfm` file):

```
object Conn: TFDConnection
  Params.Strings = (
    'Database=C:\Delphi Cookbook\BOOK\Chapter06\DATA\SAMPLES.IB'
    'User_Name=sysdba'
    'Password=masterkey'
```

```
      'DriverID=IB')
    ConnectedStoredUsage = [auDesignTime]
    Connected = True
    LoginPrompt = False
  end
  object qryPeople: TFDQuery
    Connection = Conn
    UpdateObject = updPeople
  end
  object updPeople: TFDUpdateSQL
    Connection = Conn
  end
end
```

Change the `FDConnection` parameters accordingly with your machine.

15. Now, we have to configure some data access stuff.

16. Double-click on `qryPeople`; the component editor will show up. Write the query `SELECT * FROM PEOPLE`, and click **Execute**. Hold the window open. This will be the query used to generate all the CRUD statements.

17. If you have correctly connected `qryPeople.UpdateObject` to `updPeople`, you should see an **UpdateSQL Editor** button on the right-hand side of the component editor form:

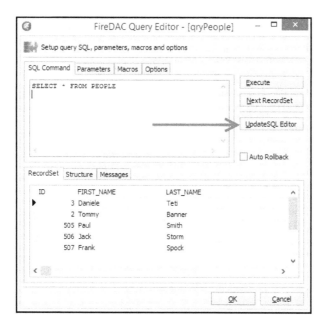

Figure 6.11: The qryPeople component editor showing the SQL and the button to configure TFDUpdateSQL that's linked to qryPeople

18. Click on the **UpdateSQL Editor** button and you will get another component editor. This time, it is related to the `updPeople` component.

19. Select the fields, as shown in the the following screenshot, and click **Generate SQL** and **OK**. Now, your `updPeople` component has been configured with all the SQL statements needed to correctly update the `PEOPLE` table:

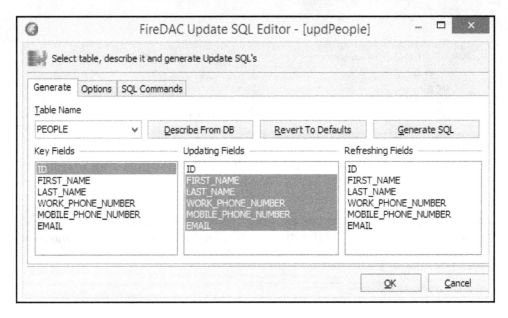

Figure 6.12: The updPeople component editor used to configure the INSERT, UPDATE, and DELETE SQL statements

20. Now, we have to create the methods used to CRUD records. Go to the `PeopleModuleU.pas` code view. Declare the following method in the class public section:

```
public
  procedure CreatePerson(APerson: TPerson);
  procedure DeletePerson(AID: Integer);
  procedure UpdatePerson(APerson: TPerson);
  function GetPersonByID(AID: Integer): TPerson;
  function FindPeople(ASearchText: String;
                      APage: Integer): TObjectList<TPerson>;
  function GetPeople: TObjectList<TPerson>;
end;
```

21. Hit *Ctrl* + *Shift* + *C* to autogenerate method bodies and fill them in with the following code:

```
procedure TPeopleModule.CreatePerson(APerson: TPerson);
var
  InsCommand: TFDCustomCommand;
begin
  InsCommand := updPeople.Commands[arInsert];
  TFireDACUtils.ObjectToParameters(InsCommand.Params, APerson,
'NEW_');
  InsCommand.Execute;
  APerson.ID := Conn.GetLastAutoGenValue('gen_people_id');
end;

procedure TPeopleModule.DeletePerson(AID: Integer);
var
  DelCommand: TFDCustomCommand;
begin
  DelCommand := updPeople.Commands[arDelete];
  DelCommand.ParamByName('OLD_ID').AsInteger := AID;
  DelCommand.Execute;
end;

function TPeopleModule.FindPeople(ASearchText: String; APage:
Integer)
  : TObjectList<TPerson>;
var
  StartRec, EndRec: Integer;
begin
  Dec(APage); // page 0 => 1, 9, page 1 => 10, 19, page 3 => 20, 29
  StartRec := (10 * APage) + 1;
  EndRec := StartRec + 10 - 1;
  qryPeople.Open('SELECT * FROM PEOPLE WHERE ' +
    'FIRST_NAME CONTAINING :SEARCH_TEXT_1 OR ' +
    'LAST_NAME CONTAINING :SEARCH_TEXT_2 OR ' +
    'EMAIL CONTAINING :SEARCH_TEXT_3 ' + 'ORDER BY LAST_NAME,
FIRST_NAME ' +
    Format('ROWS %d TO %d', [StartRec, EndRec]), [ASearchText,
ASearchText,
    ASearchText]);
  Result := qryPeople.AsObjectList<TPerson>;
end;

function TPeopleModule.GetPersonByID(AID: Integer): TPerson;
begin
  qryPeople.Open('SELECT * FROM PEOPLE WHERE ID = :ID', [AID]);
  Result := qryPeople.AsObject<TPerson>;
end;
```

```
function TPeopleModule.GetPeople: TObjectList<TPerson>;
begin
  qryPeople.Open;
  Result := qryPeople.AsObjectList<TPerson>;
end;

procedure TPeopleModule.UpdatePerson(APerson: TPerson);
var
  UpdCommand: TFDCustomCommand;
begin
  UpdCommand := updPeople.Commands[arUpdate];
  TFireDACUtils.ObjectToParameters(UpdCommand.Params, APerson,
'NEW_');
  UpdCommand.ParamByName('OLD_ID').AsInteger := APerson.ID;
  UpdCommand.Execute;
end;
```

22. These methods will be called by the controller, capping data retrieved by the HTTP request. As you can see, the CRUD methods do not have references to the HTTP environment, or to JSON objects, or whatever is related to the particular environment. These methods, and the whole class itself, can be used everywhere, even in a classic client/server application. Remember that the dependencies between the classes should be reduced as much as they can. More on this in the *How it works* section.

23. Add the MVCFramework.FireDAC.Utils and MVCFramework.DataSet.Utils units in the implementation uses clause of TPersonModule.

24. Just one more thing to do in the TPersonModule: create the event handler called **OnBeforeConnect** on the TFDConnection and write the following code (adapt it to point to the correct database path on your system):

```
procedure TPeopleModule.ConnBeforeConnect(Sender: TObject);
begin
  inherited;
  Conn.Params.Values['Database'] := '..\..\..\DATA\SAMPLES.IB';
end;
```

25. We're about to finish. Go back to WebModuleU.pas and create the **OnCreate** event handler. Here, we have to configure the DelphiMVCFramework starting point. It is really simple, just two lines of code:

```
procedure TwmMain.WebModuleCreate(Sender: TObject);
begin
  FMVC := TMVCEngine.Create(Self);
  FMVC.AddController(TPeopleController);
end;
```

26. The FMVC variable must be declared in the private section of the class and you have to add the `PeopleControllerU` unit in the `implementation uses` clause.

27. Now, your project should compile. If not, check the dependencies between all the units.

28. After running the project, you get a sad console window that informs you that an HTTP server is running on port `8080`. Launch a browser (this is better if it's Google Chrome or Mozilla Firefox) and request the following URL: `http://localhost:8080/people`.

29. Your browser should show all the data that's available in the `PEOPLE` table as a JSON array of JSON objects:

Figure 6.13: The JSON array of JSON objects returned by the HTTP call from the browser

30. If you want to try something different, get a valid person ID from the list of `PEOPLE` (look for `"ID": <some number>` in the JSON stream) and append it to the URL after a slash. This should be the effect:

Figure 6.14: The JSON object representing a single person returned by the HTTP call from the browser

How it works...

Wow! This recipe is very long! However, it summarizes all the concepts already seen in the previous recipes, so it's worth it.

The application is organized into three layers:

- Controller (TPeopleController):
    - Takes care of all the machinery needed to deserialize the JSON data into Delphi objects
    - Coordinates the job with the Table module
- Table Data Gateway (TPeopleModule):
    - Handles all the persistence needs
    - Gets objects and persists them
    - Retrieves datasets and converts them into objects
- Business Objects (TPerson):
    - Implements all the business logic required by the domain problem. In this sample, we don't have business logic, but if present, it should be inside the TPerson class.

When an HTTP request arrives at the server, the DMVCFramework router starts to find a suitable controller using the MVCPath attributes defined on all its controllers. When a matching controller and action is found, the request and response objects are packed in a TWebContext object and passed to the selected action. Here, we can read information from the request and build the response accordingly.

All the action methods look like the following:

- Read information from the HTTP request
- Invoke some methods on the TPersonModule instance
- Build the response for the client

Let's take a look at the following action used to create a new person:

```
[MVCPath]
[MVCHTTPMethod([httpPOST])]
[MVCConsumes('application/json')]
procedure CreatePerson;
. . .
procedure TPeopleController.CreatePerson;
var
  Person: TPerson;
```

```
begin
  //read information from the request
  Person := Context.Request.BodyAs<TPerson>;
  try
    //invoke some methods on the TPeopleModule instance
    FPeopleModule.CreatePerson(Person);
    //build the response for the client
    Context.Response.Location := '/people/' + Person.ID.ToString;
    Render(201, 'Person created');
  finally
    Person.Free;
  end;
end;
```

What's that `Context.Response.Location` line for? One of the RESTful features is the use of hypermedia controls. The point of hypermedia controls is that they tell us what we can do next, and the URI of the resource we need to manipulate to do it. Instead of having to know where to GET our newly created person, the hypermedia controls in the response tell us where to get the new person.

Another interesting action is mapped to the POST `/people/searches`. Here's the code:

```
[MVCPath('/searches')]
[MVCHTTPMethod([httpPOST])]
[MVCConsumes('application/json')]
procedure SearchPeople;
. . .
procedure TPeopleController.SearchPeople;
var
  Filters: TJSONObject;
  SearchText, PageParam: string;
  CurrPage: Integer;
begin
  //read informations from the requests
  Filters := TSystemJSON.StringAsJSONObject(Context.Request.Body);
  if not Assigned(Filters) then
    raise Exception.Create('Invalid search parameters');
  SearchText := TSystemJSON.GetStringDef(Filters, 'TEXT');
  if (not TryStrToInt(CTX.Request.Params['page'], CurrPage))
                            or (CurrPage < 1) then
    CurrPage := 1;
  //call some method on the TPeopleModule
  Render<TPerson>(FPeopleModule.FindPeople(SearchText, CurrPage));
  //prepare the response (also if render has been already called)
  Context.Response.CustomHeaders.Values['dmvc-next-people-page'] :=
    Format('/people/searches?page=%d', [CurrPage + 1]);
  if CurrPage > 1 then
```

```
      Context.Response.CustomHeaders.Values['dmvc-prev-people-page'] :=
        Format('/people/searches?page=%d', [CurrPage - 1]);
  end;
```

This action is a bit longer, but the three steps are still clearly defined. This action executes a search on the `people` table using a pagination mechanism. The URL to get the next and the previous page are returned, along with the response in the headers `dmvc-next-people-page` and `dmvc-prev-people-page`. So, the client doesn't have to know which kind of call to do to get the second page, but can simply navigate through the returned information.

One last note about the `TPersonModule` is that it heavily uses the `DataSet` helpers introduced in the *Serializing a dataset to JSON and back* recipe. Look at the following code that's used to get a person by ID:

```
function TPeopleModule.GetPersonByID(AID: Integer): TPerson;
begin
  qryPeople.Open('SELECT * FROM PEOPLE WHERE ID = :ID', [AID]);
  //uses the dataset helper to convert a record to an object
  Result := qryPeople.AsObject<TPerson>;
end;
```

It could not be simpler! Also, the method to create a new person is made really simple by using some of the `TFireDACUtils` methods:

```
procedure TPeopleModule.CreatePerson(APerson: TPerson);
var
  InsCommand: TFDCustomCommand;
begin
  //gets the Insert statement contained in the TFDUpdateSQL
  InsCommand := updPeople.Commands[arInsert];
  //Maps the object properties to the command parameters
  TFireDACUtils.ObjectToParameters(InsCommand.Params, APerson, 'NEW_');
  //execute the statement
  InsCommand.Execute;
  //retrieve the last assigned ID
  APerson.ID := Conn.GetLastAutoGenValue('gen_people_id');
end;
```

There's more...

What a huge topic we covered in this recipe! To test the RESTful service that you will develop from now on, you can use the `RESTDebugger.exe` program provided within since Delphi XE5 (in the `bin` folder), or the free **Postman application** (`https://www.getpostman.com/`). These tools allow you to send all the HTTP VERB requests while the browser, using only the address bar, can only issue GET requests.

Remember that if you don't know the fundamental principles of REST well, you could break all the benefits. Don't be tempted to put verbs on the URL like this: `http://server.com/people/create` or `http://server.com/people/get`.

This is not REST, it is a sort of remote procedure call (it is not bad, but it is something else; it's not REST).

Also, be coherent with the HTTP VERB used. All the HTTP methods must be idempotent except for POST and PATCH. So, if your request is executed once or repeated two, three, or a thousand times, it should not change the system further.

Read this article for a good overview on idempotence in HTTP: `http://restcookbook.com/HTTP%20Methods/idempotency/`.

Controlling the remote application using UDP

What's UDP? The **User Datagram Protocol**, or **UDP**, is a connectionless protocol used by everyone every day, but it seems that not too many people know about it. However, it can be really useful to solve particular network problems. Like TCP, UDP works at the transport layer TCP/IP model but it has a very different usage. Compared to TCP, UDP is a simpler message-based connectionless protocol. Connectionless protocols do not set up a dedicated end-to-end connection. Instead, communication is achieved by transmitting information in one direction from the source to the destination without verifying the readiness or state of the receiver. However, one primary benefit of UDP over TCP is the application to the **Voice over Internet Protocol** (**VoIP**), where latency and jitter are the primary concerns. It is assumed in VoIP UDP that the end users provide any necessary real-time confirmation that the message has been received.

Here are some features of UDP:

- **Unreliable**: When a message is sent, it cannot be known whether it will reach its destination; it could get lost along the way. In UDP, there is no concept of acknowledgment, retransmission, or timeout.
- **Not ordered**: If two or more messages are sent to the same recipient, the order in which they arrive is not deterministic and cannot be predicted.
- **No congestion control**: UDP itself does not avoid congestion, and it's possible for high-bandwidth applications to trigger congestion collapse, unless they implement congestion control measures at the application level.

Getting ready

In this recipe, we'll use UDP to autoconfigure an application in a LAN. Let's say you have some classic client/server applications (however, the same approach is valid for any type of application) in a LAN, a big LAN. Every application uses a database on a specific machine and uses internal web services. Usually, in this scenario, you have some kind of configuration stored somewhere on the client PC that is read at startup. But what if the database IP changes because something is changed on the network? Or, more generally, what if some part of the configuration is subject to change for some external reasons? If the change is only about the IP, a simple internal DNS does the job. But what about a port change? And what if something else changes? OK, I think you've got the point: you have to change the configuration on all machines (if you don't have some type of software distribution, this could be a daunting and boring task). Let's think about a well-known network service, the DHCP (http://en.wikipedia.org/wiki/Dynamic_Host_ Configuration_Protocol). When a machine with dynamic IP configuration starts, the operating system sends a broadcast on the network to ask for an IP. It doesn't know who will send the IP, and it doesn't know whether someone can reply with an IP. It doesn't know anything! In this situation, the DHCP server replies to the broadcast with the assigned IP for that machine. The machine gets its IP and can join the network. This is the same approach that we'll use in this recipe. We have a database application that doesn't know where the database it should connect to is. So, it sends a broadcast on the network saying, *Hey, I'm application X, which database I should connect to?* On the network, there is another program that we will call `ConfigDispatcher` that replies to the broadcast with the correct connection information for that specific application. So, the client reads the `ConfigDispatcher` reply and can happily connect to the correct database. No config files, no default server, no hardcoded names, but a simple autoconfiguration. Wow, this is the power of UDP.

How to do it...

This recipe is composed of two projects: the `ConfigDispatcher` and the real application. Let's start with the `ConfigDispatcher`:

1. Create a new **VCL Forms Application** and save it as `ConfigDispatcher`.
2. Drop three **TMemo** on the form and name them `MemoLog`, `MemoConfigApp1`, and `MemoConfigApp2`.
3. In the `MemoConfigApp1.Lines` property, add the following lines:

   ```
   Database=employee
   Server=localhost
   ```

   You can choose to define the connection by **FDConnectionDefs**, as shown in the following line:

   ```
   ConnectionDefName=employee
   Server=localhost
   ```

4. In the `MemoConfigApp2.Lines` property, add the following lines:

   ```
   Database=erpdb
   Server=192.168.3.4
   ```

5. In this recipe, we'll use only the first configuration, but for the sake of completeness, there is also a second (fake) configuration available that will remain unused.
6. Drop a `TidUDPServer` component and then drop three `TLabel` and arrange the form as shown in the following screenshot:

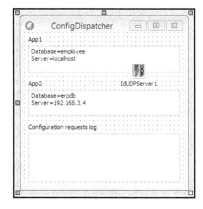

Figure 6.15: The ConfigDispatcher main form

7. Now, set the `idUDPServer1` properties as follows (this is the relevant part of the dfm form; it should not be difficult to read):

```
object IdUDPServer1: TIdUDPServer
  BroadcastEnabled = True
  DefaultPort = 8888
  Active = True
end
```

8. Now, create the `OnUDPRead` event handler for the `idUDPServer1` component and fill it in with the following code:

```
procedure TMainForm.IdUDPServer1UDPRead(AThread:
TIdUDPListenerThread;
  const AData: TIdBytes; ABinding: TIdSocketHandle);
var
  ClientCommand, ClientConfig: string;
  CommandPieces: TArray<string>;
begin
  ClientCommand := BytesToString(AData);
  MemoLog.Lines.Add(ClientCommand);
  CommandPieces := ClientCommand.Split(['#']);
  if (Length(CommandPieces) = 2) and (CommandPieces[0] =
'GETCONFIG') then
  begin
    if CommandPieces[1] = 'APP001' then
    begin
      ClientConfig := MemoConfigApp1.Lines.Text;
    end;
    if CommandPieces[1] = 'APP002' then
    begin
      ClientConfig := MemoConfigApp2.Lines.Text;
    end;
    ABinding.Broadcast(ToBytes(ClientConfig), 9999,
ABinding.PeerIP);
  end;
end;
```

9. At this time, the project doesn't compile. Add the `idGlobal` unit in the uses clause interface section and it should.

10. The `ConfigDispatcher` is finished. Let's start the `ClientDBApplication`.

11. Add to the project group a new **VCL Forms Application** (navigate to **ProjectGroup** | **Add New Project** | **VCL Forms Application**).

12. Save the new project as `ClientDBApplication` and give a meaningful name to the form.

13. Drop the following components on the main form and set their properties as follows:

```
object FDConnection1: TFDConnection
  Params.Strings = (
    'User_Name=sysdba'
    'Password=masterkey'
    'Protocol=TCPIP'
    'DriverID=IB')
  ConnectedStoredUsage = [auDesignTime]
  LoginPrompt = False
end
object FDQuery1: TFDQuery
  Connection = FDConnection1
  SQL.Strings = ('select * from customer')
end
object DataSource1: TDataSource
  DataSet = FDQuery1
end
object FDPhysIBDriverLink1: TFDPhysIBDriverLink
end
object FDGUIxWaitCursor1: TFDGUIxWaitCursor
end
object Timer1: TTimer
  Interval = 3000
end
object IdUDPServer1: TIdUDPServer
  DefaultPort = 9999
  Active = True
end
```

14. Drop a **TDBGrid** and a **TDBNavigator** and hook them up to `DataSource1`.

15. Now, if you try to activate `FDQuery1`, you should see the query data in the grid.

16. Double-click on `Timer1` and fill in the **OnTimer** event with the following code:

```
procedure TMainFormClient.Timer1Timer(Sender: TObject);
begin
  Caption := 'Waiting for configuration...';
  IdUDPServer1.Broadcast(
              ToBytes('GETCONFIG#APP001'), 8888);
end;
```

17. Include the `idGlobal` unit in the user interface clause.

18. Now, create the **OnUDPRead** event handler for the `idUDPServer1` component and fill it in with the following code:

```
procedure TMainFormClient.IdUDPServer1UDPRead(AThread:
TIdUDPListenerThread;
  const AData: TIdBytes; ABinding: TIdSocketHandle);
var
  ServerConfig: TStringList;
  i: Integer;
begin
  Timer1.Enabled := False;
  try
    Caption := 'Configuration OK...';
    ServerConfig := TStringList.Create;
    try
      ServerConfig.Text := BytesToString(AData);
      for i := 0 to ServerConfig.Count - 1 do
      begin
        FDConnection1.Params.Values[ServerConfig.Names[i]] :=
          ServerConfig.ValueFromIndex[i];
      end;
    finally
      ServerConfig.Free;
    end;
    FDConnection1.Open;
    FDQuery1.Open;
    Caption := 'Connected';
  except
    Caption := 'Wrong configuration or cannot connect';
    Timer1.Enabled := true;
  end;
end;
```

19. Now, check that the InterBase service is started on your machine. If it's not started, start it.

20. Run the `ConfigDispatcher` without debugging and then run the `ClientDBApplication`. After 3 seconds, you should see the data in the grid. The configuration has been requested with a broadcast to the `ConfigDispatcher`, and has then been parsed, understood, and used to connect to the database.

21. You can try to start the `ClientDBApplication` first, wait 6 seconds, and then start the `ConfigDispatcher`. It just works.

How it works...

This is a long recipe but the behavior is really simple. The `ConfigDispatcher` uses the two memos to maintain the strings to send to the client that requests a specific configuration. When a client requests a configuration, the server receives a command string similar to the following:

GETCONFIG#APP001

It parses the command and replies to the client with the contents of one of the memos. For `APP001`, it sends the `MemoConfigApp1` content while for `APP002`, it sends the `MemoConfigApp2` content. That's it, the `ConfigDispatcher` job is finished.

The client is simple, too. On initiation, it waits 3 seconds, is configured in the timer, and asks for a configuration. If some data arrives on the **UDPServer**, the `UDPRead` event handler is called. The code disables the timer, reads the data sent by the `ConfigDispatcher`, and tries to use it to configure its database connection. If the configuration is correct, the `ClientDBApplication` connects to its database; if not, the timer is re-enabled and after 3 seconds, another configuration request is broadcast and the cycle goes on until the client is able to connect.

As you can see, the applications talk to each other without any kind of predefined knowledge or configuration. This is the power of UDP!

There's more...

Network programming and network protocols are a really large topic. As a software developer, you have to be aware, if not yet of the possibilities that the standard networking infrastructure offers you.

The UDP protocol allows you to create strange applications that find and talk to each other using broadcasts. You could even create a complex application protocol based on UDP to remotely control some running applications. In the chapter devoted to mobile programming (Chapter 9, Recipe 5), there is another example of the power of UDP.

Using app tethering to create a companion app

App tethering is one of the main features introduced in RAD Studio XE6 and since then it has been improved in security and functionality. App tethering allows you to connect applications to exchange in a so-called serverless mode. In other words, it gives your applications the ability to interact with other applications running either on the same machine or on a remote machine without using a server because the applications communicate directly with each other.

App tethering features do not depend on a specific transport or protocol, and new protocols can be implemented using the app tethering API. Currently, app tethering can work using a network or Bluetooth Classic adapter using the same code.

To enable an application to use app tethering, only two components are required:

- `TTetheringManager`: Used to discover other applications that are using app tethering
- `TTetheringAppProfile`: Used to define the actions and data that your application shares with other applications previously paired using the *TTetheringManager*

The app tethering technology roughly follows the Bluetooth Classic model, where there are a set of Bluetooth devices that are able to interact with each other, and each application exposes a set of profiles that are usable by other applications.

One of the strengths of this technology is that it is completely independent of the platform where the resultant application runs. You can use app tethering to connect a VCL application to a mobile app running on Android or iOS, or between a FireMonkey mac OS X application and an iOS app, or even more, a VCL Windows service to a FireMonkey desktop application. I think you've got the point: you can use app tethering to create an application network that's able to make your applications more usable.

App tethering is designed to develop so-called **companion apps**. What's a companion app? Well, a companion app is an app that's designed to make another application more usable. Let's say you developed a media center running on an Android TV or on a PC. You can play videos and music, but how can you control the player while you are on the sofa? You need a remote controller! Using app tethering, you can create a companion app running on your phone that is able to control the media center to play and stop a video, to go forward, or to go to the next video. The remote controller is the typical companion app of your media center.

Getting ready

There are some nice examples of app tethering on the internet and some others have been provided by Embarcadero, but in this recipe, we're going to talk about a completely new app. We'll develop a presenter assistant (I've just coined this term!). What's a presenter assistant? Well, during training or while I'm talking at conferences or while I'm presenting something to an audience, I use a lot of slides. So, I run my Microsoft PowerPoint presentation (or `http://www.openoffice.org/` Impress) and talk over the slides about, for example, the new Delphi features. In the last few years, I have being using a presenter pointer, which allows me to go to the next slide easily without going back to the PC and pressing the space key (because I walk a lot during a presentation, I'm usually too far from the PC to go back for each slide). A presenter assistant is a small device with two buttons: **Next** and **Previous**. However, I love to talk about programming (and Delphi) so much that often I run out of time. Here's the idea for this recipe: a presenter assistant app running on my Android smartphone that allows me to go to the next slide, to the previous slide, and to also display how many minutes I have before the end of the speech. Here's the presenter assistant app while doing its job:

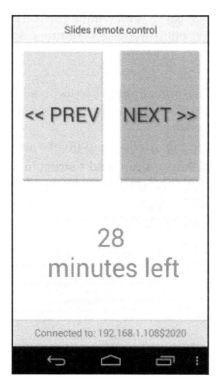

Figure 6.16: The Presenter Assistant app running on my Android phone

The Presenter VCL application is in charge of mimicking a keyboard key press when the mobile app sends the proper commands and sends the remaining minutes to the mobile app every 5 seconds (we don't need a clock; an update every five seconds is enough). Here's the VCL application:

Figure 6.17: The VCL application that will control the desktop application showing the slides

In the app tethering model, there isn't a server and a client. There is an application (or an app) that connects to other apps, but then the two, or more, apps are peers.

Each application can:

- **Share resources**: When other apps subscribe to a shared resource, every time the shared resource changes, all the subscribed apps are notified following the publish/subscribe model.
- **Share actions**: An app can discover and invoke actions published by other apps.
- **Send strings**: One of the apps can send a string to one of the other apps. The string can contain anything, which includes a complex JSON object.
- **Send streams**: One of the apps can send a stream to one of the other apps. The stream can also contain binary data such as an image or an MP3 file.

The presenter assistant we're talking about is very simple. The mobile app needs to send two strings to the desktop application. The first is when we need the next slide and the second is when we need the previous slide. The VCL application running on the PC needs to publish a resource showing the remaining minutes.

How it works...

Open the project group in `Chapter6\RECIPE08`. There are two projects: `Presenter.dproj` (the VCL application) and `PresenterRemote.dproj` (the Android app).

Let's start by showing you how the application works. Run the Presenter application, then run the `PresenterMobile` app on your phone and press **Connect**. If your phone is connected to the same network of your PC, you should be able to connect and see something like **Connected to: 192.168.1.101$2020** on your phone. This means that the mobile app is connected to the VCL application listening on port 2020. Now, go to your desktop, write an integer number in `TSpinEdit`, and press **Start Speech**. The application goes to the taskbar. Now, open Microsoft PowerPoint with a presentation (or another program which is sensible to the left and right arrow; the Delphi source code editor is also good) and repeatedly press the left or right button in the mobile app. You should see the slides (or the cursor) moving.

The following schema shows the communication between the mobile app and the VCL application after the discovering and pairing phases:

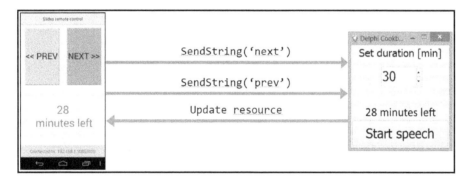

Figure 6.18: The communication between the mobile app and the VCL application

When the `PresenterMobile` app sends the `NEXT` command (using `SendString`), the Presenter application receives it and sends a `VK_RIGHT` Windows keyboard event. Sending a Windows keyboard event, the application is miming a user who is using the keyboard, so the key sent is intercepted by the window that has got the focus at that moment (just as a normal keyboard works). If in the foreground Microsoft PowerPoint (or `http://openoffice.org/` Impress) is showing a presentation, you get the next slide (because if you hit the right arrow during a presentation, you go to the next slide). It's the same for the `PREV` command, which in turn sends a `VK_LEFT` key to Microsoft PowerPoint.

The relevant part of this message exchange is as follows:

```
const
  NEXT_SLIDE = Ord(VK_RIGHT);
  PREV_SLIDE = Ord(VK_LEFT);
  DEFAULT_MINUTES = 30;

procedure SendKey(const C: Word);
var
  kb: TInput;
begin
  kb.Itype := INPUT_KEYBOARD;
  kb.ki.wVk := C;
  kb.ki.wScan := MapVirtualKey(C, 0);
  kb.ki.dwFlags := 0;
  SendInput(1, kb, SizeOf(kb));
  kb.ki.dwFlags := KEYEVENTF_KEYUP;
  SendInput(1, kb, SizeOf(kb));
end;

procedure TMainForm.TetheringAppProfile1ResourceReceived(const Sender:
TObject;
  const AResource: TRemoteResource);
var
  Cmd: string;
begin
  Caption := AResource.Value.AsString;
  if AResource.Hint.Equals('cmd') then
  begin
    Cmd := AResource.Value.AsString;
    if Cmd.Equals('prev') then
      SendKey(PREV_SLIDE)
    else if Cmd.Equals('next') then
      SendKey(NEXT_SLIDE);
  end;
end;
```

How about the connection between the applications? For this app, I've used the **Group** feature of app tethering. As you know, there are two ways to connect your applications:

- Define two applications as belonging to the same group and use automatic discovering and pairing. This approach is very simple, but not as flexible.
- Obtain a list of discovered applications and then request to pair with specific applications. This approach is more flexible, but requires a bit of work.

Considering the scenario, I've used the `Group` property and the autoconnect feature. Here's the code under the **Connect** button in the mobile app:

```
procedure TMainForm.btnConnectClick(Sender: TObject);
begin
  TetheringManager1.AutoConnect(2000);
end;
```

In order for the `AutoConnect` function to work properly, both of the `TetheringAppProfile` components must have the same value in the group property. In our case, the value is `com.danieles.presenters`.

Also, the **Presenter** application shares a `Resource` with **PresenterMobile**. In order to automatically subscribe to the resource update notification, the resource name must be the same on all of the paired apps:

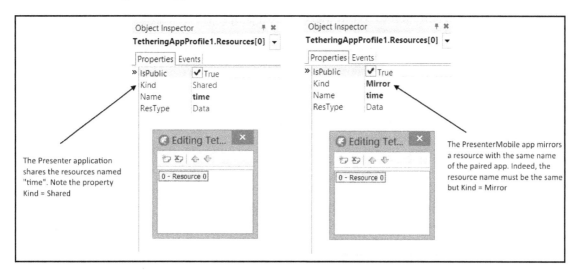

Figure 6.19: The resource configuration

Having this configuration, you can simply update the value of the resource in the desktop application using the following code:

```
TetheringAppProfile1.Resources.
  FindByName('time').Value := MinutesLeft;
```

Updating the local resource 'time' causes an update to the remote resource with the same name, and the following event handler is executed on the mobile app:

```
procedure TMainForm.TetheringAppProfile1Resources0ResourceReceived
  (const Sender: TObject; const AResource: TRemoteResource);
begin
  lblMinutes.Text := AResource.Value.AsString +
sLineBreak + ' minutes left';
end;
```

There's more...

App tethering is a nice technology. It is not a replacement for a server, but a good tool to easily create companion applications. Here's some documentation about it:

- A fast introduction to app tethering with Delphi XE6: `https://www.youtube.com/watch?v=oeMQdvxi560`
- App tethering with RAD Studio 10 Seattle (by Al Mannarino): `https://www.youtube.com/watch?v=da0-e38XYrk`
- Using app tethering: `http://docwiki.embarcadero.com/RADStudio/en/Using_App_Tethering`
- Adding app tethering to your application: `http://docwiki.embarcadero.com/RADStudio/en/Adding_App_Tethering_to_Your_Application`
- Connecting to remote applications using app tethering: `http://docwiki.embarcadero.com/RADStudio/en/Connecting_to_Remote_Applications_Using_App_Tethering`
- Sharing and running actions on remote applications using app tethering: `http://docwiki.embarcadero.com/RADStudio/en/Sharing_and_Running_Actions_on_Remote_Applications_Using_App_Tethering`
- Sharing data with remote applications using app tethering: `http://docwiki.embarcadero.com/RADStudio/en/Sharing_Data_with_Remote_Applications_Using_App_Tethering`
- Fun with Delphi XE6 app tethering and barcodes: `http://fixedbycode.blogspot.it/2014/04/fun-with-delphi-xe6-app-tethering-and.html`

Creating DataSnap Apache modules

One of the most awaited Delphi features by server-side Delphi developers is the support for the building of Apache web server modules. Since Delphi XE6, Delphi has been able to generate Apache modules, and this is very good news! The most recent Apache Versions, are 2.0, 2.2, and 2.4, are supported.

An Apache module is only compatible with the specific version for which it has been compiled. So, be sure about the Apache version you have to deploy your module before you create the project. However, it's possible to change the target Apache version by just changing a unit name.

Getting ready

In this recipe, we'll create a very simple REST service, with only one method returning a list of people. The service will be built using the Embarcadero DataSnap framework and the service itself will be packaged as an Apache web server module. The real goal of this recipe is to show you how to use the Delphi strength in creating an Apache module, and a very light introduction to DataSnap.

How to do it...

This recipe requires some steps, so here's the list:

1. The Apache HTTP Server (`"httpd"`) is a project of the Apache Software Foundation and has been the most popular web server on the internet since April 1996. On Windows, one of the recommended binary distributions is maintained by the Apache Lounge community. Go to `http://www.apachelounge.com/download/` and download the most updated 2.4.x version as a ZIP file. In this recipe, we'll use the Win32 version, so download that one please.

2. Unzip the Apache distribution in a folder named `Apache24` (for example, `C:\DEV\Apache24`).

3. The Apache main configuration is contained in the `C:\DEV\Apache24\conf\httpd.conf` file. Open it with a good text editor. This file contains all the main configuration and includes a bounce of other configuration files. Configuring Apache is trivial; however, in this recipe, we'll configure it to let it run our module. Let's start with a very basic configuration; however, the `http.conf` syntax can be complex, so pay attention to the following steps.

4. Look for `ServerRoot`. Currently, it should look as follows:

   ```
   Define SRVROOT "c:/Apache24"
   ```

5. Change the folder name to `C:/DEV/Apache24`. Warning: we're using / as a folder separator and not \. Also, don't terminate the folder name with a trailing slash. Now, the line should look as follows:

   ```
   Define SRVROOT "c:/DEV/Apache24"
   ```

6. Look for `ServerName`. The `ServerName` directive gives the name and port that the server uses to identify itself. Currently, the line is commented:

   ```
   #ServerName www.example.com:80
   ```

7. Just after the commented line, add the following:

   ```
   ServerName localhost:80
   ```

8. Let's test whether our Apache server is correctly configured. Open a command prompt, go to the `C:\DEV\Apache24` folder, and launch the following command:

   ```
   bin\httpd.exe
   ```

15. Errors will be printed on the standard output. If no errors have been printed, launch a browser and navigate to `http://localhost`; you should get a white page with **It works!** text on it. If so, your Apache installation is running correctly. Now, Apache is running in application mode. It is possible to install it as a service with a simple command that we'll see later.

16. As a warning, we are configuring Apache just to run our modules. It is not configured to be exposed on the internet. So, please read the documentation about the configuration carefully or ask an Apache expert before letting your server go into the wild!

17. Terminate Apache by pressing *Ctrl + C*, leave the command prompt for a moment, and go back to Delphi.

18. Let's create our DataSnap WebBroker project as an Apache 2.4 module.

19. Navigate to **File** | **New** | **Other** then navigate to **Delphi Projects** | **DataSnap Server** | **DataSnap WebBroker Application**.

20. The first wizard asks you for which platform the application will be created: in this case, we leave things as they are (we will see the delpoy on Linux in the next chapter), so click **Next**.

21. The next wizard asks which kind of project we're about to create; select Apache dynamic link module and press **Next**, which is shown in the following screenshot:

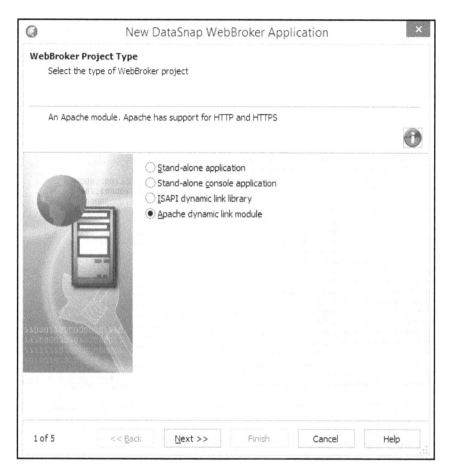

Figure 6.20: The DataSnap Wizard—we choose the Apache module option

16. Then, the wizard asks which Apache version our module will be built for. Select **Apache version 2.4**, name it `datasnap_module`, and press **Next**:

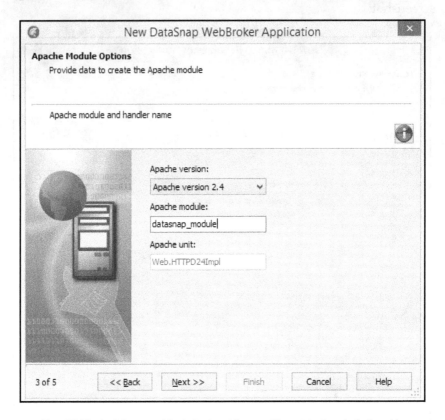

Figure 6.21: The wizard allows you to define the Apache module name and the target Apache version for the module

17. On the next screen, the wizard asks about the functionalities that we need to include in our DataSnap module. Leave the defaults and press **Next**:

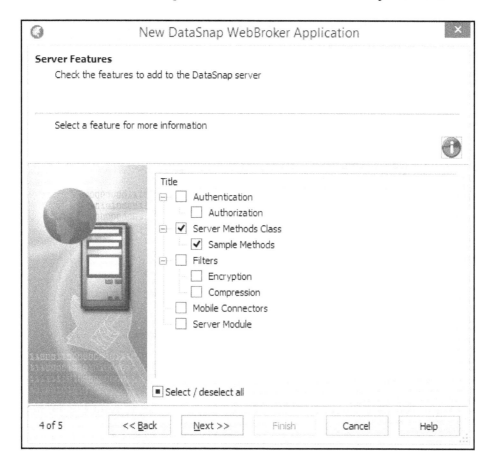

Figure 6.22: Let the wizard include some sample methods in the DataSnap module

18. At the next screen, select `TDataModule` and press **Finish**:

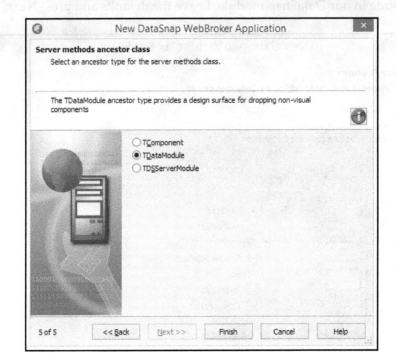

Figure 6.23: Using the TDataModule as ancestor class, we got a design surface without the overhead of the IAppServer interface that we won't use

19. Delphi has created a complete Apache 2.4 module containing a DataSnap REST server. WOW! Now, let's add some features to it.

20. Save the project using the default names.

21. Open `ServerMethodsUnit1.pas`, show the designer, and drop on it a `TFDConnection` and a `TFDQuery`. Connect the `TFDQuery` to the `TFDConnection`, and configure the `TFDConnection` to point at the sample database in the `DATA` folder contained in this recipe. The connection configuration parameters should be similar to the following:

```
Database=C:\DEV\Chapter06\CODE\RECIPE09\DATA\SAMPLES.IB
User_Name=sysdba
Password=masterkey
DriverID=IB
```

22. Go to the code editor and declare the following method in the public section of the `TDataModule`:

```
public
    . . . //other methods
    function GetEmployees: TJSONArray;
end;
```

23. Press *Ctrl + Shift + C* to implement the method body and fill it with the following code:

```
function TServerMethods1.GetEmployees: TJSONArray;
begin
    FDQuery1.Open('SELECT * FROM PEOPLE');
    Result := FDQuery1.AsJSONArray;
end;
```

24. Go to the `implementation uses` clause and add the `MVCFramework.DataSet.Utils` unit (it is a unit contained in the `DelphiMVCFramework` project that we'll use to do standard DataSet serialization).

25. Build the project. Now, our Apache module is ready, but how can we test and debug it? First, we have to put the compiled DLL in the right place. To allow Apache to load our module, it is useful to have it at the same level as the built-in modules. Navigate to **Project | Options | Delphi compiler** and write in the `Output` directory section the `C:\DEV\Apache24\modules\` path, as shown in the following screenshot, and then press **OK**:

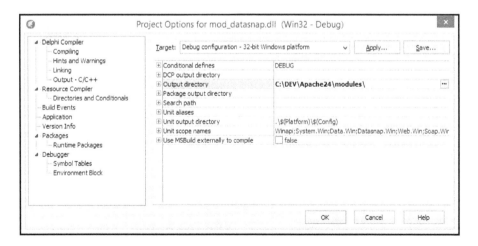

Fig 6.24: Configuring the project output directory to compile directly where Apache looks for modules

26. Compile the project and go back to the `httpd.conf` file.

27. Look for the `LoadModule` string in the file. You will find a lot of lines with this directive and many of them are commented. Just after the last `LoadModule` line (it doesn't matter whether it is commented or not), add the following lines and save the file:

```
LoadModule datasnap_module modules/mod_datasnap.dll
<Location /api>
    SetHandler mod_datasnap-handler
</Location>
```

28. Now, go back to the command prompt. Go to the `C:\DEV\Apache24` folder and launch the following command:

```
bin\httpd.exe
```

29. Go to a browser, and navigate to the URL `http://localhost/api/datasnap/rest/TServerMethods1/getemployees`; you should get the DataSnap JSON response from the Apache module just created.

30. How do we debug our module? Terminate Apache by pressing *Ctrl + C* from the command line and go back to Delphi.

31. Navigate to **Run** | **Parameters**, configure the values shown as follows, and press **OK**:

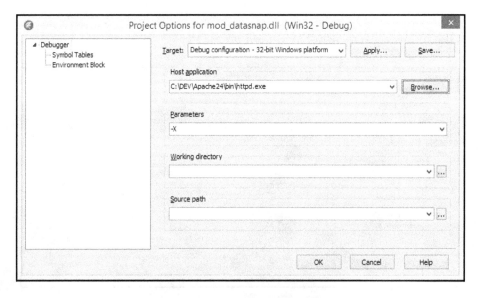

Figure 6.25: Let's set up the debugger to debug the module; note the -X parameter passed to the httpd.exe executable

32. Now, Delphi will start Apache for us and we'll be able to debug the module as with any Delphi program. The `-X` parameter passed to `httpd.exe` launches Apache in debug mode with only one worker, so Delphi doesn't need to debug the web server spawned processes.

33. Run the project. Apache will silently start, launched by Delphi, and our module will be loaded by the `httpd.exe` process. Now, we are able to debug the module using breakpoints and all the ordinary stuff.

How it works...

Apache is configured to load our module. The source code of that module is opened in the Delphi IDE. When Delphi compiles the DLL module, it writes it to where Apache looks for it. Just after the compilation, Delphi launches Apache in application mode with the `-X` parameters (avoiding the spawned process). Apache loads the DLL as configured in the `httpd.conf` file, and Delphi attaches its debugger to the `httpd.exe` process and to its `datasnap_module.dll`. This approach is valid for any DLL that is loaded at runtime by some other software, and is also still valid for every WebBroker program compiled as an Apache module or ISAPI DLL.

There's more...

There are a lot of concepts in this recipe. Here are some links for those who want to go deeper. DataSnap is a complex and powerful framework from Embarcadero that's able to create TCP/IP and HTTP/S servers. I have held many training sessions about it, and suggest that you give it a try. It is also present in Delphi Professional versions:

- *DataSnap Overview and Architecture*: `http://docwiki.embarcadero.com/RADStudio/en/DataSnap_Overview_and_Architecture`

- Tutorial on *Using a DataSnap Server with an Application*: `http://docwiki.embarcadero.com/RADStudio/en/Tutorial:_Using_a_DataSnap_Server_with_an_Application`

- Tutorial on *Using a REST DataSnap Server with an Application*: `http://docwiki.embarcadero.com/RADStudio/en/Tutorial:_Using_a_REST_DataSnap_Server_with_an_Application`

- And now something about the Apache web server: Apache HTTP server security tips: `http://httpd.apache.org/docs/current/misc/security_tips.html`

After you have configured and secured your Apache web server, you can install it as a Windows service using the following command line:

```
.\bin\httpd.exe -k install -n "My DataSnap Server"
```

And to uninstall it:

```
.\bin\httpd.exe -k uninstall -n "My DataSnap Server"
```

In this way, you can package a customized Apache distribution to deploy and run your custom modules. I do it very often with my services that need to be published on the internet, because Apache is stronger and more secure compared to the Delphi built-in web server based on INDY (to each their own).

However, even though in this recipe we have used a dedicated Apache installation to host our module, you can also use an already deployed instance, and often you will do so. The deployment process is the same—copy your module in some path that's accessible from the web server, change the `httpd.conf` file to load your module, and restart the server. That's it.

Creating WebBroker Apache modules

As we have already said WebBroker is a technology that's been available since Delphi 4 to help create web server applications, exposing an HTTP/HTTPS interface. It is a very thin layer on top of HTTP/S, but I love it because it doesn't try to do a lot of things but remains at a low level, allowing you to implement the rest of the architecture as you need. Therefore, Embarcadero used WebBroker as a framework to create DataSnap and EMS. It is very important to know this, because by knowing it, you have all the power to create web things such as HTTP/S services, HTML interfaces, and so on.

In this recipe, we'll create a 64-bit WebBroker Apache module, install it in a custom Apache 2.4 distribution, and secure the server by configuring HTTPS access. Internally, our application uses DelphiMVCFramework, but all the steps are still valid for other frameworks or no frameworks at all (apart from WebBroker). Let's start!

Getting ready

We need to get an Apache distribution. You can use the one provided by Apache Lounge (`https://www.apachelounge.com/`). In this recipe, we'll use the 64-bit Apache, so go to the website and download the latest 2.4 version at 64 bit as a ZIP package. Now, follow steps 1 to 13 of the previous recipe to set up Apache.

You should be able to run Apache from the command line with the following command without any errors:

```
C:\DEV\Apache24> bin\httpd -X
```

How to do it...

Let's create the Apache module. We'll create a fake management system with a list of users. There is one resource that's accessible with two HTTP verbs: GET and POST:

1. Navigate to **File | New | Other**.
2. Navigate to **Delphi Projects | WebBroker | Web Server Application**.
3. The first wizard asks you for which platform the application will be created: in this case, we leave things as they are (we will see the deploy on Linux in the next chapter), so click **Next.**
4. In the resultant modal dialog, select Apache dynamic link module and click **Next**.
5. In the combo box, select **Apache version 2.4** and write `peoplemanager_module` as the module name.
6. Click **Finish.**
7. Save the project using the following file names:
   - mod_peoplemanager.dproj
   - WebModuleU.pas

7. Add a new unit to the project, save it as `SampleControllerU.pas`, and write the following code:

   ```
   unit SampleControllerU;

   interface

   uses
     MVCFramework, MVCFramework.Commons;
   ```

```delphi
type

  [MVCPath('/')]
  [MVCDoc('Just a sample controller')]
  TSampleController = class(TMVCController)
  public
    [MVCPath('/users')]
    [MVCHTTPMethods([httpGET])]
    [MVCDoc('Returns the users list')]
    procedure GetUsers;

    [MVCPath('/users')]
    [MVCHTTPMethods([httpPOST])]
    [MVCConsumes('application/json')]
    [MVCDoc('Creates a new user')]
    procedure CreateUser;
  end;

implementation

uses System.JSON;

{ TSampleController }

procedure TSampleController.CreateUser;
begin
  // just a fake, we don't create any user.
  // simply echo the body request as body response
  if Context.Request.ThereIsRequestBody then
  begin
    Context.Response.StatusCode := HTTP_STATUS.Created;
    Render(Context.Request.Body)
  end
  else
    raise EMVCException.Create(HTTP_STATUS.BadRequest, 'Expected
JSON body');
end;

procedure TSampleController.GetUsers;
var
  LJObj: TJSONObject;
  LJArray: TJSONArray;
begin
  LJArray := TJSONArray.Create;

  LJObj := TJSONObject.Create;
  LJObj.AddPair('first_name', 'Daniele').AddPair('last_name',
'Teti')
```

```
      .AddPair('email', 'd.teti@bittime.it');
    LJArray.AddElement(LJObj);

    LJObj := TJSONObject.Create;
    LJObj.AddPair('first_name', 'Peter').AddPair('last_name',
  'Parker')
      .AddPair('email', 'pparker@dailybugle.com');
    LJArray.AddElement(LJObj);

    LJObj := TJSONObject.Create;
    LJObj.AddPair('first_name', 'Bruce').AddPair('last_name',
  'Banner')
      .AddPair('email', 'bbanner@angermanagement.com');
    LJArray.AddElement(LJObj);

    Render(LJArray);
  end;

  end.
```

8. Re-open the web module.

9. Create an event handler for the **OnCreate** event, and fill it with the following code:

```
    FMVCEngine := TMVCEngine.Create(self);
    FMVCEngine.AddController(TSampleController);
```

10. Declare the `FMVCEngine` private variable as `TMVCEngine`.

11. Include `MVCFramework` in the interface uses clause.

12. Include `SampleControllerU` in the `implementation uses` clause.

13. Save all.

14. Right-click on the **Target Platform** node in the **Project Manager**. Select **Add Platform** and `64 bit Windows` in the dialog.

15. Be sure that the 64-bit Windows node is selected in the **Project Manager** because we have to build the module for a 64-bit Apache.

16. Navigate to **Project** | **Options** and select **All Configuration – All Platform** in the upper combo box.

17. Then, click on the **Delphi Compiler** node and set `C:\DEV\Apache24\modules\` as the output directory.
    1. Click **OK**, save all, and build. Now, your DLL should be compiled in `C:\DEV\Apache24\modules\mod_peoplemanager.dll`.

18. Now, open the `C:\DEV\Apache24\conf\https.conf` file with your preferred text editor.

19. Look for the `LoadModule` string in the file. You will find a lot of lines with this directive and many of them are commented. Just after the last `LoadModule` line (it doesn't matter whether it is commented or not), add the following lines and save the file:

```
LoadModule peoplemanager_module modules/mod_peoplemanager.dll
<Location /people>
  SetHandler mod_peoplemanager-handler
</Location>
```

20. Remember this location because it is a temporary configuration; we have to remove the `<location>` node before the end.

21. Enable `ssl_module`, by removing the character # from this line:

    20. `#LoadModule ssl_module modules/mod_ssl.so`

21. Now, go back to the command prompt. Go to the `C:\DEV\Apache24` folder and launch the following command:

    bin\httpd.exe -X

23. If some errors come up, double-check the previous steps.

24. Open a web browser (Google Chrome or Mozilla Firefox are good choices) and write down the following URL: `http://localhost/people/users`.

25. You should see a JSON array showing some user data, as shown in the following screenshot:

Figure 6.26: The resource users as JSON array; this particular formatting is because of the extension JSON View for Google Chrome

Now, the Apache module works! What we still have to do is configure **HTTPS** access to this service. Remember that HTTP is a textual protocol, so every byte you send or receive from an HTTP server is not encrypted and can be sniffed by anyone with basic networking knowledge and a good sniffer. So, let's make this server secure!

To configure HTTPS on Apache, you need to generate the certificate files. OpenSSL command-line tools can do the job, and we have a copy of OpenSSL in the Apache\bin folder. Here's the list:

1. Shut down Apache if it is still running. You should still be in the C:\DEV\Apache folder; if not, go there please.
2. Enter the bin folder and write the following command:

   ```
   set OPENSSL_CONF=C:\DEV\Apache24\conf\openssl.cnf
   ```

3. This command sets the environment variable used by the openssl.exe executable and should contain the full path of openssl.cnf. Check whether the openssl.cnf file is actually in your Apache distribution.
4. Now, we have to actually generate the certificate, which is composed of two files: the certificate file and the private key file.
5. Execute the following command line:

   ```
   openssl req -x509 -nodes -days 365 -newkey rsa:2048 -keyout
   delphicookbook.key -out delphicookbook.crt
   ```

6. You will be prompted to enter your organizational information and a common name. The common name should be the fully qualified domain name for the site you are securing (www.mydomain.com), or just empty in this case. You can leave the email address, challenge password, and optional company name blank. When the command has finished running, it will create two files: a delphicookbook.key file and a delphicookbook.crt self-signed certificate file, valid for 365 days. Copy these files into the C:\DEV\Apache24\conf folder.
7. Open your Apache configuration file (conf\httpd.conf) in a text editor.
8. Remove the following lines, because we know that the module works but we don't need it running in HTTP:

```
<Location /people>             #REMOVE
SetHandler mod_peoplemanager-handler    #REMOVE
</Location>          #REMOVE
```

9. Search for line Listen 80 and add the following lines just after it:

```
Listen 443
<VirtualHost *:443>
  ServerName localhost
  SSLEngine on
  SSLCertificateFile "conf/delphicookbook.crt"
  SSLCertificateKeyFile "conf/delphicookbook.key"
  <Location /people>
  SetHandler mod_peoplemanager-handler
  </Location>
</VirtualHost>
```

10. Save the changes and exit the text editor.
11. Run Apache with the usual command (`bin/httpd -X`).
12. Now, you should be able to access the `/people/users` URI using only the HTTPS protocol; let's check it.
13. Open a browser and go to `http://localhost/people/users`.
14. You should see a big **Not Found**; good! The service is no longer accessible via the HTTP protocol.
15. Now, change the address and write `https://localhost/people/users`.
16. You should see our JSON array:
    - The browser may tell you that the certificate is not verified. We know it is because it's a self-signed certificate and no Certification Authority has been contacted to obtain it. So, you can accept the certificate without a problem.

17. Our WebBroker Apache module is running behind an HTTPS secured web server and is ready to rock!
18. How can we debug our module? Terminate Apache by pressing *Ctrl* + *C* from the command line and go back to Delphi.
19. Navigate to **Run** | **Parameters**, and configure the values, as shown in figure 6.24.
20. Now, Delphi will start Apache for us and we'll be able to debug the module as for any Delphi program. The `-X` parameter passed to `httpd.exe` launches Apache in debug mode with only one worker so that Delphi doesn't need to debug the web server spawned processes.
21. Run the project. Apache will silently start, launched by Delphi, and our module will be loaded by the `httpd.exe` process. Now, we are able to debug the module using breakpoints and all the ordinary stuff.

How it works...

The project is quite simple. Using `DelphiMVCFramework`, we defined one resource which supports two verbs, GET and POST, with the following meanings:

URI	VERB	DESCRIPTION
`/users`	GET	Retrieves a JSON array of JSON objects representing fake users
`/users`	POST	Doesn't create anything, but echoes the JSON request body as the body response

There's more...

A lot of topics here! Now, we understand how to create Apache modules and how to make them secure using HTTPS. Warning! Don't think about publishing this server to the internet without a proper hardening by a skilled person. This configuration is just a minimal HTTPS setup; the server could still be vulnerable at some other point:

- If you want to know more about Apache and HTTPS, check out the following article: `https://httpd.apache.org/docs/2.4/ssl/ssl_howto.html`
- Curious about OpenSSL? Check out the project's site: `https://www.openssl.org/`
- Here's a WebBroker framework introduction: `http://docwiki.embarcadero.com/RADStudio/en/Using_Web_Broker_Index`
- To get support on DelphiMVCFramework, check out the following Facebook group: `https://www.facebook.com/groups/delphimvcframework/`
- Alternatively, if you need professional support, email `dmvcframework@bittime.it`.

Using native HTTP(S) client libraries

The RTL provides two components that you can use to send HTTP requests to servers and handle their responses:

- `TNetHTTPClient`
- `TNetHTTPRequest`

Alternatively, as we saw in Chapter 3, *Knowing Your Friends – The Delphi RTL*, you can use an instance of THTTPClient to manage your HTTP requests.

Why use these components instead of good old TidHTTP from the INDY suite? The reasons have been explained in Chapter 3, *Knowing Your Friends – The Delphi RTL*, in the *Delphi RTL* section. However, in this recipe we'll use the new HTTP client to show how much of the deployment is simplified, also in mobile apps, using these new components instead of the INDY ones, at least for HTTP communications.

Long story short, Embarcadero developed a native HTTP client library that is not based on INDY or OpenSSL, but that relies on the OS API to implement HTTP protocol. So, when Microsoft, Apple, or Google release a new security patch, your application is already updated. Great! You simply rely on the OS security infrastructure and don't depend on the OpenSSL DLLs anymore!

Getting ready

In this recipe, we'll see a simple but complete cross-platform HTTPS client that's able to connect to the following:

- An HTTPS service using a valid certificate provided by a certification authority
- An HTTPS service that uses a self-signed certificate

The TNetHTTPClient component does a great job of integrating with the underlying OS to provide a uniform development and deployment experience to the Delphi developer. This project reuses the unit AsyncTask.pas developed in *Recipe 6* of Chapter 5, *The Thousand Faces of Multithreading*. Let's see how it works.

How it works...

Open the project in Chapter06\CODE\RECIPE11\XPlatNativeHTTPClient.dproj. The main form is similar to the following:

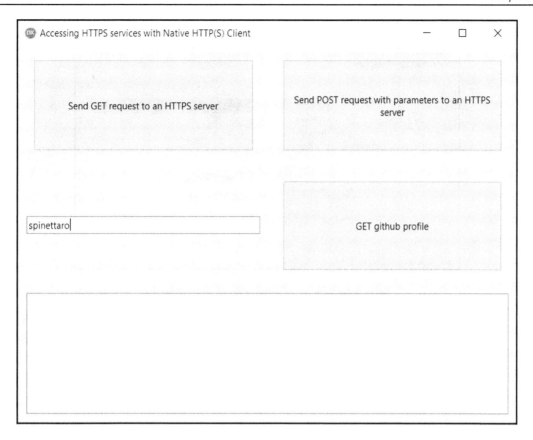

Figure 6.27: The cross-platform HTTPS client running on Windows 10

The two upper buttons send a request to the Apache HTTPS service developed in the previous recipe, while the third button sends a GET request to the GitHub API to get information about the username written in the edit. While the GitHub HTTPS contains a valid certificate provided by a recognized certification authority, the local service uses a self-signed certificate.

All the HTTPS requests are executed by a data module that contains an instance of `TNetHTTPClient`.

Let's check out the THTTPDM data module:

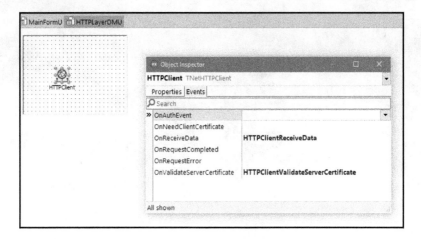

Figure 6.28: The data module in charge of actually sending the HTTPS requests

This data module contains all the methods used by the main form to do the actual HTTPS requests, plus some other helpers required to make asynchronous requests a bit simpler.

Here's the code:

```
unit HTTPLayerDMU;

interface

uses
  System.SysUtils, System.Classes, System.Net.URLClient,
System.Net.HttpClient,
  System.Net.HttpClientComponent;

type
  THTTPDM = class(TDataModule)
    HttpClient: TNetHTTPClient;
    procedure HTTPClientValidateServerCertificate(const Sender: TObject;
      const ARequest: TURLRequest; const Certificate: TCertificate;
      var Accepted: Boolean);
    procedure HTTPClientReceiveData(const Sender: TObject;
      AContentLength, AReadCount: Int64; var Abort: Boolean);
  private
    FReadCount: UInt64;
    FCertificate: TCertificate;
    procedure Clear;
  public
```

```
    function Get(const URL: String): IHTTPResponse;
    function Post(const URL: String; BodyRequest: TStream; Headers:
TNetHeaders)
      : IHTTPResponse;
    property ReadCount: UInt64 read FReadCount;
    property Certificate: TCertificate read FCertificate;

  type
    TResponseData = record
      Response: IHTTPResponse;
      ReadedBytes: UInt64;
      Certificate: TCertificate;
      function HeadersAsStrings: TArray<String>;
    end;
  end;

implementation

{ %CLASSGROUP 'FMX.Controls.TControl' }

{$R *.dfm}

function THTTPDM.Get(const URL: String): IHTTPResponse;
begin
  Clear;
  Result := HttpClient.Get(URL);
end;

procedure THTTPDM.HTTPClientReceiveData(const Sender: TObject;
  AContentLength, AReadCount: Int64; var Abort: Boolean);
begin
  FReadCount := AReadCount;
end;

procedure THTTPDM.HTTPClientValidateServerCertificate(const Sender:
TObject;
  const ARequest: TURLRequest; const Certificate: TCertificate;
  var Accepted: Boolean);
begin
  Accepted := (Certificate.Start <= Now) and (Certificate.Expiry >= Now);
  FCertificate := Certificate;
end;

function THTTPDM.Post(const URL: String; BodyRequest: TStream;
  Headers: TNetHeaders): IHTTPResponse;
begin
  Clear;
  Result := HttpClient.Post(URL, BodyRequest, nil, Headers);
```

```
  end;

procedure THTTPDM.Clear;
begin
  FReadCount := 0;
end;

{ THTTPDM.TResponseData }

function THTTPDM.TResponseData.HeadersAsStrings: TArray<String>;
var
  Pair: TNameValuePair;
begin
  Result := [];
  for Pair in Response.Headers do
  begin
    Insert(Pair.Name + ':' + Pair.Value, Result, MaxLongInt);
  end;
end;

end.
```

This code is quite simple and is just a high level wrapper for TNetHTTPClient. How does the main form use this? Let's check out the main form code:

```
unit MainFormU;

interface

uses
  System.SysUtils, System.Types, System.UITypes, System.Classes,
  System.Variants,
  FMX.Types, FMX.Controls, FMX.Forms, FMX.Graphics, FMX.Dialogs,
  FMX.ScrollBox, FMX.Memo, FMX.Controls.Presentation,
  FMX.StdCtrls, FMX.Layouts, HTTPLayerDMU, FMX.Edit;

type
  TMainForm = class(TForm)
    GridPanelLayout1: TGridPanelLayout;
    btnGet: TButton;
    btnPost: TButton;
    Layout1: TLayout;
    mmResponse: TMemo;
    EditGithubUser: TEdit;
    btnGithub: TButton;
    procedure btnGetClick(Sender: TObject);
    procedure btnPostClick(Sender: TObject);
    procedure btnGithubClick(Sender: TObject);
```

```
private
  { Private declarations }
  procedure UpdateGUI(const Value: THTTPDM.TResponseData);
public
  { Public declarations }
end;

var
  MainForm: TMainForm;

implementation

uses
  System.JSON.Writers, System.JSON.Builders, REST.Types, System.Threading,
  AsyncTask, FMX.Ani, System.Net.HttpClient,
  System.Net.URLClient;

{$R *.fmx}

procedure TMainForm.btnGetClick(Sender: TObject);
const
  URL = 'https://192.168.1.103/people/users';
begin
  (Sender as TControl).Enabled := False;
  Async.Run<THTTPDM.TResponseData>(
    function: THTTPDM.TResponseData
    var
      LHTTPReq: THTTPDM;
      LResp: IHTTPResponse;
    begin
      LHTTPReq := THTTPDM.Create(nil);
      try
        LResp := LHTTPReq.Get(URL);
        if LResp.StatusCode <> 200 then
        begin
          raise Exception.CreateFmt('Error %d: %s', [LResp.StatusCode,
            LResp.StatusText]);
        end;
        Result.ReadedBytes := LHTTPReq.ReadCount;
        Result.Certificate := LHTTPReq.Certificate;
      finally
        LHTTPReq.Free;
      end;
      Result.Response := LResp;
    end,
    procedure(const Value: THTTPDM.TResponseData)
    begin
      UpdateGUI(Value);
```

```
      (Sender as TControl).Enabled := True;
    end);

end;

procedure TMainForm.btnGithubClick(Sender: TObject);
const
  URL = 'https://api.github.com/users/%s';
var
  LGithubuser: String;
begin
  LGithubuser := EditGithubUser.Text;
  (Sender as TControl).Enabled := False;
  Async.Run<THTTPDM.TResponseData>(
    function: THTTPDM.TResponseData
    var
      LHTTPReq: THTTPDM;
      LResp: IHTTPResponse;
    begin
      LHTTPReq := THTTPDM.Create(nil);
      try
        LResp := LHTTPReq.Get(Format(URL, [LGithubuser]));
        if LResp.StatusCode <> 200 then
        begin
          raise Exception.CreateFmt('Error %d: %s', [LResp.StatusCode,
            LResp.StatusText]);
        end;
        Result.ReadedBytes := LHTTPReq.ReadCount;
        Result.Certificate := LHTTPReq.Certificate;
      finally
        LHTTPReq.Free;
      end;
      Result.Response := LResp;
    end,
    procedure(const Value: THTTPDM.TResponseData)
    begin
      UpdateGUI(Value);
      (Sender as TControl).Enabled := True;
    end);

end;

procedure TMainForm.btnPostClick(Sender: TObject);
const
  URL = 'https://192.168.1.103/people/users';
begin
  (Sender as TControl).Enabled := False;
  Async.Run<THTTPDM.TResponseData>(
```

```
function: THTTPDM.TResponseData
var
  LJSONStream: TStringStream;
  LJSONWriter: TJsonWriter;
  LStreamWriter: TStreamWriter;
  LJSONObjectBuilder: TJSONObjectBuilder;
  LHeaders: TNetHeaders;
  LHTTPReq: THTTPDM;
  LResp: IHTTPResponse;
begin
  LHTTPReq := THTTPDM.Create(nil);
  try
    LJSONStream := TStringStream.Create;
    try
      LStreamWriter := TStreamWriter.Create(LJSONStream);
      try
        LJSONWriter := TJsonTextWriter.Create(LStreamWriter);
        try
          LJSONObjectBuilder := TJSONObjectBuilder.Create(LJSONWriter);
          try
            LJSONObjectBuilder.BeginObject.Add('first_name', 'Daniele')
              .Add('last_name', 'Spinetti').Add('email',
'd.spinetti@bittime.it')
              .EndObject;
            LJSONWriter.Flush;
            LJSONStream.Position := 0;
            LHeaders := [TNetHeader.Create('content-type',
              CONTENTTYPE_APPLICATION_JSON)];
            LResp := LHTTPReq.Post(URL, LJSONStream, LHeaders);
            if LResp.StatusCode <> 201 then
            begin
              raise Exception.CreateFmt('Error %d: %s',
[LResp.StatusCode,
                LResp.StatusText]);
            end;
          finally
            LJSONObjectBuilder.Free;
          end;
        finally
          LJSONWriter.Free;
        end;
      finally
        LStreamWriter.Free;
      end;
    finally
      LJSONStream.Free;
    end;
    Result.ReadedBytes := LHTTPReq.ReadCount;
```

```
          Result.Certificate := LHTTPReq.Certificate;
        finally
          LHTTPReq.Free;
        end;
        Result.Response := LResp;
      end,
      procedure(const Value: THTTPDM.TResponseData)
      begin
        UpdateGUI(Value);
        (Sender as TControl).Enabled := True;
      end);
  end;


  procedure TMainForm.UpdateGUI(const Value: THTTPDM.TResponseData);
  begin
    mmResponse.Lines.Clear;
    if not Value.Certificate.Subject.IsEmpty then
    begin
      mmResponse.Lines.Add('** Certificate Validity: from ' +
        DateToStr(Value.Certificate.Start) + ' to ' +
        DateToStr(Value.Certificate.Expiry));
      mmResponse.Lines.Add(sLineBreak + '** Certificate Subject: ' +
        Value.Certificate.Subject.Replace(sLineBreak, ', '));
    end
    else
    begin
      mmResponse.Lines.Add(sLineBreak + '** Certificate is not self-signed');
    end;
    mmResponse.Lines.Add(sLineBreak + '** Total bytes read: ' +
      Value.ReadedBytes.ToString);
    mmResponse.Lines.Add(sLineBreak + '** Headers: ' + sLineBreak +
      String.Join(sLineBreak, Value.HeadersAsStrings));
    mmResponse.Lines.Add(sLineBreak + '** Content charset: ' + sLineBreak +
      Value.Response.ContentCharSet);
    mmResponse.Lines.Add(sLineBreak + '** Response Status: ' + sLineBreak +
      Value.Response.StatusCode.ToString + ': ' + Value.Response.StatusText);
    mmResponse.Lines.Add(sLineBreak + '** Response body: ' + sLineBreak +
      Value.Response.ContentAsString);
  end;

end.
```

Quite a lot of code, but there are few things to understand. The most important is the wrapping of `TNetHTTPClient` inside the data module. We want to isolate the event handlers needed to accept the self-signed certificates and all the helpers. Then, in the main form, we can simply use the simpler interface offered by the wrapper.

All the data retrieved by the request, plus ReadBytes and Certificate, are packaged in a `TResponseData` record. All the requests simply return the `IHTTPResponse` returned by the `TNetHTTPClient`. The requests are asynchronous, so that we can use the code on Windows and on mobile platforms. `THTTPDM` could also be more specialized with specific methods that completely hide the internals of the system, but in this case, our objective is not to do this. Please don't use the `btnPostClick` event handler. Also, it is not really needed here; I'm using it to show you how to generate JSON text using the `TJSONBuilder` object to achieve a small memory requirement, which is especially useful on mobile, but also in servers and in general, when your JSON may be big.

Now, try to run this project on Windows. Did it work? Try to click on the buttons and check the certificate information returned for a self-signed certificate (the one from our server) and for a proper certificate (the one from GitHub). Now, select **Android** or **iOS** as the **Target Platform** and try to run the project as a mobile app, as shown in the following screenshot (remember to change the hardcoded IP address to yours):

Figure 6.29: The HTTPS client running on Android

As you can see, it simply works! You don't need to deploy OpenSSL DLL or object code, like it was usual to do with INDY. Moreover, I quite love this integration between the OS HTTP layer and mine; it makes a lot of things simpler and is completely cross-platform.

There's more...

The TNetHTTP* components are a must if you want to build HTTP-enabled applications, especially on mobile, so you have to know them. Play with them, use them; if you have HTTP-enabled apps using INDY, consider switching to TNetHTTPClient; you will simplify deployment and development and take advantage of the deeper integration with the OS.

Here's a well-written document about the TNetHTTP* components directly from Embarcadero's WikiDoc:

- Using an HTTP client: http://docwiki.embarcadero.com/RADStudio/en/Using_
  an_HTTP_Client

Logging like a pro using LoggerPro

The developer's life is difficult, very often we have to diagnose problems and understand why an application does not start, or perform reverse engineering tasks on the functioning of parts of the system.

The first source to be consulted for each troubleshooting operation are logs, simple files that track errors and particular actions performed by the system, such as changing a password, a user's login, or an error message from an application.

These files are extremely important, so it is essential to choose the right tool. In this recipe, we will analyze the LoggerPro framework, which is a very valid tool for the cause in question.

Getting ready

This recipe uses LoggerPro, a Delphi open source framework defined as a modern and pluggable logging framework for Delphi. You can find the project code here: https://
github.com/danieleteti/loggerpro.

To download LoggerPro, go to the project website and clone the repository using a Git client (TortoiseGit, for example: `https://tortoisegit.org/`). You can also use the command-line version and then use the following command:

```
git clone https://github.com/danieleteti/loggerpro loggerpro
```

Or, you can use Delphi to directly download the repository. Navigate to **File** | **Open from version control** | **Git**. Then, in the window that appears, write the following information and click **OK**:

Figure 6.30: The dialog used to download LoggerPro project from its GitHub repository

Now the integrated Git client will clone the repository, downloading all the necessary files. At the end of this process, the wizard asks about which project we want to open. Click **Cancel** and **Close** the dialog. The LoggerPro files have been downloaded in `C:\DEV\Delphi\Libs\loggerpro`; configure the Delphi library path to point there.

This recipe is also configured to log through SysLog protocol, and a valid tool for Window OS is this one: `https://www.solarwinds.com/free-tools/kiwi-free-syslog-server`. If you want to be up and running for this protocol, go to the link, download the free edition, and install it (the only advice I can give you during installation is to install it as a service). No more configuration is needed to be up and running for this recipe, but only start it after installation.

> *"In computing, SysLog is a standard for message logging. It allows separation of the software that generates messages, the system that stores them, and the software that reports and analyzes them. Each message is labeled with a facility code, indicating the software type generating the message, and assigned a severity label."*

> *– Wikipedia*

How to do it...

We will start from the project created in the first recipe, take that project and clone it in your file system; this will be the starting point.

The goal is to write a log mechanism that does the following:

- Writes all Log levels to file and through the **SysLog** protocol
- Sends emails in case of error

Perform the following steps:

1. First, we have to define a configuration file called `LoggerProConfig.pas`. In this file, we define the appenders, the log level, and we provide a log interface that's ready for use:

```
unit LoggerProConfig;

interface

uses
  LoggerPro;

const
  LOG_TAG = 'PHONEBOOK_SERVER';

function Log: ILogWriter;

implementation

uses
  LoggerPro.FileAppender,
  LoggerPro.EMailAppender,
  LoggerPro.OutputDebugStringAppender,
  LoggerPro.UDPSyslogAppender,
  System.SysUtils,
  idSMTP, System.IOUtils,
  IdIOHandlerStack, IdSSL,
  IdSSLOpenSSL, IdExplicitTLSClientServerBase;

var
  _Log: ILogWriter;

function Log: ILogWriter;
begin
  Result := _Log;
end;
```

```
function GetSMTP: TidSMTP;
begin
  Result := TidSMTP.Create(nil);
  try
    Result.IOHandler :=
TIdSSLIOHandlerSocketOpenSSL.Create(Result);
    Result.Host := 'smtp.gmail.com';
    Result.Port := 25;
    Result.UseTLS := TIdUseTLS.utUseImplicitTLS;
    Result.AuthType := satDefault;
    Result.Username := 'daniele.spinetti@bittime.it';
    if not TFile.Exists('config.txt') then
      raise Exception.Create
        ('Create a "config.txt" file containing the password');
    Result.Password := TFile.ReadAllText('config.txt'); //
'<yourpassword>';
  except
    Result.Free;
    raise;
  end;
end;


procedure SetupLogger;

const
  LOG_LEVEL = TLogType.Debug;

var
  lEmailAppender: ILogAppender;
begin
  lEmailAppender := TLoggerProEMailAppender.Create(GetSMTP,
    'LoggerPro<daniele.spinetti@bittime.it>',
'd.spinetti@bittime.it');
  lEmailAppender.SetLogLevel(TLogType.Error);
  _Log := BuildLogWriter([TLoggerProFileAppender.Create,
lEmailAppender,
    TLoggerProOutputDebugStringAppender.Create,
    TLoggerProUDPSyslogAppender.Create('127.0.0.1', 514 //
UDPClientPort.Value
    , 'COMPUTER', 'd.spinetti', 'PhoneBookServer', '0.0.1', '',
True, False)], nil, LOG_LEVEL);
end;


initialization

SetupLogger;

end.
```

2. Next, redefine the `WebModuleU.pas` in order to use the newly configured log interface (in the following code, only the changed parts are shown):

```
unit WebModuleU;

interface

uses System.SysUtils, System.Classes, Web.HTTPApp, Data.DBXJSON,
  FireDAC.Stan.Intf, FireDAC.Stan.Option,
  FireDAC.Stan.Error, FireDAC.UI.Intf, FireDAC.Phys.Intf,
FireDAC.Stan.Def,
  FireDAC.Stan.Pool, FireDAC.Stan.Async,
  FireDAC.Phys, FireDAC.Stan.Param, FireDAC.DatS,
FireDAC.DApt.Intf,
  FireDAC.DApt, FireDAC.Phys.IBBase, FireDAC.Phys.IB,
  Data.DB, FireDAC.Comp.DataSet, FireDAC.Comp.Client,
FireDAC.Moni.Base,
  FireDAC.Moni.FlatFile, FireDAC.Moni.Custom, System.JSON,
  FireDAC.Phys.IBDef, FireDAC.VCLUI.Wait,
MVCFramework.Serializer.JSON;

type
  TwmMain = class(TWebModule)
    Connection: TFDConnection;
    FDPhysIBDriverLink1: TFDPhysIBDriverLink;
    cmdInsertPerson: TFDCommand;
    qryPeople: TFDQuery;
    cmdUpdatePerson: TFDCommand;
    WebFileDispatcher1: TWebFileDispatcher;
    procedure wmMainDefaultHandlerAction(Sender: TObject; Request:
TWebRequest;
      Response: TWebResponse; var Handled: Boolean);
    procedure wmMainwaGetPeopleAction(Sender: TObject; Request:
TWebRequest;
      Response: TWebResponse; var Handled: Boolean);
    procedure wmMainwaDeletePersonAction(Sender: TObject; Request:
TWebRequest;
      Response: TWebResponse; var Handled: Boolean);
    procedure wmMainwaSavePersonAction(Sender: TObject; Request:
TWebRequest;
      Response: TWebResponse; var Handled: Boolean);
    procedure ConnectionBeforeConnect(Sender: TObject);
    procedure WebModuleCreate(Sender: TObject);
    procedure WebModuleDestroy(Sender: TObject);
    procedure WebModuleException(Sender: TObject; E: Exception;
      var Handled: Boolean);
    procedure WebModuleBeforeDispatch(Sender: TObject; Request:
TWebRequest;
```

```
      Response: TWebResponse; var Handled: Boolean);
  private
    FSerializer: TMVCJSONSerializer;
    procedure PrepareResponse(AJSONValue: TJSONValue;
      AWebResponse: TWebResponse);
  end;

var
  WebModuleClass: TComponentClass = TwmMain;

implementation

{$R *.dfm}

uses
  MVCFramework.DataSet.Utils, {this unit comes from
delphimvcframework project}
  MVCFramework.Serializer.Commons,
  System.RegularExpressions,
  System.IOUtils,
  System.StrUtils,
  LoggerProConfig;


procedure TwmMain.wmMainwaSavePersonAction(Sender: TObject;
  Request: TWebRequest; Response: TWebResponse; var Handled:
Boolean);
var
  InsertMode: Boolean;
  LastID: Integer;
  HTTPFields: TStrings;
  JPeople: TJSONObject;

  procedure CheckNoEmpty(const AName, AValue: String);
  const
    NO_EMPTY_FIELD: array of string = ['FIRST_NAME', 'LAST_NAME',
'EMAIL'];
  begin
    if not MatchText(AName, NO_EMPTY_FIELD) then
      exit;
    if AValue.IsEmpty then
      raise Exception.CreateFmt('Field %s cannot be empty',
[AName]);
  end;

  procedure MapStringsToParams(AStrings: TStrings; AFDParams:
TFDParams);
  var
```

```
    i: Integer;
  begin
    for i := 0 to HTTPFields.Count - 1 do
    begin
      if AStrings.ValueFromIndex[i].IsEmpty then
      begin
        CheckNoEmpty(AStrings.Names[i].ToUpper,
AStrings.ValueFromIndex[i]);
        AFDParams.ParamByName(AStrings.Names[i].ToUpper).Clear();
      end
      else
        AFDParams.ParamByName(AStrings.Names[i].ToUpper).Value :=
          AStrings.ValueFromIndex[i];
    end;
  end;

begin
  HTTPFields := Request.ContentFields;
  InsertMode := HTTPFields.IndexOfName('id') = -1;
  if InsertMode then
  begin
    MapStringsToParams(HTTPFields, cmdInsertPerson.Params);
    cmdInsertPerson.Execute();
    LastID := Connection.GetLastAutoGenValue('GEN_PEOPLE_ID');
  end
  else
  begin
    MapStringsToParams(HTTPFields, cmdUpdatePerson.Params);
    cmdUpdatePerson.Execute();
    LastID := HTTPFields.Values['id'].ToInteger;
  end;

  // execute query and prepare response
  qryPeople.Open('SELECT * FROM PEOPLE WHERE ID = ?', [LastID]);
  try
    JPeople := TJSONObject.Create;
    FSerializer.DataSetToJSONObject(qryPeople, JPeople,
      TMVCNameCase.ncLowerCase, []);
    PrepareResponse(JPeople, Response);
  finally
    qryPeople.Close;
  end;
end;

procedure TwmMain.ConnectionBeforeConnect(Sender: TObject);
begin
  Connection.Params.Values['Database'] :=
    TPath.GetDirectoryName(WebApplicationFileName) +
```

```
  '\..\..\DATA\SAMPLES.IB';
end;

procedure TwmMain.WebModuleAfterDispatch(Sender: TObject; Request:
TWebRequest;
  Response: TWebResponse; var Handled: Boolean);
begin
  Log.DebugFmt('Response ok for request <%s>', [Request.PathInfo],
LOG_TAG);
end;

procedure TwmMain.WebModuleBeforeDispatch(Sender: TObject; Request:
TWebRequest;
  Response: TWebResponse; var Handled: Boolean);
begin
  Log.DebugFmt('Start processing of request <%s> - <%s>',
    [Request.PathInfo, Request.Method], LOG_TAG);
end;

procedure TwmMain.WebModuleException(Sender: TObject; E: Exception;
  var Handled: Boolean);
begin
  Log.Error(E.Message, LoggerProConfig.LOG_TAG);
end;

end.
```

How it works...

First, in WebModuleU.pas, WebModuleBeforeDispatch, WebModuleAfterDispatch, and WebModuleException events were implemented. In the first two events, a log (with DEBUG level) was been inserted that allows tracking the requests before and after their execution: in this case, the log is written to file and through the SysLog. In WebModuleException, a log with an ERROR level has been inserted, so when an exception occurs, in addition to writing on the previously viewed appenders, you also send an email with the exception message (its configuration is written in LoggerProConfig.pas).

To allow raising an exception (in an easy way), the CheckNoEmpty method has been added, which checks that the FIRST_NAME, LAST_NAME, and EMAIL fields are not left empty, in which case it raises an exception that will be reported in the WebModuleException event.

The log writers and all the appenders are asynchronous.

Hitting *F9*, you should see a console window informing you that a server has started. Open the browser and point it to `http://localhost:8080`. If you try to create a person, leaving all the fields empty, an exception will raise. Now, if you go into the file system at the same level as the executable, you will find the log file, as shown in the following screenshot:

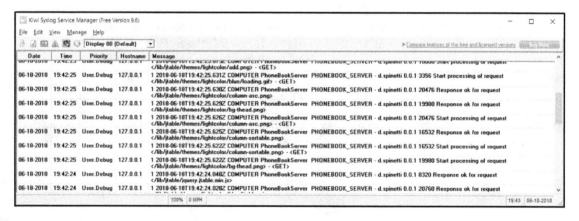

Figure 6.31: Server file log after initial request

Furthermore, if you have correctly configured your email settings, you should also find a new message in the inbox and if you have installed a SysLog server as suggested in *Getting Started*, you will see something like this on your **SysLog Service Manager**:

Figure 6.32: Syslog service manager after the initial request

There's more...

What a fantastic framework! But the good news does not end here—**LoggerPro,** through a modular and extensible design, lets you output to multiple logging targets by defining an appender—an object that implements how the log operation should take place. The framework contains the following built-in log appenders:

Appender	Description
File (`TLoggerProFileAppender`)	To write the log in the file
Console (`TLoggerProConsoleAppender`)	To write the log in the console output
Output Debug String (`TLoggerProOutputDebugStringAppender`)	To write the log in the Output Debug String
VCL Memo (`TVCLMemoLogAppender`)	To write the log in a **TMemo** component
VCL ListView (`TVCLListViewAppender`)	To write the log in a **TListView** component
Redis Appender with LogsViewer (`TLoggerProRedisAppender`)	To aggregate logs from different instances on a single Redis instance
Email Logger (`TLoggerProEMailAppender`)	To send an email as a log
SysLog Logger (`TLoggerProUDPSyslogAppender`-RFC 5424 compliant)	To write a log through SysLog protocol

LoggerPro is also very useful in client applications, so I recommend going to see the samples contained in the project repository, where you can find all the implementations of the various appenders we have just described.

See also

- The following are some links to deepen the topics covered:
    - LoggerPro project, samples, and documentation: `https://github.com/danieleteti/loggerpro`
    - SysLog RFC 5424: `https://tools.ietf.org/html/rfc5424`

Linux Development

7

In this chapter, we will cover the following recipes:

- Creating Linux TCP/IP servers
- How to correctly handle Linux signals
- To fork or not to fork? a.k.a. How to build a modern Linux daemon?
- Building a TCP/IP Linux server and daemonizing it
- Building a RESTful server for Linux
- Building a complete RESTful server with database access and web client interface
- Creating WebBroker Apache modules for Linux

Introduction

In this chapter, we'll look at how to develop Linux applications using Delphi. Delphi, from Version 10.2 Tokyo, officially allows you to develop applications for the Linux platform. Using your own Delphi instance from a Windows machine, you can create a 64-bit Linux application and deploy it to the Linux machine. However, initial support does not allow you to create visual applications. To be clear, it is not possible to create VCL or FireMonkey applications—officially, there is no support, but there are frameworks that allow you to do so. Here is a list of the types of applications that are available for the Linux platform:

- Console application
- EMS package (RAD Server)
- DataSnap
- DataSnap WebBroker
- WebBroker
- Dynamic-link libraries
- DUnitX project

Looking at the types of apps available, the recipes in this chapter are focused on console applications and server-side technology.

An installation of RAD Studio on a Windows PC and another machine with the Linux platform (even a virtual machine is okay) is needed to develop Linux applications. The recipes in this chapter require a working development configuration for your PC to talk with the Linux machine. A detailed tutorial on how to properly configure your system for this purpose can be found on the Embarcadero DocWiki: `http://docwiki.embarcadero.com/RADStudio/en/Linux_Application_Development`. There is also a video on the *Embarcadero Technologies* channel on YouTube: `https://www.youtube.com/watch?v=1-nnF5o3bhU`.

> The official supported distributions are **Ubuntu** and **Red Hat Enterprise**. The other distributions are not directly supported, but you can always make an attempt based on the similarity of the target Linux distribution.

If you also need to configure your Linux machine, a great tutorial can be found in Craig Chapman's blog: `http://chapmanworld.com/2016/12/29/configure-delphi-and-redhat-or-ubuntu-for-linux-development/`.

> ATTENTION!! By default, no launcher application is used. This means that if you run an application, it will be used in the Terminal window of the running instance of the Platform Assistant. If you want to open a separate Terminal window for running and debugging your application, you have to enable the **Use Launcher Application** checkbox (you can find it in **Project | Options | Debugger**). You can find more info at `http://docwiki.embarcadero.com/RADStudio/Tokyo/en/Debugger`.

The Linux distribution that I have used and recommend you use is **Ubuntu 18.04 LTS** (`http://releases.ubuntu.com/18.04/`).

Creating Linux TCP/IP servers

Linux has always been a highly appreciated operating system, especially for the server-side. So what better start for development on Linux? Creating a small TCP/IP server.

Getting ready

In this recipe, we're going to create a TCP/IP server that listens on a port for incoming requests from a TCP client. These client requests can then be processed at the server, and a response can be sent back to the client.

The server we are going to build will listen for client commands. It will always return the reverse string of the command passed, unless you pass the `quit` command; in that case, it will disconnect the client.

This recipe is very simple, but it will be of great help in taking the first steps in development for Linux.

INDY is a fantastic library, appreciated and widely used by Delphi developers. INDY offers client and server components for using internet protocols, such as **TCP**, **UDP**, **Echo**, **FTP**, **HTTP**, **Telnet**, and many others. The good news is that it is also available for the Linux platform!

In this recipe, we're going to use `TIdTCPServer` of INDY components. `TIdTCPServer` is a component that encapsulates a complete multithreaded **Transmission Control Protocol (TCP)** server.

How to do it...

Let's see how to do this:

1. Let's start by creating a new console application by performing the following steps: **File | New | Other | Console Application**.
2. The idea is to insert the `TIdTCPServer` in a `TDataModule` in order to give a closer implementation to a real environment. So, add a new **Data Module** to the project and name it `ServerModuleU`. Right-click on **Project**, then click on **Add | New | Delphi Files | Data Module**.
3. Add the `TIdTCPServer` component (from the **Tool Palette**) on the **Data Module** surface.
4. We want the server listening on port `8888`, so by **Object Inspector**, set the **DefaultPort** property of `TIdTCPServer` to `8888`.

5. We have to manage the connection to the server to show a welcome message. Select the component. In the **Events** tab of **Object Inspector**, double-click on the **OnConnect** event to define this event handler and enter the following code:

```
procedure TServerModule.IdTCPServer1Connect(AContext: TIdContext);
var
  lHandler: TIdIOHandlerSocket;
begin
  lHandler := AContext.Connection.Socket;
  lHandler.WriteLn('WELCOME TO THE BEST TCPSERVER EVER');
end;
```

6. Now we have to define the **OnExecute** event handler to handle the commands sent by clients (in accordance with the description in the introduction). Select the component, and in the **Events** tab of **Object Inspector**, double-click on the **OnExecute** event to define this event handler. Then enter the following code:

```
procedure TServerModule.IdTCPServer1Execute(AContext: TIdContext);
var
  lHandler: TIdIOHandlerSocket;
  lCmd: string;
  lResp: string;
begin
  lHandler := AContext.Connection.Socket;
  lCmd := lHandler.ReadLn;
  lCmd := lCmd.ToLower;
  if lCmd = 'quit' then
    lResp := 'quit'
  else
    lResp := System.StrUtils.ReverseString(lCmd);

  lHandler.WriteLn('RECEIVED: ' + lCmd);
  if lResp = 'quit' then
    AContext.Connection.Disconnect
  else
    lHandler.WriteLn(lResp);
end;
```

7. Define two public methods in `TDataModule`, `Start`, and `Stop` to allow the activation and deactivation of the server:

```
TServerModule = class(TDataModule)
  ..

  ..
public
  procedure Start;
  procedure Stop;
```

```
end;

...

procedure TServerModule.Start;
begin
  IdTCPServer1.Active := True;
end;

procedure TServerModule.Stop;
begin
  IdTCPServer1.Active := False;
end;
```

8. In the program unit (.dpr), add a procedure called `Main` to instantiate the `TServerModule` component and start the TCP/IP server inside an infinite loop:

```
procedure Main;
begin
  // instantiate the ServerModule
  lServerModule := TServerModule.Create(nil);
  // START TCP/IP server
  lServerModule.Start;
  while true do
  begin
    Sleep(1000);
    Log('Loop');
  end;
  Log('Stopping');
  // STOP TCP/IP server
  lServerModule.Stop;
  Log('Stopped');
  lServerModule := nil;
end;
```

9. Call the `Main` procedure inside the main `begin end` as follows:

```
begin
  try
    Log('Started');
    Main;
    Log('Finished');
  except
    on E: Exception do
      Log('ERROR> ' + E.ClassName + ': ' + E.Message);
  end;

end.
```

10. You can find the `Log` function in the `LogU.pas` file, it is in the `Commons` folder of the code recipes. So if you want to use it, add the unit to your project and in the `uses` section

How it works...

Before we run the program, remember to do the following:

1. Set the target platform to **Linux**.
2. Ensure you have a Linux machine with the **Platform Assistant Server** started up (you can find more info about this in the *Introduction* section of this recipe).
3. Enable the **Use Launcher Application** checkbox (you can find more info about this in the *Introduction* section of this recipe) to run it in a new Terminal.
4. Now you can run the program. Hit *F9*, or **Run | Run**.
5. Once launched, on the Linux machine you will see a new Terminal open; the TCP/IP server is running!
6. You've just developed and deployed an application for the Linux system, don't you feel proud of yourself? This is just a small demo to take the first steps into the Linux world:

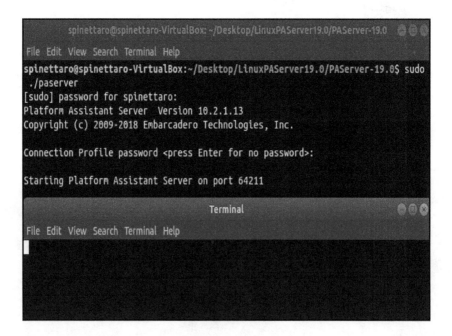

Figure 7.1: TCP/IP server running

7. Now, you can interact with it! You can use a simple Telnet program (or whatever you want) to test it and interact with commands, as shown in the following screenshot:

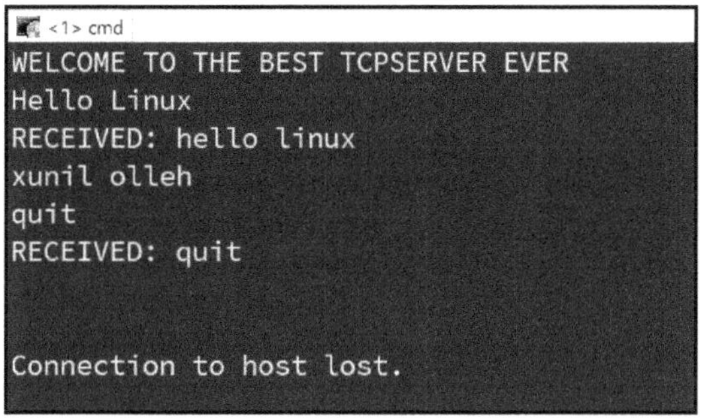

Figure 7.2: TCP/IP server in action

The application is quite simple—always return the reverse string of the command passed, unless you pass the `quit` command. In that case, it will disconnect the client.

There's more...

In this unit, we have used the `log` function made available by the `Commons/LogU.pas` unit. This function is excellent for logging into Linux, as it points directly to the `/var/log/` path.

How to correctly handle Linux signals

Signals in Linux are a very important concept to understand. A robust program needs to handle signals. This is because signals are a way to deliver asynchronous events to the application.

A signal is nothing but a way of communicating a message from one process to another. These messages are popularly known as **notifications**, which the receiving process is free to process, ignore, or leave up to the OS to take default action. For example, a notification can be to kill the receiving process, or let it know that it has accessed an invalid memory area, or it can be a notification of availability of a resource, and so on. So, you can see that signals are a way for **inter-process communication** (**IPC**). Signals are important because they drive some of the most popular programming and system administration activities, so they're also important when integrating your applications with the system.

It is very useful to know how to handle Linux signals, and that is exactly what we'll be doing in this recipe.

Getting ready

In this recipe, I'll show you how to correctly handle Linux signals; in particular, the classics SIGINT, SIGKILL, SIGSTOP and the user-defined signal, USR1.

In Linux, a program can handle signals in the following three ways (from https://www.thegeekstuff.com/2012/03/linux-signals-fundamentals/):

- The signal can be ignored. By ignoring we mean that nothing will be done when the signal occurs. Most of the signals can be ignored, but signals generated by hardware exceptions like divide by zero, if ignored can have weird consequences. Also, a couple of signals cannot be ignored.
- The signal can be caught. When this option is chosen, then the process registers a function with the kernel. This function is called by the kernel when that signal occurs. If the signal is non-fatal for the process, then in that function the process can handle the signal properly, or otherwise it can choose to terminate gracefully.
- Let the default action apply. In a scenario where a program just doesn't care for a signal, the default action corresponding to that particular signal is taken by the OS kernel.

If a process needs to handle certain signals, the process has to register a signal handling function to the kernel. The following code is the prototype of a signal handling function:

```
procedure SignalHandler(SigNum: Integer); cdecl;
```

The signal handler procedure accepts a signal number corresponding to the signal that needs to be handled.

To make sure that the signal handler function is registered to the kernel, the signal handler function pointer is assigned to the sa_handler field property of a record of type sigaction_t. This record is then passed as a second argument to the sigaction function (declared in Posix.Signal.pas), as shown in the following code:

```
procedure SignalHandler(SigNum: Integer); cdecl;
begin
  Log('SIGNAL RECEIVED: ' + SigNum.ToString);
end;

var
  lAction: sigaction_t;
```

```
begin
  lAction._u.sa_handler := SignalHandler;
  sigaction(SIGINT, @lAction, nil))
```

The `sigaction` function returns 0 on success or `SIG_ERR` (−1) on error. For example, if you cannot catch the signal. This is the case for `SIGKILL` and `SIGSTOP`.

How to do it...

Perform the following steps:

1. Create a new **Console Application** performing these actions: **File | New | Other | Console Application**.

2. In the `uses` section, add the units `Posix.Unistd` and `Posix.Signal`.

3. Declare the signal handler procedure to catch the signals:

```
procedure SignalHandler(SignalNo: Integer); cdecl;
begin
  if SignalNo = SIGINT then
    WriteLn('Received SIGINT');
  if SignalNo = SIGUSR1 then
    WriteLn('Received SIGUSR1');
  if SignalNo = SIGKILL then
    WriteLn('Received SIGKILL');
  if SignalNo = SIGSTOP then
    WriteLn('Received SIGSTOP');
end;
```

4. In the console's main section, add the following code inside the autogenerated `try...except`:

```
lAction._u.sa_handler := SignalHandler;

if sigaction(SIGINT, @lAction, nil) = SIG_ERR then
  WriteLn('Can''t catch SIGINT');
if sigaction(SIGUSR1, @lAction, nil) = SIG_ERR then
  WriteLn('Can''t catch SIGUSR1');
if sigaction(SIGKILL, @lAction, nil) = SIG_ERR then
  WriteLn('Can''t catch SIGKILL');
if sigaction(SIGSTOP, @lAction, nil) = SIG_ERR then
  WriteLn('Can''t catch SIGSTOP');

while True do
begin
  WriteLn(TimeToStr(now) + ': Hey, I''m alive...');
```

```
      Sleep(2000);
    end;
```

5. Declare the variable `lAction`:

```
var
    lAction: sigaction_t;
```

6. Set the target platform to **Linux64**.

How it works...

In the preceding code, I tried to handle four signals:

- `SIGINT`
- `SIGKILL`
- `SIGSTOP`
- `SIGUSR1` (this last one is a user-defined signal, `USR1`)

If the handler procedure catches a signal, it writes an output `Received + the signal catched` (for example, `Received SIGINT`), while every two seconds informs the user that it is alive by printing on the console, `Hey, I'm alive...`.

When we run the program, we get the following output:

Figure 7.3: Linux Signals in action

As shown in the preceding figure, the first things printed out are as follows:

- `Can't catch SIGKILL`
- `Can't catch SIGSTOP`

This is because the signals SIGKILL and SIGSTOP cannot be caught or ignored; these two signals provide a way for either the root user or the kernel to kill or stop any process, in any situation.

 Did you ever wonder what makes a program aware that a breakpoint has been hit and it needs to halt the execution? Yes, it is a signal (SIGSTOP) that is sent to the program in this case. This is a problem because a SIGSTOP signal cannot be caught, otherwise you cannot debug a program.

Whenever *Ctrl + C* is pressed, a signal SIGINT is sent to the process. So, with the process running, try to press *Ctrl + C*—the default action of this signal is to terminate the process, but the signal is handled in the code, so you'll see something like this following screenshot:

```
12:19:40 PM: Hey, I'm alive...
12:19:42 PM: Hey, I'm alive...
12:19:44 PM: Hey, I'm alive...
12:19:46 PM: Hey, I'm alive...
12:19:48 PM: Hey, I'm alive...
^CReceived SIGINT
12:19:50 PM: Hey, I'm alive...
12:19:52 PM: Hey, I'm alive...
```

Figure 7.4: SIGINT signal handled

To see the USR1 signal in action, you have to use the kill command:

```
kill -USR1 <Program_PID>
```

There's more...

A signal handler function must be very careful, since processing elsewhere may be interrupted at some arbitrary point in the execution of the program. POSIX has the concept of the safe function. If a signal interrupts the execution of an unsafe function, and the handler calls an unsafe function, then the behavior of the program is undefined.

– Linux man page

So, if you don't know which functions are safe to use inside a signal handler and which functions are not, refer to that list at: `http://man7.org/linux/man-pages/man7/signal-safety.7.html` and make sure you are not doing anything dangerous.

If you really want your signal handler to be safe and portable, consider not doing anything inside it except modifying a global interlocked variable. For more on this scenario, I've left you a bonus recipe project called *LinuxSignalsSafe*.

See also

As mentioned several times, Linux signals are a very important concept, so I want to leave several links to deepen this topic:

- **Signals introduction**: `http://man7.org/linux/man-pages/man7/signal.7.html`
- **Signal handling**: `http://man7.org/linux/man-pages/man2/signal.2.html`
- **Sigaction**: `http://man7.org/linux/man-pages/man2/sigaction.2.html`
- **Signal-safety**: `http://man7.org/linux/man-pages/man7/signal-safety.7.html`

How to build a modern Linux daemon

Each Linux system has a bunch of processes running. Some processes have the goal to run for a long time on the system in the background: daemons. A **daemon** (or service) is a particular type of process that runs in the background and is designed to run autonomously, with little or no user intervention.

Many programs use daemons to solve current scenarios: a web server to provide static content for web pages, email service like SMTP, and so on.

You must use a specific set of rules in a given order to create a daemon in Linux. To understand how daemons works in Linux, we have to see them in action. They give you incredible flexibility and power, and for this reason they are one of the fundamental building blocks in Linux.

Getting ready

Throughout this recipe, I'll show you how to build a Linux daemon by transforming a process into a daemon with the classic method of using the `fork()` function, or in the modern method supported by the `sistemd` process.

How to do it...

As mentioned earlier, transforming a process into a daemon uses a specific set of rules in a given order:

1. First, we have to fork off the parent process and let it terminate autonomously.
2. According to the requirement of the daemon, we have to change the file mode mask (umask).
3. Then, ensure that the process runs in a new session and has a new group. So, we will have to create a unique session ID (setsid).
4. Next, ensure the current working directory points to a valid path (chdir).
5. Close stdin, stdout, and stderr.
6. Let the main logic of the daemon process run.

Next, perform the following steps:

1. Create a new **Console Application** by performing these actions, **File | New | Other | Console Application**.
2. Declare two constants that represent the exit code in case of success or failure:

```
const
    EXIT_FAILURE = 1;
    EXIT_SUCCESS = 0;
```

3. Implement a Log procedure to log information, as shown in the following code. In this way, we are logging in the same path of application. It will be useful then, to understand what the internal processes of the app are doing:

```
procedure Log(const Value: String);
var
   SW: TStreamWriter;
begin
   SW := TStreamWriter.Create(TFile.Open(getpid.ToString,
TFileMode.fmAppend,
      TFileAccess.faWrite));
   SW.BaseStream.Seek(0, TSeekOrigin.soEnd);
   SW.WriteLine(DateTimeToStr(now) + ': ' + Value);
   SW.Close;
end;
```

4. Fork off the parent process:

```
// Fork off the parent process
Log('2. BEFORE FORK ');
```

```
ProcessID := fork();
if (ProcessID < 0) then
   raise Exception.Create('Error. Cannot fork the process');
// If we got a good ProcessID, then let parent process terminate
autonomously.
if (ProcessID > 0) then
begin
   Log('3. INTO FORK - PARENT PROCESS - EXIT SUCCESS');
   Halt(EXIT_SUCCESS);
end;
```

5. It's necessary to modify the file mask (`umask`) to ensure that they can be written or read correctly. The function used to do that is the `umask()` function. To have full access to the files generated by the daemon, we have to set the `umask()` to 0 (octal value):

    ```
    // Change the file mode mask
    umask(0);
    ```

6. The following code will place the server in a new process group and session, and detach its controlling Terminal (no user intervention):

    ```
    // Create a new SID for the child process
    SID := setsid();
    if (SID < 0) then
       raise Exception.Create('Error in creating a new SID');
    ```

7. Now, it's necessary to change the current working directory, and we can do this by using the `chdir()` function. This is because many Linux distributions do not follow the **Linux Filesystem Hierarchy Standard**. Drawing attention to this, the only path that certainly exists and is valid is the root (`/`):

    ```
    // Change the current working directory
    chdir('/');
    ```

8. Now we have to close the three standard streams—`STDIN`, `STDOUT`, and `STDERR`. These file descriptors turn out to be redundant and a potential security flaw, since by definition a daemon can not use the Terminal:

    ```
    // Close out the standard file descriptors
    for idx := sysconf(_SC_OPEN_MAX) downto 0 do
       __close(idx);
    ```

9. A daemon's main code is typically inside an infinite loop. Here is where the application logic should be defined:

```
// the Daemon Loop
I := 0;
while true do
begin
  // Do some task here ...
  Log('LOOPING...');
  // wait 500 milliseconds
  sleep(500);
  Inc(I);
  if I >= 100 then
    break;
end;
```

10. Remember, before exiting the application, it is necessary to set the output value on success, as demonstrated in the following code:

```
ExitCode := EXIT_SUCCESS;
```

11. Putting all the points we've discussed and turning them into code, we have the following:

```
var
  ProcessID: Integer;
  SID: Integer;
  I: Integer;

begin
  Log('1. START');
  try
    // Fork off the parent process
    Log('2. BEFORE FORK ');
    ProcessID := fork();
    if (ProcessID < 0) then
      raise Exception.Create('Error. Cannot fork the process');
    // If we got a good PID, then let parent process terminate
autonomously.
    if (ProcessID > 0) then
    begin
      Log('3. INTO FORK - PARENT PROCESS - EXIT SUCCESS');
      Halt(EXIT_SUCCESS);
    end;

    Log('4. AFTER FORK');
    // Change the file mode mask
    umask(0);
```

```
// Create a new SID for the child process
SID := setsid();
if (SID < 0) then
  raise Exception.Create('Error in creating a new SID');

// Change the current working directory
chdir('/');

// Close out the standard file descriptors
for idx := sysconf(_SC_OPEN_MAX) downto 0 do
  __close(idx);

// the Daemon Loop
I := 0;
while true do
begin
  // Do some task here ...
  Log('LOOPING...');
  // wait 500 milliseconds
  sleep(500);
  Inc(I);
  if I >= 100 then
    break;
end;
Log('6. END WHILE');
ExitCode := EXIT_SUCCESS;
Log('7. END - EXIT SUCCESS');
except
  on E: Exception do
  begin
    Log(E.Message);
    Writeln(E.ClassName, ': ', E.Message);
    ExitCode := EXIT_SUCCESS;
  end;
end;
```

12. Always remember to set the target platform to **Linux64**, and have a **PAServer** up and running before you run the Linux application.

How it works...

In Linux, there are three different ways to start a daemon:

- By the system
- By a user in a Terminal
- By using a script

But historically, there is only one primary method to create a new process—this is done by using the `fork()` function; the function takes its name from this bifurcation process. So, let's analyze the `fork()` function closely because it is the key topic of this recipe:

```
pid := fork();
if (pid < 0) then
  raise Exception.Create('Error in forking the process');
// If we got a good PID, then we can KILL the parent process.
if (pid > 0) then
begin
  Log('3. INTO FORK - PARENT PROCESS - EXIT SUCCESS');
  Halt(EXIT_SUCCESS);
end;
// Daemon execution here...
```

The `fork()` function is used to create a new process. Once executed, we will have two processes—**parent** and **child**—that need to be distinguished. This can be done by analyzing the returned value of the `fork()` function. The `fork()` function returns:

- `PID = 0`: If it is the child process
- `PID > 0`: If it is the parent
- `PID < 0`: If it is an error

When writing a daemon, you will have to code as defensively as possible.

The interesting thing to understand is that after the `fork()` function call, there are two processes that run separately—child and parent. For this reason, it is important to determine, based on the return of the `fork()` function, whether we are in the context of parent or child (daemon). In the first case, we have to end the process gracefully (with the `HALT()` function and successful exit code), while in the second case, the child process continues the execution from here on out in the code (the `Daemon` loop, referring to the preceding code).

If you run the application, after it is terminated, you can see the logs generated by the application. I voluntarily entered numbered logs to allow you to better understand the fork mechanism, process separation, and their execution.

The rest of the code is well explained in the *How do it...* section until the final `while...true` loop, where within the simple logic of this recipe is applied.

There's more...

There is another way to create daemons, or services, in this case, through the use of `systemd`.

systemd

> *systemd is a suite of basic building blocks for a Linux system. It provides a system and service manager that runs as PID 1 and starts the rest of the system. systemd provides aggressive parallelization capabilities, uses socket and D-Bus activation for starting services, offers on-demand starting of daemons, keeps track of processes using Linux control groups, maintains mount and automount points, and implements an elaborate transactional dependency-based service control logic. systemd supports SysV and LSB init scripts, and works as a replacement for sysvinit. Other parts include a logging daemon, utilities to control basic system configuration like the hostname, date, locale, and a list of logged-in users and running containers and virtual machines, system accounts, runtime directories and settings, and daemons to manage simple network configuration, network time synchronization, log forwarding, and name resolution.*
>
> *– Wiki ArchLinux*

Please note that although `systemd` has become the default init system for many Linux distributions, it isn't implemented universally across all distributions. As you go through this recipe, if your Terminal outputs the error `systemctl` is not installed, then it is likely that your machine has a different init system installed. The distribution chosen as a reference for this chapter, **Ubuntu 18.04 LTS**, has it.

To examine and control `systemd`, a command is made available—`systemctl`. Here, some of its commands need to manage the services. When using `systemctl`, you generally have to specify the complete name of the unit file, including its suffix, for example, `consoled.service`. For services, if you do not specify the suffix, `systemctl` will assume `.service`. For example, `consoled` and `consoled.service` are equivalent):

- List running units:

    ```
    $ systemctl
    ```

- Start a unit immediately:

    ```
    $ systemctl start unit
    ```

- Stop a unit immediately:

    ```
    $ systemctl stop unit
    ```

- Restart a unit:

    ```
    $ systemctl restart unit
    ```

- Ask a unit to reload its configuration:

    ```
    $ systemctl reload unit
    ```

- Show the status of a unit, including whether it is running or not:

    ```
    $ systemctl status unit
    ```

- Check whether a unit is already enabled or not:

    ```
    $ systemctl is-enabled unit
    ```

- Enable a unit to be started on bootup:

    ```
    $ systemctl enable unit
    ```

- Enable a unit to be started on bootup and start immediately:

  ```
  $ systemctl enable --now unit
  ```

- Disable a unit from starting during bootup:

  ```
  $ systemctl disable unit
  ```

- Reload `systemd`, scanning for new or changed units:

  ```
  $ systemctl daemon-reload
  ```

You can find these commands in the following Wikipedia link: `https://wiki.archlinux.org/index.php/systemd`. To interact with the commands made available by `systemd`, a resource (such as, in our case, a daemon or service) must have defined a specific file called **unit.** A unit is a standardized representation of system resources. Where you read unit in the preceding commands, it means just that.

Let's get to the point. How can you use `systemd` to run a console application like a daemon (service)? Follow these steps:

1. Define the unit file for consoled (`consoled.service`) in `/etc/systemd/system/consoled.service`:

   ```
   [Unit]
   Description=Example of simple TCPIP daemon program

   [Service]
   Type=simple
   ExecStart=/home/spinettaro/PAServer/scratch-dir/d.spinetti-
   ubuntudev/consoled

   [Install]
   WantedBy=multi-user.target
   ```

2. Use `systemd` to start it:

   ```
   $ sudo systemctl start consoled.service
   ```

3. If you want to run the service at startup, you can use the following commands:

   ```
   $ sudo systemctl enable consoled.service
   $ sudo systemctl start consoled.service

   or

   $ systemctl enable --now consoled.service
   ```

In the code of this recipe, I have also left the unit `consoled.service` as previously written.

See also

Here are some links about daemons and the `fork` process:

- Introduction to daemons: `https://wiki.archlinux.org/index.php/daemons`
- Detailed explanation of the `fork` method: `http://man7.org/linux/man-pages/man2/fork.2.html`

Here are some links to the `systemd` documentation, its commands, and how to structure the units:

- `https://wiki.archlinux.org/index.php/Systemd`
- `https://access.redhat.com/documentation/en-us/red_hat_enterprise_linux/7/html/system_administrators_guide/sect-managing_services_with_systemd-unit_files`

Building a TCP/IP Linux server and daemonizing it

In the previous recipes, we have seen how to create a TCP/IP server, manage the Linux signals, and how to create a Linux daemon. Through this newly acquired knowledge, in this new recipe we will be able to build a TCP/IP server that includes all the necessary considerations to be daemonized.

Getting ready

We will start from the project created in the first recipe, take that project and clone it in your filesystem; it will be the starting point.

In the code of this project, we should add the signal management and the parts necessary to be compatible with `systemd`, so that it can be daemonized.

Let's start!

How to do it...

Perform the following steps:

1. Open the project just cloned.
2. Rename it in `tcpipserver_daemon`; you can do it by right-clicking on the project (in the project manager) and then choosing **Rename.**
3. Add `Posix.Signal` in the `uses` section.
4. We have to introduce the signal handling: `SIGINT` and `SIGTERM` are the two signals that we're going to manage. The first represents the handle of a user hitting *Ctrl + C* on console, and the second represents the handle of the `systemctl` stop service. In both cases, we have to terminate the application. So, let's start by implementing the signal handler procedure:

```
procedure SigHandler(SignalNo: Integer); cdecl;
begin
  Log('SIGNAL : ' + SignalNo.ToString);
  TInterlocked.Exchange(lTerminated, 1);
end;
```

5. Now we have to update the `Main` function by entering the code for signal handling:

```
procedure Main;
begin
  lTerminated := 0;
  lServerModule := TServerModule.Create(nil);
  lServerModule.Start;
  lAction._u.sa_handler := SigHandler;
  sigaction(SIGINT, @lAction, nil); // sent by the console hitting
ctrl+c
  sigaction(SIGTERM, @lAction, nil); // sent by "systemctl stop
dtservice"
  while lTerminated = 0 do
  begin
    Sleep(1000);
    Log('Loop');
  end;
  Log('Stopping');
  lServerModule.Stop;
  Log('Stopped');
  lServerModule := nil;
end;
```

6. Also, the `var` section must be updated:

```
var
  lServerModule: TServerModule;
  lAction: sigaction_t;
  lTerminated: Int64;
```

7. Define the unit file for `systemctl` and save it as `tcpserver_daemon.service` (remember to replace `{your_user}` and `{your_machine}` with your data) and put the unit file in `etc/systemd/system/tcpserver_daemon.service`):

```
[Unit]
Description=Example of simple TCPIP server daemon

[Service]
Type=simple
ExecStart=/home/{your_user}/PAServer/scratch-
dir/{your_machine}/tcpserver_daemon

[Install]
WantedBy=multi-user.target
```

How it works...

Now you can start the server via `systemctl`:

```
$ systemctl start tcpserver_daemon.service
```

The server is started! Great work! You can now check its status:

```
$ systemctl status tcpserver_daemon.service
```

You can make a further check on its operation by looking at the logs in
`/var/logs/tcpserver_daemon.log`, or you can interact with it. You can use a simple
Telnet program (or whatever you want) to test it and interact with commands, as shown in
the following figure:

Figure 7.5: TCP/IP server daemonized in action

The application will always return the reverse string of the command passed, unless you
pass the `quit` command. In that case, it will disconnect the client.

Since we have managed the `SIGTERM` signal, we can proceed to the `stop` command to
understand how the app behaves:

```
$ systemctl start tcpserver_daemon.service
```

The server is now stopped, and you can check this by typing the `status` command, as
show in the following code:

```
$ systemctl status tcpserver_daemon.service
```

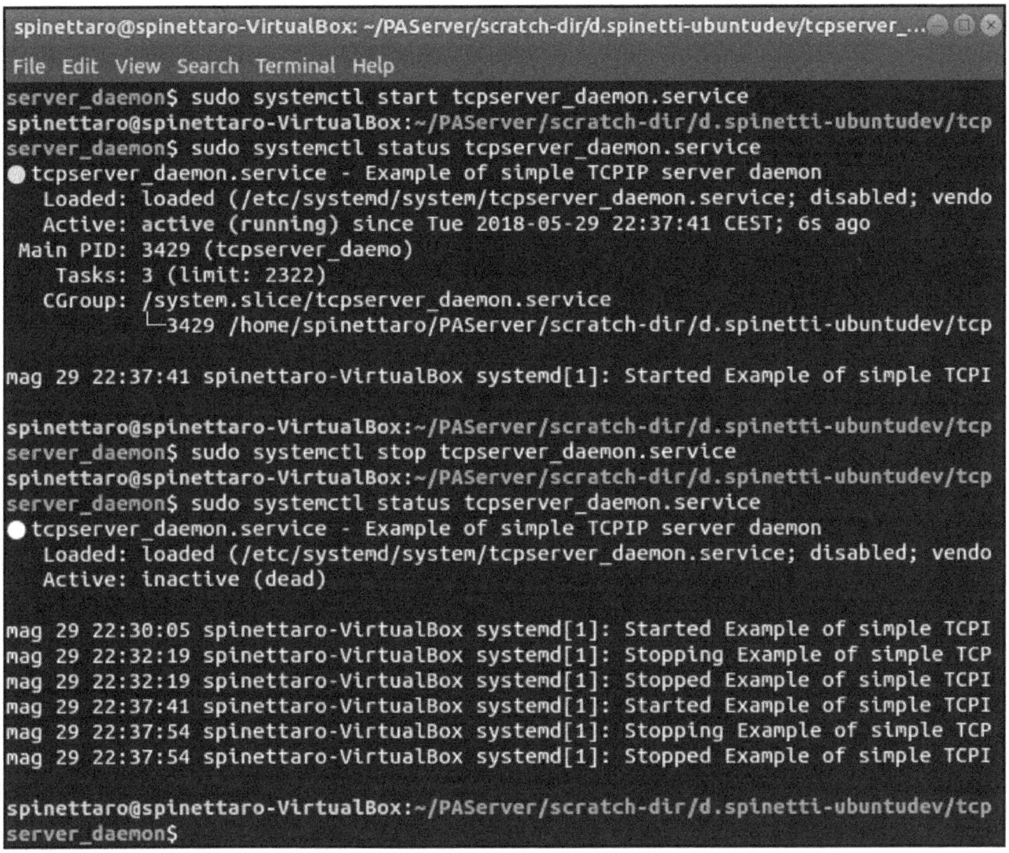

Figure 7.6: Start, Status and Stop of TCP Server

You can also check the server works and the signal **TERM** management, by reading logs in `/var/logs/tcpserver_daemon.log`.

There's more...

This server can also be daemonized another way: using the fork mechanism.

I left you a bonus recipe about this: `TCPIP_SERVER_FORK`.

I want to explain some steps about the procedure. I simply took the code skeleton related to the fork mechanism explained in Recipe 3 of this chapter, and replaced the main loop with the Main procedure of this recipe:

```
begin
  Log('1. START');
  try
    // Fork off the parent process
    Log('2. BEFORE FORK ');
    ProcessID := fork();
    if (ProcessID < 0) then
      raise Exception.Create('Error. Cannot fork the process');
    // If we got a good ProcessID, then let parent process terminate
autonomously.
    if (ProcessID > 0) then
    begin
      Log('3. INTO FORK - PARENT PROCESS - EXIT SUCCESS');
      Halt(EXIT_SUCCESS);
    end;

    Log('4. AFTER FORK');
    // Change the file mode mask
    umask(0);

    // Create a new SID for the child process
    SID := setsid();
    if (SID < 0) then
      raise Exception.Create('Error in creating a new SID');

    // Change the current working directory
    // ATTENTION -- for demo purpose comment the line below to
    // make sure that the log files are written to the same level of
application file
    chdir('/');

    // Close out the standard file descriptors
    for idx := sysconf(_SC_OPEN_MAX) downto 0 do
      __close(idx);

    // Daemon - specific initialization goes here

    // the Daemon Loop
    Main; // main procedure of TCP SERVER
    Log('6. END WHILE');
    ExitCode := EXIT_SUCCESS;
    Log('7. END - EXIT SUCCESS');
  except
    on E: Exception do
```

```
begin
  Log(E.Message);
  Writeln(E.ClassName, ': ', E.Message);
  ExitCode := EXIT_SUCCESS;
  end;
  end;
end;
```

Once deployed and started on the Linux machine, the server's operation is the same.

As an additional operation, you can test that the server is interrupted with a SIGINT through the following command:

```
$ sudo kill -s SIGINT <pid_process>
```

Building a RESTFul server for Linux

It is time to understand how to develop a RESTful server for the most widespread OS for server-side. This is what I will show you in this recipe.

Getting ready

This recipe uses DMVC, a Delphi open source framework based on WebBroker that allows you to create powerful RESTful web services. You can find the project code here: https://github.com/danieleteti/delphimvcframework.

Version 3.0.0 of DMVC has brought several innovations, and a very useful one is that it is not necessary to download the Git repository to use it. Just download the latest version (https://github.com/danieleteti/delphimvcframework/releases) as a .zip file and you'll be okay.

Once you have downloaded the zip file, unzip the file and put it in a folder of your filesystem. Next, you have to configure your IDE to find the DMVC units:

Navigate to **Tools** | **Options** | **Environment Options** | **Delphi Options** | **Library**

Then, click the **...** on the **Library Path** edit, and add the following paths one by one (change C:\DEV\DMVCFramework to the appropriate path on your machine):

- C:\DEV\DMVCFramework\sources
- C:\DEV\DMVCFramework\lib\loggerpro

This recipe uses many DMVC features, and could be a little confusing if you don't know the basics of REST and DMVC. If so, please read the following documentation before going ahead:

- Building Web Services the REST Way: `http://www.xfront.com/REST-WebServices.html`
- RESTful web services: The basics: `https://www.ibm.com/developerworks/webservices/library/ws-restful/`
- The latest information about DelphiMVCFramework is available in the Developer Guide: `https://danieleteti.gitbooks.io/delphimvcframework/content/`
- The Developer Guide is also available as a PDF here: `https://www.gitbook.com/download/pdf/book/danieleteti/delphimvcframework`

A valuable resource for DelphiMVCFramework is its samples, so please check the `\Samples` folder in the project root folder. From this point onward, I'll not repeat concepts and information already explained in the previously mentioned articles, so read them with care.

In this recipe, we are going to build a simple web book shop.

How to do it...

Perform the following steps:

1. Navigate to **Delphi Project** | **Web Broker** | **Web Server Application**.
2. The wizard asks you to specify the type of platform for the application. Obviously check **Linux** and click **Next**.
3. Now the wizard asks you what type of web server application you want to create. This demo will be built as a console application; however, you can take advantage of the flexibility of **WebBroker** and add another type of application, for instance an **Apache** module, which we will see in the last recipe of this chapter. At this point, select **Stand-alone console application** and click **Next**.
4. The wizard proposes a TCP port where the service will listen. Click on the **Test** port. If the test port succeeds, use it; otherwise, change the port until the test passes. In this recipe, port `8080` is used.
5. Click **Finish**.

6. Save all. Name the project `BookManager.dproj`, server constants `ConstantsU.pas`, and the WebModule `WebModuleU.pas`.

7. We start from the business object classes. This web service will manage books, so let's create a new unit and declare the following class:

```
TBook = class(TObject)
  public
    property ID: Integer;
    property TITLE: String;
    property AUTHOR: String;
    property NUMBER_OF_PAGES: Integer;
    property YEAR: Integer;
    property PLOT: String;
  end;
```

8. Hit *Ctrl + Shift + C* to autocomplete the declaration; save the file as `BookBO.pas` in a folder named `BusinessObjects`.

9. Now it is time to create a DelphiMVCFramework controller. Create a new unit, name it `BooksControllerU.pas`, and save it in the `Controllers` folder.

10. Fill `BooksControllerU.pas` with the following code:

```
unit BooksControllerU;

interface

uses MVCFramework, BooksModuleU, MVCFramework.Commons;

type

  [MVCPath('/books')]
  TBooksController = class(TMVCController)
  private
    FBookModule: TBookModule;
  protected
    procedure OnAfterAction(Context: TWebContext;
      const AActionNAme: string); override;
    procedure OnBeforeAction(Context: TWebContext; const
AActionNAme: string;
      var Handled: Boolean); override;
  public
    [MVCPath]
    [MVCHTTPMethod([httpGET])]
    procedure GetBooks;
    [MVCPath('/($id)')]
    [MVCHTTPMethod([httpGET])]
```

```
      procedure GetBookByID(id: Integer);
      [MVCPath]
      [MVCHTTPMethod([httpPOST])]
      [MVCConsumes('application/json')]
      procedure CreateBook;
      [MVCPath('/($id)')]
      [MVCHTTPMethod([httpPUT])]
      [MVCConsumes('application/json')]
      procedure UpdateBook(id: Integer);
      [MVCPath('/($id)')]
      [MVCHTTPMethod([httpDELETE])]
      procedure DeleteBook(id: Integer);
    end;

implementation

uses
  BookBO, SysUtils, System.JSON, System.Math;

procedure TBooksController.CreateBook;
var
  Book: TBook;
begin
  Book := Context.Request.BodyAs<TBook>;
  FBookModule.CreateBook(Book);
  Context.Response.Location := '/books/' + Book.ID.ToString;
  Render(201, 'Book created');
end;

procedure TBooksController.UpdateBook(id: Integer);
var
  Book: TBook;
begin
  Book := Context.Request.BodyAs<TBook>;
  try
    Book.ID := id;
    FBookModule.UpdateBook(Book);
    Render(200, 'Book updated');
  finally
    Book.Free;
  end;
end;

procedure TBooksController.DeleteBook(id: Integer);
begin
  FBookModule.DeleteBook(id);
  Render(204, 'Book deleted');
end;
```

```delphi
procedure TBooksController.GetBookByID(id: Integer);
var
  Book: TBook;
begin
  Book := FBookModule.GetBookByID(id);
  if Assigned(Book) then
    Render(Book, false)
  else
    Render(404, 'Book not found');
end;

procedure TBooksController.GetBooks;
begin
  Render<TBook>(FBookModule.GetBooks, false);
end;

procedure TBooksController.OnAfterAction(Context: TWebContext;
  const AActionNAme: string);
begin
  inherited;
  FBookModule.Free;
end;

procedure TBooksController.OnBeforeAction(Context: TWebContext;
  const AActionNAme: string; var Handled: Boolean);
begin
  inherited;
  FBookModule := TBookModule.Create(nil);
end;

end.
```

11. All our RESTful interface is implemented in this unit. Now we have to write the part that actually accesses the database. Also in this recipe, we'll use the **Table Data Gateway** design pattern (http://martinfowler.com/eaaCatalog/tableDataGateway.html). In this recipe, we will not cover the part related to DB access, so books data will be stored in an objects list as a unit variable.

12. Let's create our TDG using a **Data Module**. Add a new **Data Module**, name it `BooksModule`, and save it into the `Modules` folder as `BooksModuleU.pas`.

13. Fill the `BooksModuleU.pas` with the following code:

```
unit BooksModuleU;

interface

uses
  System.SysUtils, System.Classes, BookBO,
System.Generics.Collections;

type
  TBookModule = class(TDataModule)
  public
    procedure CreateBook(ABook: TBook);
    procedure UpdateBook(ABook: TBook);
    procedure DeleteBook(AID: Integer);
    function GetBookByID(AID: Integer): TBook;
    function GetBooks: TObjectList<TBook>;
  end;

implementation

{$R *.dfm}

var
  BooksData: TObjectList<TBook>;

  { TBookModule }

procedure TBookModule.CreateBook(ABook: TBook);
begin
  BooksData.Add(ABook);
end;

procedure TBookModule.DeleteBook(AID: Integer);
var
  LBook: TBook;
begin
  LBook := GetBookByID(AID);
  BooksData.Extract(LBook).Free;
end;

function TBookModule.GetBookByID(AID: Integer): TBook;
var
  LBook: TBook;
```

```
begin
  Result := nil;
  for LBook in BooksData do
    if LBook.ID = AID then
      Exit(LBook);
end;

function TBookModule.GetBooks: TObjectList<TBook>;
begin
  Result := BooksData;
end;

procedure TBookModule.UpdateBook(ABook: TBook);
var
  LBook: TBook;
begin
  LBook := GetBookByID(ABook.ID);

  LBook.TITLE := ABook.TITLE;
  LBook.AUTHOR := ABook.AUTHOR;
  LBook.NUMBER_OF_PAGES := ABook.NUMBER_OF_PAGES;
  LBook.YEAR := ABook.YEAR;
  LBook.PLOT := ABook.PLOT;

end;

initialization

BooksData := BookBO.GenerateRandomData;

finalization

BooksData.Free;

end.
```

14. As you can see in the preceding code, `BooksData` is filled in the initialization section by `GenerateRandomData` provided by `BookBO`. As mentioned earlier, in this recipe we will not see data access, so we will use a unit variable as data store:

```
function GenerateRandomData: TObjectList<TBook>;

  function NewBook(AID: Integer; ATitle, AAuthor: String;
    AYear, ANumberOfPages: Integer): TBook;
  begin
    Result := TBook.Create;
    Result.ID := AID;
    Result.TITLE := ATitle;
```

```
        Result.AUTHOR := AAuthor;
        Result.YEAR := AYear;
        Result.NUMBER_OF_PAGES := ANumberOfPages;
    end;

begin
    Result := TObjectList<TBook>.Create;
    Result.Add(NewBook(1, 'Divine Comedy', 'Dante Alighieri', 1321,
352));
    Result.Add(NewBook(2, 'In Search of Lost Time', 'Marcel Proust',
1908, 4500));
    Result.Add(NewBook(3, 'Ulysses', 'James Joyce', 1922, 730));
    Result.Add(NewBook(4, 'The Great Gatsby', 'F. Scott Fitzgerald',
1925, 240));
    Result.Add(NewBook(5, 'The Brothers Karamazov', ' Fyodor
Dostoevsky',
        1880, 1033));

end;
```

15. We're about to finish. Go back to `WebModuleU.pas` and create the **OnCreate** event handler. Here, we have to configure the DelphiMVCFramework starting point. Copy these lines of code:

```
procedure TwmMain.WebModuleCreate(Sender: TObject);
begin
    FMVC := TMVCEngine.Create(Self);
    FMVC.AddController(TBooksController);
end;
```

16. The `FMVC` variable must be declared in the private section of the class, and you have to add the **BooksControllerU** unit in the `implementation uses` clause.

17. Select **Linux** as **Target Platform.**

18. Now your project should compile. If not, check the dependencies between all the units.

19. If true, **Deploy** the project on the Linux machine.

How it works...

It's time to start the newly deployed server. Go into the specific deployed folder and run it as shown in the following figure:

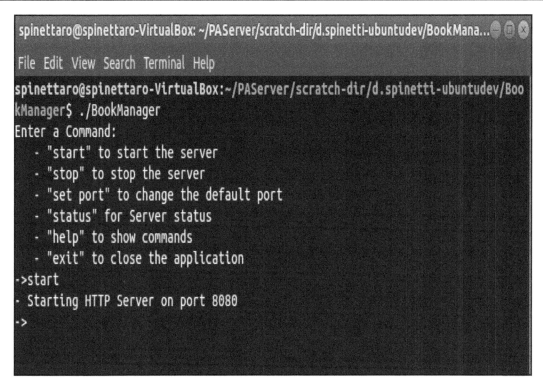

Figure 7.7: BookManager start up

As all recipes are based on DMVC, this application is organized into three layers:

- Controller (TBookController)
- Table Data Gateway (TBookModule)
- Business Objects (TBook)

We have already seen how the internal mechanisms of this kind of application work, so to not be repetitive I will only include a brief summary:

- The controller reads information from the HTTP request
- It invokes some methods on the Table Data Gateway layer (TBookModule instance)
- It builds the response for the client

You can interact with the book shop through these REST APIs:

HTTP VERB	URL	DESCRIPTION
GET	/books	Returns a JSON array containing one JSON object for each element present in the object list `BooksData`. In each object, the property name is the name of the field, while the property values are the values of the fields.
GET	/book/($id) $id is a URL parameter	Returns a JSON object representing the specific book that has ID = $id.
POST	/book	Creates a new book and adds it in the `BooksData` object list. Requires a request body containing the new book to create as a JSON object. The request content type must be `application/json`.
PUT	/book/($id)	Updates the book with ID = $id with the data passed in the request body. Requires a request body containing the new book to update as a JSON object. The request content type must be `application/json`.
DELETE	/book/($id)	Deletes book with ID = $id

Launch a browser (preferably Google Chrome or Mozilla Firefox) and request the following URL: `http://<your_linux_machine_ip>:8080/books`.

Your browser should show all the data available in the **BOOKS** list as a JSON array of JSON objects, as shown in the following screenshot:

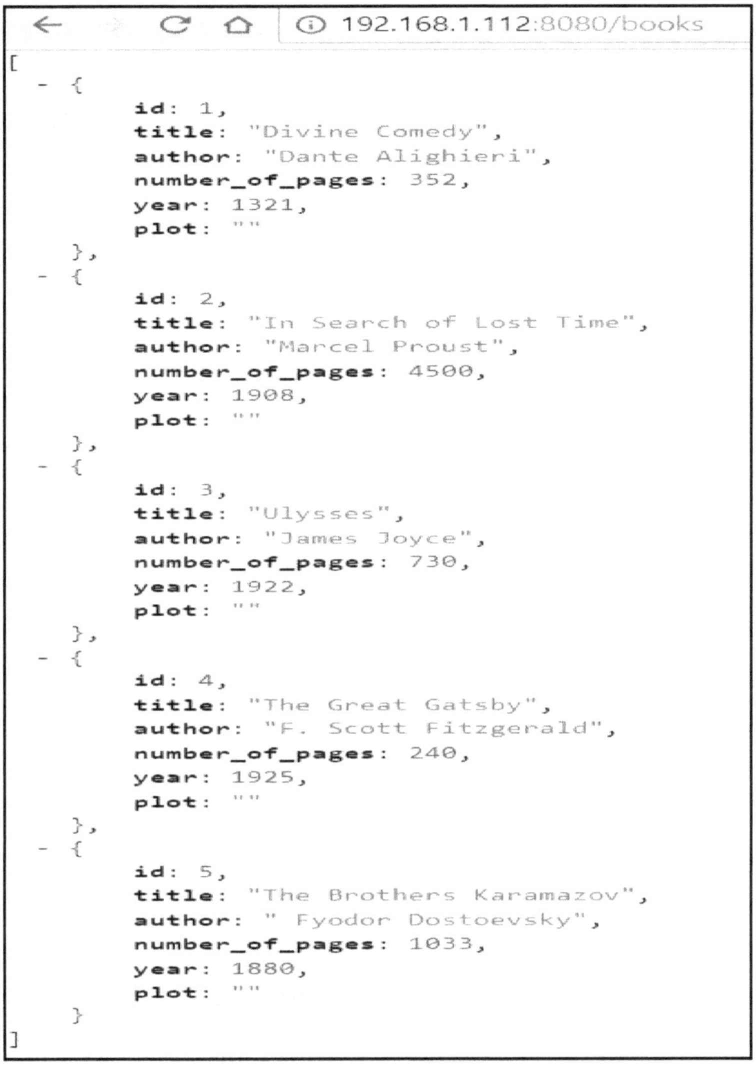


```
[
  - {
        id: 1,
        title: "Divine Comedy",
        author: "Dante Alighieri",
        number_of_pages: 352,
        year: 1321,
        plot: ""
    },
  - {
        id: 2,
        title: "In Search of Lost Time",
        author: "Marcel Proust",
        number_of_pages: 4500,
        year: 1908,
        plot: ""
    },
  - {
        id: 3,
        title: "Ulysses",
        author: "James Joyce",
        number_of_pages: 730,
        year: 1922,
        plot: ""
    },
  - {
        id: 4,
        title: "The Great Gatsby",
        author: "F. Scott Fitzgerald",
        number_of_pages: 240,
        year: 1925,
        plot: ""
    },
  - {
        id: 5,
        title: "The Brothers Karamazov",
        author: " Fyodor Dostoevsky",
        number_of_pages: 1033,
        year: 1880,
        plot: ""
    }
]
```


Figure 7.8: BookManager in action. Here the JSON array of JSON objects returned by the HTTP call from the browser.

WOW! We have created a simple RESTful server running on the Linux machine! The REST server just created does the minimum necessary to be considered as such, but as a first step in Linux development there may be. Our path is not yet complete, so get ready, because in the next recipe we are going to introduce advanced steps!

There's more...

To test the RESTful service that you will develop, you can use the `RESTDebugger.exe` program provided within Delphi XE5 (in the `bin` folder, or by clicking **Tools** | **Rest Debugger**), or the free **POSTMAN** application (`https://www.getpostman.com/`). These tools allow you to send all the `HTTP VERB` requests while the browser, using only the address bar, can only issue `GET` requests.

Building a complete RESTful server with database access and web client interface

In the previous recipe, we saw how to build a RESTful server. Our path is not yet complete; there are two basic elements to make sure that our server is complete:

- Database access
- Web interface for data viewing and manipulation

In the previous RESTful server we built, the data was managed through the use of an objects list. Now, we will see how to integrate a database access in order to give it an appropriate place. With regard to the visualization and manipulation of data, it is necessary to create a web interface to guarantee an adequate user experience for the described operations.

Getting ready

This recipe is an evolution of the previous one, so you can proceed by directly editing the code already written, or clone the previous one in another path and start with this new one.

The first thing we have to do is define our database. For this recipe, we will use MySQL—one of the most popular open source **relational database management systems (RDBMS)**. If you still do not know MySQL, this may be the right opportunity to start analyzing it (`https://www.mysql.com`).

 Example, platforms such as LAMP and WAMP incorporate MySQL for server implementation to manage dynamic websites. In addition, many of the successful **content management systems** (**CMSs**) such as WordPress, Joomla, and Drupal are born with MySQL default support.

We need to install MySQL. MySQL is available within Ubuntu's default software repositories, making it possible to install it using conventional package management tools (run this command on the Linux machine):

```
# update package list
$ sudo apt update
# install the server
$ sudo apt install mysql-server
# configure the server, this will guide you to the database configuration
through different steps
# such as user settings, password, remote access etc ...
$ sudo mysql_secure_installation
# check is the service is running
$ systemctl status mysql
```

If it has been configured well, you will see a positive result through the last command.

Being a normal `systemd` service, MySQL can be managed using the usual commands:

```
#start
sudo systemctl start mysql
```

```
#stop
sudo systemctl stop mysql
```

By default, MySQL disallows remote access to the database and binds only to localhost. To change this behavior, we have to change the main config file called `mysqld.conf`:

```
$ cd /etc/mysql/mysql.conf.d
$ sudo nano mysqld.cnf
# look for "bind-address = 127.0.0.1"
# change to "bind-address = 0.0.0.0"
# Hit CTRL+O, then HIT return key to save the file
# Hit CTRL+X (the terminal comes back)
$ sudo systemctl stop mysql
$ sudo systemctl start mysql
```

Now it's time to create a database and a remote user:

```
$ sudo mysql --user=root
mysql> create database delphicookbook;
mysql> use delphicookbook;
mysql> create user 'cookuser'@'%' identified by 'cookpass';
mysql> grant all privileges on *.* to 'cookuser'@'%' with grant option;
mysql> flush privileges;
mysql> quit
```

As a final step, we have to create a database table to store our books:

```
$ mysql --user=cookuser --password=cookpass delphicookbook
mysql> CREATE TABLE `BOOKS` (
    ->  `ID` BIGINT(20) NOT NULL AUTO_INCREMENT,
    ->  `author` VARCHAR(100) NOT NULL,
    ->  `title` VARCHAR(250) NOT NULL,
    ->  `year` INT(11) NULL DEFAULT NULL,
    ->  `number_of_pages` INT(11) NULL DEFAULT NULL,
    ->  `plot` VARCHAR(500) NULL DEFAULT NULL,
    ->   PRIMARY KEY (`ID`)
    -> );
# insert data into table
mysql> INSERT INTO `BOOKS` (`author`, `title`, `year`, `number_of_pages`)
VALUES
    ->  ('Dante Alighieri', 'Divine Comedy', 1321, 352),
    ->  ('Marcel Proust', 'In Search of Lost Time', 1908, 4500),
    ->  ('James Joyce', 'Ulysses', 1922, 730),
    ->  ('F. Scott Fitzgerald', 'The Great Gatsby', 1925, 240),
    ->  ('Fyodor Dostoevsky', 'The Brothers Karamazov', 1880, 1033);
```

One last thing before proceeding further—make sure that FireDAC is correctly configured for use with MySQL. Follow a guide link to make sure it's all okay—http://docwiki. embarcadero.com/RADStudio/en/Connect_to_MySQL_Server_(FireDAC). In particular, be sure to correctly configure Windows (for design time) and Linux (once deployed) clients. You can find the details under the sections **Windows Client Software** and **Linux Client Software**.

How to do it...

It's time to complete our RESTful server with database access and web client interface. Let's start:

1. We have to modify our TDG to access database data. Open BooksModuleU.pas and drop on the data module and link the components, as follows (this is an extract of the .dfm file):

```
object BookModule: TBookModule
  object Conn: TFDConnection
    Params.Strings = (
      'Server=192.168.1.112'
      'Database=delphicookbook'
      'User_Name=cookuser'
      'Password=cookpass'
      'DriverID=MySQL')
```

```
      LoginPrompt = False
      ConnectedStoredUsage = [auDesignTime]
      Connected = True
    end
    object updBooks: TFDUpdateSQL
      Connection = Conn
    end
    object qryBooks: TFDQuery
      Connection = Conn
      UpdateObject = updBooks
    end
  end
end
```

Change the FDConnection params accordingly with your machine.

2. Now we have to configure some data access stuff.
3. Double-click on qryBooks; the component editor shows up. Write the query "select * from books", and click **Execute**. Hold the window open. This will be the query used to generate all the CRUD statements.
4. If you have correctly connected qryBooks.UpdateObject to updBooks, you should see an **UpdateSQL Editor** button on the right side of the component editor form.
5. Click on the **UpdateSQL Editor** button and you will get another component editor. This time, it is related to the updBooks component,
6. Select **ID** in **Key Fields** and all entries except **ID** in **Updating fields**.
7. Click **Generate SQL** and **OK**.
8. Now your updBooks component has been configured with all the SQL statements necessary to correctly update the BOOKS table.
9. Now we have to create the methods used to CRUD records. Go to the BookModuleU.pas code view. Replace the previously implemented methods with the following:

MVCFramework.FireDAC.Utils.pas unit contains some FireDAC useful stuff. Take a look at it.

```
uses
  MVCFramework.FireDAC.Utils;

{ TBookModule }

procedure TBookModule.CreateBook(ABook: TBook);
```

```
var
  InsCommand: TFDCustomCommand;
begin
  InsCommand := updBooks.Commands[arInsert];
  TFireDACUtils.ObjectToParameters(InsCommand.Params, ABook,
'NEW_');
  InsCommand.Execute;
  ABook.ID := Conn.GetLastAutoGenValue('gen_book_id');
end;

procedure TBookModule.DeleteBook(AID: Integer);
var
  DelCommand: TFDCustomCommand;
begin
  DelCommand := updBooks.Commands[arDelete];
  DelCommand.ParamByName('OLD_ID').AsInteger := AID;
  DelCommand.Execute;
end;

function TBookModule.GetBookByID(AID: Integer): TDataSet;
begin
  qryBooks.Open('SELECT * FROM books WHERE ID = :ID', [AID]);
  Result := qryBooks;
end;

function TBookModule.GetBooks: TDataSet;
begin
  qryBooks.Open('SELECT * FROM books ');
  Result := qryBooks;
end;

procedure TBookModule.UpdateBook(ABook: TBook);
var
  UpdCommand: TFDCustomCommand;
begin
  UpdCommand := updBooks.Commands[arUpdate];

  TFireDACUtils.ObjectToParameters(UpdCommand.Params, ABook,
'NEW_');
  UpdCommand.Params.ParamByName('OLD_ID').AsInteger := ABook.ID;
  UpdCommand.Execute;
end;
```

10. We also have to update `BooksControllerU.pas` to manage new data returned from `TBookModule`:

```
procedure TBooksController.CreateBook;
var
  Book: TBook;
begin
  Book := Context.Request.BodyAs<TBook>;
  FBookModule.CreateBook(Book);
  Context.Response.Location := '/books/' + Book.ID.ToString;
  Render(201, 'Book created');
end;

procedure TBooksController.UpdateBook(id: Integer);
var
  Book: TBook;
begin
  Book := Context.Request.BodyAs<TBook>;
  try
    Book.ID := id;
    FBookModule.UpdateBook(Book);
    Render(200, 'Book updated');
  finally
    Book.Free;
  end;
end;

procedure TBooksController.DeleteBook(id: Integer);
begin
  FBookModule.DeleteBook(id);
  Render(204, 'Book deleted');
end;

procedure TBooksController.GetBookByID(id: Integer);
var
  BookDS: TDataSet;
begin
  MVCFramework.Logger.LogI('Requested book with id ' +
id.ToString);
  BookDS := FBookModule.GetBookByID(id);
  Render(BookDS, False, dstSingleRecord);
end;

procedure TBooksController.GetBooks;
var
  BooksDS: TDataSet;
begin
  MVCFramework.Logger.LogI('Books list request');
```

```
            BooksDS := FBookModule.GetBooks;
            Render(BooksDS, False, dstAllRecords);
        end;
```

11. To properly set the Logger (the Logger of DMVC is a writer of LoggerPro, seen in Chapter 6, *Putting Delphi on the Server*), define the logger file path, and initialize it in the initialization section of `WebModuleU.pas`:

```
function GetLogFilePath: String;
begin
  Result := '/var/log/';
end;

initialization

SetDefaultLogger(BuildLogWriter([TLoggerProFileAppender.Create(5,
2000,
  GetLogFilePath)]));
```

12. It's time to define our web client interface. In this case, the web client interface is composed of an HTML file that uses some JavaScript functionality (in particular, jQuery, a very powerful JavaScript library) to make requests to the server API and manipulate the DOM. You can find these static files in the `www` folder of this recipe. To configure DMVC so that it makes the static files available, we have to configure the document root property. Open `WebModuleU` and add the following line in the `WebModuleCreate` method:

```
FMVC.Config['document_root'] := 'www';
```

13. We have completed all the implementation steps! Now, build the project by right-clicking on **Project** | **Build**, and deploy it to the Linux machine (**Project** | **Deploy BookManager**).

14. Copy the `www` folder of this recipe into the same path of the deployed project of your Linux machine.

15. Go in the specific deployed folder and run it as shown in following figure:

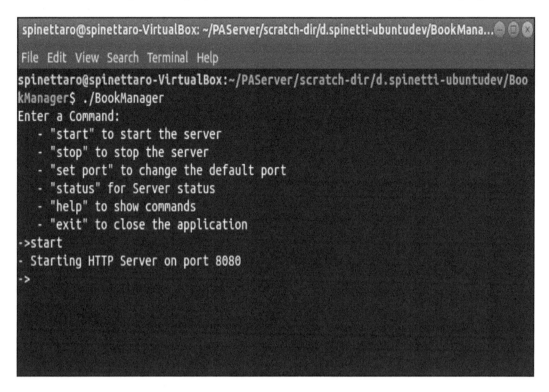

Figure 7.11: BookManager start up

16. Launch a browser, preferably Google Chrome or Mozilla Firefox, and request the following URL: `http://<your_linux_machine_ip>:8080`. Your browser should show the web client as follows:

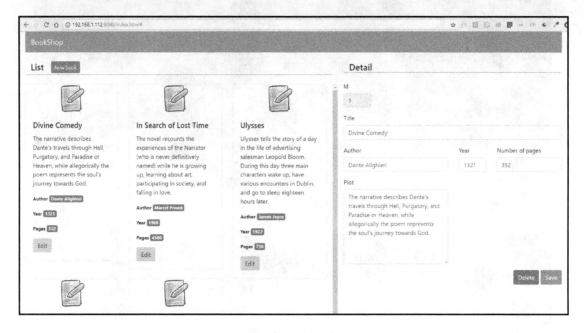

Figure 7.12: Web client interface

How it works...

This recipe is an extension of the previous recipe, so the architecture and application logic does not change:

- The application is organized in three layers:
    - Controller (`TBookController`)
    - Table Data Gateway (`TBookModule`)
    - Business Objects (`TBook`)
- Internal mechanisms perform steps as follows:
    - The controller reads information from the HTTP request
    - It invokes some methods on the Table Data Gateway layer (`TBookModule` instance)
    - It builds the response for the client

The difference this time is that we have implemented a data access, so the operations performed in the `TBookModule` (**TDG**) will change the configured database table directly.

Now you can interact with the **Book Shop** through the REST API as shown in the previous recipe, or by the web interface just implemented and as shown in *Figure 7.9*.

There's more...

In this recipe, I have also used the **Log** functions provided by the logging mechanism of DMVC. At its base, there is a particular configuration of the awesome framework seen in the previous chapter, LoggerPro. Just import the `MVCFramework.Logger.pas` and you are up and running to be able to log with several functions:

```
procedure Log(AMessage: string); overload;
procedure LogD(AMessage: string); overload;
procedure LogI(AMessage: string);
procedure LogW(AMessage: string);
procedure LogE(AMessage: string);
```

You have to choose the appropriate method based on the severity:

- **I**: Information
- **D**: Debug
- **W**: Warning
- **E**: Error.

A single clarification: since we are in a Linux environment, the correct path for the logs must be `/var/log`.

See also

The following are some links to deepen the topics dealt with:

- Official site of MySQL RDBMS: `https://www.mysql.com/it/`
- MySQL setup and testing: `https://dev.mysql.com/doc/refman/en/postinstallation.html`
- Official site of jQuery library: `https://jquery.com/`
- DMVC: `https://github.com/danieleteti/delphimvcframework`
- LoggerPro: `https://github.com/danieleteti/loggerpro`

Creating WebBroker Apache modules for Linux

In this recipe, we'll create a 64-bit WebBroker Apache module for Linux, install it in Apache 2, and secure the server by configuring HTTPS access. Internally, our application uses DelphiMVCFramework, but all the steps are still valid for other frameworks or no frameworks at all (apart from WebBroker).

 The application logic of this recipe is exactly the same as the WebBroker Apache module for Windows, so you can skip the application logic points and dwell only on the parts related to the Apache configuration on the Linux system. This is so that you can be more focused on the part related to deploy on Linux.

Let's start!

Getting ready

We need to get an Apache distribution. Apache is available within Ubuntu's default software repositories, making it possible to install it using conventional package management tools. Run these commands on the Linux machine:

1. Update the local package index to reflect the latest upstream changes:

    ```
    $ sudo apt update
    ```

2. Install the Apache2 package:

    ```
    $ sudo apt install apache2
    ```

3. Now, we have to allow traffic on port 80. Modify the Firewall settings to allow outside access to the web port as follows:

    ```
    $ sudo ufw allow 'Apache'
    ```

4. Check with the systemd init system (go to the *Fork* recipe for more info) to make sure the service is running, by typing the following code:

    ```
    $ sudo systemctl status apache2
    ```

5. From the output generated, you'll be able to understand whether the service has been started successfully. However, the best way to test this is to request a page from Apache (use a browser to navigate on `http://your_machine_ip`):

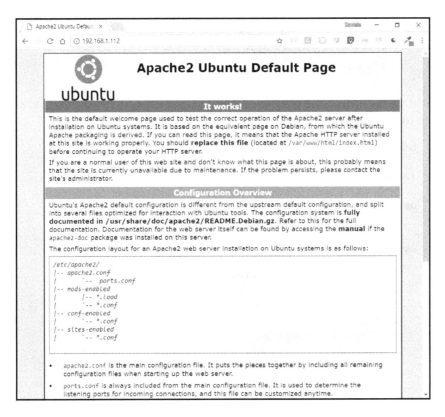

Figure 7.13 : Apache works!

As you can see from the preceding image, we have successfully installed and started Apache!

How to do it...

Let's create the Apache module. We'll create a fake management system with a list of users. There is one resource accessible with two HTTP verbs: **GET** and **POST**:

1. Navigate to **File | New | Other**.
2. Navigate to **Delphi Projects | WebBroker | Web Server Application**.
3. In the resultant modal dialog, select **Linux** and click **Next**.

4. In the next modal dialog, select the Apache dynamic-link module and click **Next**.

5. In the combo box, select Apache Version 2.4 and write `peoplemanager_module` as the module name.

6. Click **Finish**.

7. Save the project using the following filenames:
   - `mod_peoplemanager.dproj`
   - `WebModuleU.pas`

8. Add a new unit to the project, save it as `SampleControllerU.pas`, and write the following code:

```
unit SampleControllerU;

interface

uses
  MVCFramework, MVCFramework.Commons;

type

  [MVCPath('/')]
  [MVCDoc('Just a sample controller')]
  TSampleController = class(TMVCController)
  public
    [MVCPath('/users')]
    [MVCHTTPMethods([httpGET])]
    [MVCDoc('Returns the users list')]
    procedure GetUsers;

    [MVCPath('/users')]
    [MVCHTTPMethods([httpPOST])]
    [MVCConsumes('application/json')]
    [MVCDoc('Creates a new user')]
    procedure CreateUser;
  end;

implementation

uses System.JSON, MVCFramework.Serializer.JSON;

{ TSampleController }

procedure TSampleController.CreateUser;
begin
  // just a fake, we don't create any user.
  // simply echo the body request as body response
```

```
  if Context.Request.ThereIsRequestBody then
  begin
    Context.Response.StatusCode := HTTP_STATUS.Created;
    Render(Context.Request.Body);
  end
  else
    raise EMVCException.Create(HTTP_STATUS.BadRequest, 'Expected
JSON body');
end;

procedure TSampleController.GetUsers;
var
  LJObj: TJSONObject;
  LJArray: TJSONArray;
begin
  LJArray := TJSONArray.Create;

  LJObj := TJSONObject.Create;
  LJObj.AddPair('first_name', 'Daniele').AddPair('last_name',
'Spinetti')
    .AddPair('email', 'd.spinetti@bittime.it');
  LJArray.AddElement(LJObj);

  LJObj := TJSONObject.Create;
  LJObj.AddPair('first_name', 'Peter').AddPair('last_name',
'Parker')
    .AddPair('email', 'pparker@dailybugle.com');
  LJArray.AddElement(LJObj);

  LJObj := TJSONObject.Create;
  LJObj.AddPair('first_name', 'Bruce').AddPair('last_name',
'Banner')
    .AddPair('email', 'bbanner@angermanagement.com');
  LJArray.AddElement(LJObj);

  Render(LJArray);
end;

end.
```

9. Reopen the web module.
10. Create an event handler for the **OnCreate** event, and fill it with the following code:

```
FMVCEngine := TMVCEngine.Create(self);
FMVCEngine.AddController(TSampleController);
```

11. Declare the FMVCEngine private variable as TMVCEngine.

12. Include the MVCFramework in the `interface uses` clause.
13. Include `SampleControllerU` in the implementation uses clause.
14. Save all.
15. On the **Target Platform** node in the **Project Manager**, select **64-bit Linux**.
16. Be sure that the Linux 64-bit machine and the **PAServer** are up and running, because we have to build the module for a **64-bit Linux Apache.**

17. Let's build this. Right-click on **Project** and **Build.**
18. Then, deploy the project to the server by clicking on **Project | Deploy** `libmod_peoplemanager.so`.
19. Move onto Linux.
20. Move the file `/home/{your_user}/PAServer/scratch-dir/{your_configuration_name}/projname/libmod_peoplemanager.so` just deployed to this new folder, `/etc/apache2/modules`:

```
$ sudo mkdir /etc/apache2/modules
$ sudo mv /home/{your_user}/PAServer/scratch-
dir/{your_configuration_name}/{project_name}/libmod_peoplemanager.s
o /etc/apache2/modules
```

21. In the next two steps, we have to configure Apache so that it loads `peoplemanager_module`.
22. We have to configure the load file to enable `libmod_peoplemanager.so`, so create and edit the following file on the Linux machine:

```
$ touch /etc/apache2/mods-enabled/libmod_peoplemanager.load
$ nano /etc/apache2/mods-enabled/libmod_peoplemanager.load
# this is the content of the above file
LoadModule peoplemanager_module
/etc/apache2/modules/libmod_peoplemanager.so
```

You can find the right `{module_name}`, to pass in `LoadModule` (as `peoplemanager_module` in the preceding configuration), in the export section of the `.dpr` file of this project.

23. We have to configure the file, so create and edit the following file on the Linux machine:

```
$ touch /etc/apache2/mods-enabled/libmod_peoplemanager.conf
$ nano /etc/apache2/mods-enabled/libmod_peoplemanager.conf
# this is the content of the above file
<Location /people>
```

```
    SetHandler libmod_peoplemanager-handler
    Require all granted
</Location>
```

The `/people` means that we will be able to reach our module at
http://<ip_address>/people/.

24. To enable our module, we only have to restart Apache:

    ```
    $ sudo systemctl restart apache2
    ```

25. Open a web browser, preferably Google Chrome or Mozilla Firefox, and write
    down the following URL:
    `http://<your_linux_machine_ip>/people/users`. The result is shown in
    the following screenshot:

Figure 7.14: The resource users as JSON array; this particular formatting is
because of the extension JSON View for Google Chrome

26. Open a web browser, preferably Google Chrome or Mozilla Firefox, and write
    down the following URL: `http://localhost/people/users`.

Now the Apache module works! What we still have to do is to configure **HTTPS** access to
this service. Remember that HTTP is a textual protocol, so every byte you send or receive
from an HTTP server is not encrypted, and can be sniffed by anyone with a basic
networking knowledge and a good sniffer. So, let's make this server secure!

To configure HTTPS on Apache, you need to generate the certificate files. OpenSSL command-line tools can do the job, and we already have this tool available. Here's what to do:

1. We have to actually generate the certificate, which is composed of two files: the certificate file and the private key file.

2. Execute the following command-line (`opnessl` installed by default on Ubuntu):

   ```
   $ openssl req -x509 -nodes -days 365 -newkey rsa:2048 -keyout
   delphicookbook.key -out delphicookbook.crt
   ```

3. You will be prompted to enter your organizational information and a common name. The common name should be the fully qualified domain name for the site you are securing (`www.mydomain.com`), or, in this case, just empty. You can leave the email address, challenge password, and optional company name blank. When the command has finished running, it will create two files: a `delphicookbook.key` and a `delphicookbook.crt` self-signed certificate file, valid for 365 days.

4. Next, we have to move these files:

   ```
   $ sudo mv delphicookbook.crt /etc/ssl/certs
   $ sudo mv delphicookbook.key /etc/ssl/private
   ```

5. We are in Linux, so it's important that we adjust the key file permissions. Let's change our key file permissions accordingly, giving the owner reading and writing permissions and read-only privileges for the group:

   ```
   $ sudo chown root:ssl-cert /etc/ssl/private/delphicookbook.key
   $ sudo chmod 640 /etc/ssl/private/delphicookbook.key
   ```

6. Now we have to enable the SSL Apache module. We do it by using the **a2enmod** command:

   ```
   $ sudo a2enmod ssl
   ```

7. We are almost there. Now it's time to modify the SSL virtual host. Go to `/etc/apache2/sites-available/default-ssl.conf` and set it as follows:

   ```
   <VirtualHost *:443>
   ...
   ...
     # Enable ssl engine
     SSLEngine on

     SSLCertificateFile /etc/ssl/certs/delphicookbook.crt
   ```

```
SSLCertificateKeyFile /etc/ssl/private/delphicookbook.key

...
...
...
</VirtualHost>
```

8. We have to enable the SSL site. We do this by putting the `default-ssl` in the site available creating a symbolic link:

```
$ sudo ln -s /etc/apache2/sites-available/default-ssl.conf
/etc/apache2/sites-enabled/000-default-ssl.conf
```

9. We have to adjust the firewall rules to enable Apache on SSL port (`443`). So run this command:

```
$ sudo ufw allow 'Apache Secure'
```

10. OK, it's done. Restart Apache (check the command in the previous step if you don't remember it), and you will be able to access the `/people/users` URI by using the HTTPS protocol:

Figure 7.15: The resource users as JSON array accessed with the HTTPS protocol

11. You should now see our JSON array. The browser may tell you that the certificate is not verified. We know it is because it's a self-signed certificate and no certification authority has been contacted to obtain it. So you can accept the certificate without a problem.

12. Our WebBroker Apache module is running behind an HTTPS-secured web server, and is ready to rock!

Now, a good idea could be to keep non-ssl connection (HTTP), but permanently redirect it to your SSL Virtual Host (HTTPS).

How it works...

The project is quite simple. Using DelphiMVCFramework, we defined one resource that supports two verbs, **GET** and **POST**, with the following meanings:

URI	VERB	DESCRIPTION
/users	GET	Retrieves a JSON array of JSON objects representing fake users
/users	POST	Doesn't create anything, but echoes the JSON request body as the body response

There's more...

Now we understand how to create Apache modules for the Linux system and how to make them secure using HTTPS. Warning! Don't attempt to publish this server to the internet without a proper hardening by a skilled person. This configuration is just a minimal HTTPS setup; the server could still be vulnerable in some other parts.

See also

The following are some helpful links:

- If you want to know more about Apache and HTTPS, check out the following article: `https://httpd.apache.org/docs/2.4/ssl/ssl_howto.html`
- Curious about OpenSSL? Check out the project's site: `https://www.openssl.org/`
- Here's a WebBroker framework introduction: `http://docwiki.embarcadero.com/RADStudio/en/Using_Web_Broker_Index`
- DelphiMVCFramework project: `https://github.com/danieleteti/delphimvcframework`
- DelphiMVCFramework group support: `https://www.facebook.com/groups/delphimvcframework/`

8
Riding the Mobile Revolution with FireMonkey

In this chapter, we will cover the following recipes:

- Taking a photo, applying effects, and sharing it
- Using TListView to show and search local data
- Using SQLite databases to handle a to-do list
- Do not block the main thread!
- Using a styled TListView to handle long lists of data
- Customizing the TListView
- Taking a photo and location and sending it to a server continuously
- Talking with the backend using HTTPS
- Making a phone call from your app
- Tracking the application's life cycle
- Building your own SMS sending service with the REST API

Introduction

In this chapter, we will look at how to develop mobile apps using Delphi. The recipes in this chapter require a working development configuration of your PC and, in the case of iOS, your Mac, to talk with an Android or iOS device. A detailed tutorial on how to properly configure your system for this purpose can be found on the Embarcadero DocWiki. To develop and deploy an app for iOS, you require an Apple computer and an actual iOS device, while to develop and deploy for Android, you only need to have the device. There is also an emulator in the SDK where you can deploy an app but, currently, it is very slow; if you really want to develop for Android, having an actual device where deploying is faster than using an emulator is recommended.

Visit the following links for more information and relevant documentation that will help you to configure different environments:

- **For Android configuration**: The *Set Up Your Development Environment on Windows PC (Android)* documentation can be found at `http://docwiki.embarcadero.com/RADStudio/en/Mobile_Tutorial:_Set_Up_Your_Development_Environment_on_Windows_PC_(Android)`.
- **For iOS configuration**: The *Set Up Your Development Environment on the Mac (iOS)* documentation can be found at `http://docwiki.embarcadero.com/RADStudio/en/Mobile_Tutorial:_Set_Up_Your_Development_Environment_on_the_Mac_(iOS)`.
- **For Windows configuration**: The *Set Up Your Development Environment on Windows PC (iOS)* documentation can be found at `http://docwiki.embarcadero.com/RADStudio/en/Mobile_Tutorial:_Set_Up_Your_Development_Environment_on_Windows_PC_(iOS)`.

Taking a photo, applying effects, and sharing it

This recipe will introduce the mobile development world using a simple app that shows you how to take a photo directly from the camera or from the photo library, apply some effects to it, and then share it using one of the installed apps on the device.

Getting ready

This recipe makes extensive use of Delphi actions. Actions are an implementation of the GoF command design pattern and have been an important tool for the Delphi developer since the initial versions of Delphi. You can use them as much as you like. In the mobile era, actions are even more important and useful. Indeed, actions can be used to execute common tasks such as taking a photo with the camera, getting a photo from the library, or sharing some content with other apps. Here's how our app will look:

Figure 8.1: The Photo with Effects app, with buttons on the top; three out of the four buttons are bound to standard actions

How to do it...

Now, we are about to create our first FireMonkey mobile app. Let's start:

1. Create a new mobile app by navigating to **File | New | Multi-Device Application – Delphi.**

2. Select the **Header/Footer** template and click on **OK.**

3. In the upper-left corner of the form designer, there is a combo box which allows you to select the OS style used by the form designer to show the form. Select **Android** from the drop-down menu.

4. The IDE has just created a base for us. Name the MainForm form and let's add our logic and adapt the UI.

5. Select the **HeaderLabel** label and change its **Text** property to **Photos with Effects.**

6. Select the **TToolbar** component named **Footer** and delete it.

7. Now, drop a **TPanel** component and align it to the **Top** so that it is just below the header.

8. Add four buttons to the **TPanel** component that you just dropped. Align three of them to the left-hand side and the other one to the right-hand side. Now, starting from the left-hand side, set the following values for their **StyleLookup** property:
    - **cameratoolbutton**
    - **organizetoolbutton**
    - **composetoolbutton**
    - **actiontoolbutton**

9. Now, the buttons should look like the ones in the previous screenshot.

10. Drop a **TImage** component in the center of the form and align it to **Client**. This component will be our main working area.

11. Set **TImage.MarginWrapMode** to **Fit**.

12. Drop a **TListView** component in the center of the form, make it a bit wider, and name it lvEffects. This list view will be used to show the available effects to the user.

13. Select the lvEffects control and set
    ItemAppearanceObjects.ItemObjects.Accessory.Visible = false.

14. Drop a **TActionList** component, double-click on it, and then, from the little menu button on the left-hand side, click on **New Standard Action** (or you can use *Ctrl + Ins*).

15. From the resultanting window, select TTakePhotoFromCameraAction and click on **OK**. Repeat the process and add the TTakePhotoFromLibraryAction and TShowShareSheetAction actions. Note that these actions are actually invisible components with properties and events, just like a persistent field in a dataset. In a few moments, we will go back to these components to customize their default behaviors.

16. Starting from the left-hand side, connect the following actions to the buttons placed in the **TPanel** component at the top.

17. Set the first button, Action = TakePhotoFromCameraAction1.

18. Set the second button, Action = TakePhotoFromLibraryAction1.

19. Do not assign an action to the third button; instead, name it btnEffects.

20. Set the fourth button, `Action = ShowShareSheetAction1`.

21. In the app, there will be a mechanism to dynamically load the available effects inspecting the **TFilterEffect** descendants placed on the form. So we can simply drop some effects on the form and the app will automatically load them in a list, allowing the user to use them. Drop the following effects on the form: **TEmbossEffect, TRadialBlurEffect, TContrastEffect, TColorKeyAlphaEffect, TInvertEffect, TSepiaEffect, TTilerEffect, TPixelateEffect, TToonEffect, TPencilStrokeEffect, TRippleEffect, TWaveEffect, TWrapEffect,** and **TInnerGlowEffect**.

22. Now, we have to write some code. In the `private` section of the `TMainForm` class, declare the following instance members:

```
private
  FItemsEffectsMap: TDictionary<Integer, TFilterEffect>;
  FUndoEffectsList: TObjectStack<TFilterEffect>;
  FUndoEffectItem: TListViewItem;
  FTopWhenShown: Extended;
  procedure LoadPhoto(AImage: TBitmap);
  procedure RecalcMenuPosition;
  procedure RemoveCurrentEffect(ARemoveFromList: boolean);
  function EffectNameByClassName(const AClassName: String): String;
```

23. Hit *Ctrl + Shift + C* to create empty methods and fill them in with the following code:

```
procedure TMainForm.LoadPhoto(AImage: TBitmap);
begin
  Label1.Text := '';
  RemoveCurrentEffect(False);
  FUndoEffectsList.Clear;
  Image1.Bitmap.Assign(AImage);
end;

procedure TMainForm.RecalcMenuPosition;
begin
  FTopWhenShown := ClientHeight / 2 - lvEffects.Height / 2;
  lvEffects.Height := ClientHeight / 2;
  lvEffects.Position.X := ClientWidth / 2 - lvEffects.Width / 2;
end;

procedure TMainForm.RemoveCurrentEffect(ARemoveFromList:
  boolean);
begin
  if FUndoEffectsList.Count = 0 then
    Exit;
```

```
      Image1.RemoveObject(FUndoEffectsList.Peek);
      if ARemoveFromList then
        FUndoEffectsList.Pop;
      Image1.Repaint;
    end;

    function TMainForm.EffectNameByClassName(
      const AClassName: String): String;
    begin
      Result := AClassName.Substring(1);
      Result := TRegEx.Replace(Result, '[A-Z]', ' $0').TrimLeft;
    end;
```

24. To compile this code, add System.Generics.Collections in the uses interface section and add System.RegularExpressions in the uses implementation section. Build the project just to ensure that everything is alright.

25. Now, create the OnCreate event handler for the form and add the following code:

```
    procedure TMainForm.FormCreate(Sender: TObject);
    var
      eff: TFmxObject;
      lbi: TListViewItem;
    begin
      FItemsEffectsMap := TDictionary<Integer, TFilterEffect>.Create;
      FUndoEffectsList := TObjectStack<TFilterEffect>.Create(False);
      lvEffects.Position.Y := -lvEffects.Height;
      lvEffects.BeginUpdate;
      try
        FUndoEffectItem := lvEffects.Items.Add;
        FUndoEffectItem.Text := 'Undo';

        for eff in Children do
        begin
          //if it's an effect, add it to the listview
          //and to the dictionary. Use the classname
          //to create a friendly name
          if eff is TFilterEffect then
          begin
            lbi := lvEffects.Items.Add;
            lbi.Text := EffectNameByClassName(eff.ClassName);
            FItemsEffectsMap.Add(lbi.Index, TFilterEffect(eff));
          end;
        end;
      finally
        lvEffects.EndUpdate;
```

```
  end;
  lvEffects.ApplyStyleLookup;
end;
```

26. Now, create the `FormResize` and `FormShow` event handlers. In the `body` section of these event handlers, call the `RecalcMenuPosition` procedure.

27. Select the list view and create the **OnItemClick** event handler. This event will be called when the user selects an effect from the list. Now, we have to remove, with an animation, the list from the form and apply the effect. Fill the event handler with this code:

```
procedure TMainForm.lvEffectsItemClick(const Sender:
  TObject; const AItem: TListViewItem);
begin
  TAnimator.AnimateFloatDelay(
    lvEffects, 'Position.Y', -lvEffects.Height, 0.3, 0.1,
    TAnimationType.&In, TInterpolationType.Back);

  if AItem = FUndoEffectItem then
  begin
    //undo and revert to the previous one
    RemoveCurrentEffect(true);
    if FUndoEffectsList.Count > 0 then
      Image1.AddObject(FUndoEffectsList.Peek);
  end
  else
  begin
    // apply new effect
    RemoveCurrentEffect(False);
    FUndoEffectsList.Push(FItemsEffectsMap[AItem.Index]);
    Image1.AddObject(FUndoEffectsList.Peek);
  end;
end;
```

28. Now, we have to create something that is able to show the list of available effects when the user needs to apply one of them. The effect list will drop down from the top of the form with a little bouncing effect, and will go away in the same way (but in a reversed manner).

29. Create `btnEffects` on the `click` event handler and fill it in with the following code:

```
procedure TMainForm.btnEffectsClick(Sender: TObject);
begin
  if FUndoEffectsList.Count = 0 then
    FUndoEffectItem.Text := '<No effect to undo>'
  else
```

```
        FUndoEffectItem.Text := '[Undo ' +
          EffectNameByClassName(
        FUndoEffectsList.Peek.ClassName) + ']';

      TAnimator.AnimateFloat(lvEffects, 'Position.Y',
        FTopWhenShown, 0.4, TAnimationType.Out,
        TInterpolationType.Back);
    end;
```

30. We have to customize the actions' behavior. Double-click on `TActionList1`, select the `ShowShareSheet1` action, create the **OnBeforeExecute** event handler, and then fill it in with the following code:

```
procedure TMainForm.ShowShareSheetAction1BeforeExecute(Sender:
TObject);
begin
  if FUndoEffectsList.Count > 0 then
  begin
    //actually apply the effect to the bitmap
    FUndoEffectsList.Peek.ProcessEffect(nil,
      Image1.Bitmap, 0);
  end;
  ShowShareSheetAction1.Bitmap.Assign(Image1.Bitmap);
end;
```

31. Create the `OnDidFinishTaking` event handler for the `TakePhotoFromCameraAction1` and `TakePhotoFromLibraryAction1` actions and fill both with the following code:

```
procedure TMainForm
  .TakePhotoFromCameraAction1DidFinishTaking(
    Image: TBitmap);
begin
  LoadPhoto(Image);
end;

procedure TMainForm
  .TakePhotoFromLibraryAction1DidFinishTaking(
    Image: TBitmap);
begin
  LoadPhoto(Image);
end;
```

32. Select an available target in the **Project Manager** window (in your phone or an available emulator in the case of Android) and run the app.

Tap the first button from the left-hand side and take a photo. The image should be placed in the main area. Tap on the `btnEffects` button, and you should see the list view falling from the top to allow you to choose effects. The first item should be **<No effect to undo>**. Select an effect, let's say, **Contrast Effect**, and see how the effect is applied to the photo. Tap `btnEffect` again, and you should see the first item saying, **[Undo Contrast Effect]**. Play with the app by adding effects and using the **undo** features to sequentially go back to the beginning. Note that the effects can not be added on top of each other (so you cannot have **Emboss** along with **Blur** applied at the same time), but must be applied individually. When you are satisfied with the result, tap on the button on the right-hand side to share the photo with the applied effects using an installed app:

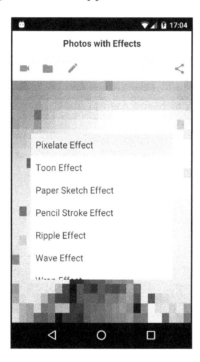

Figure 8.2: A photo taken from the camera with the Pixelate effect applied; the menu is visible and ready to apply another effect

How it works...

When launched, the app loads the available effects, inspecting all the **TEffectFilter** descendants placed on the form, and stores the component reference in a dictionary indexed with the **ListItem** index in the list. To create a friendly effect name for the UI, the effect's class name is used. Indeed, all the effect classes have the typical Pascal case naming convention (just like everything else in Delphi) and the `EffectNameByClassName` method uses a regular expression to make a string such as `TRadialBlurEffect` into something like the radial blur effect. To do this, the initial `T` is removed and then it is used as a regular expression and the words are split, as shown in the following code:

```
function TMainForm.EffectNameByClassName(const AClassName: String): String;
begin
  Result := AClassName.Substring(1);
  Result := TRegEx.Replace(Result, '[A-Z]', ' $0').TrimLeft;
end;
```

Another nice feature that's been implemented is the **Undo** stack. Each time a new effect is applied to the image, the current one is pushed onto the stack. So, when you tap on **Undo <current effect>**, the current effect is removed and the top of the stack is used to retrieve the last effect. With this approach, which is used in multiple scenarios, we can go back to the beginning without losing any steps.

The last note goes to the share functionality. The effects are applied, adding the related components to the image child controls list. Following the parenting relation, FireMonkey performs all of the drawing jobs; however, the image itself is not transformed, with only its visual representation being effected. Now, if you try to read the bitmap that's contained by the **TImage** control programmatically, the image is not "effected" and you get the original image. So, how do we actually apply the effect to the image? Check the `ShowShareSheetAction1BeforeExecute` event handler:

```
procedure TMainForm.ShowShareSheetAction1BeforeExecute(Sender: TObject);
begin
  if FUndoEffectsList.Count > 0 then
  begin
    FUndoEffectsList.Peek.ProcessEffect(nil, Image1.Bitmap, 0);
  end;
  ShowShareSheetAction1.Bitmap.Assign(Image1.Bitmap);
end;
```

As you can see, the `effect` component has a `ProcessEffect` method that actually takes an image and applies the transformation to it. In this case, the effect is not only visually applied but is actually applied. So, when you share the effected image, the image is really affected.

There's more...

Many concepts have been covered in this first mobile recipe. As you will see, the base approach to mobile development is no different from a normal FireMonkey application. This is an extraordinary feature of FireMonkey: one framework for all platforms. If you are good at FireMonkey, you are at least 80% good at all the supported platforms. However, in the mobile scope, all things get a bit slower and more difficult due to platform limits and the inherently slower `edit/run/test` loop.

To get more information about effects, you can check out the following articles:

- `http://docwiki.embarcadero.com/RADStudio/en/FireMonkey_Image_Effects`
- `http://docwiki.embarcadero.com/RADStudio/en/Applying_FireMonkey_Image_Effects`

To get more information about regular expressions as they are implemented in Delphi, you can check out the following articles:

- `http://docwiki.embarcadero.com/RADStudio/en/Regular_Expressions`
- `http://docwiki.embarcadero.com/CodeExamples/en/RTL.RegExpressionFMX_Sample`

To get some information about the Command Design Pattern and the other 22 fundamental patterns, you can read the classic book, *Design Patterns: Elements of Reusable Object-Oriented Software*, by *Erich Gamma, Richard Helm, Ralph Johnson, and John Vlissides, Addison Wesley Professional* (`http://www.amazon.com/Design-Patterns-Elements-Reusable-Object-Oriented/dp/0201633612`).

Using TListView to show and search local data

In many cases with mobile apps, data is read from remote servers and then stored locally to make them available even without an internet connection. In this recipe, you'll see how to read and write to a file as well as how to show and search that data in a `TListView`.

Getting ready

This recipe is short and simple, but it is really useful because the concepts exposed are reusable and allow you to gain confidence with some very important best practices. The final aspect of the app is shown in the following screenshot. Note that the **Delete** button is only visible when an item is selected:

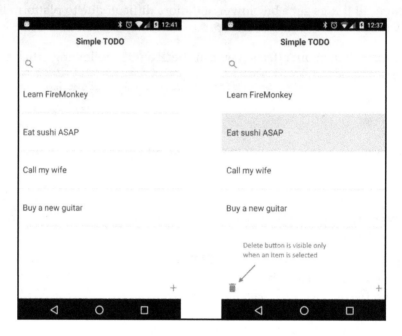

Figure 8.3: The Simple TODO app; when an item is selected, the Delete button is visible

How to do it...

1. Create a new mobile app by navigating to **File** | **New** | **Other...** | **Delphi Projects** | **Multi-Device Application**.
2. Choose the **Header/Footer** template and click on **OK**.
3. As soon as Delphi creates the project template, save all the files with the following names:
   - Save the project as `SimpleTODO.dproj`
   - Save the form as `MainFormU.pas`

4. Drop a **TListView** component on the form and set the following properties (the relevant properties are extracted from the `MainFormU.fmx` file):

```
object ListView1: TListView
  Align = Client
  ItemAppearance.ItemHeight = 80
  ItemAppearanceObjects.ItemObjects.Text.WordWrap = True
ItemAppearanceObjects.ItemObjects.Accessory.Visible =
False
  SearchVisible = True
end
```

5. Drop a **TActionList** component on the form and add two actions. Name them `acNew` and `acDelete`.

6. Create the **OnExecute** event handler for the two actions using the following code:

```
procedure TMainForm.acDeleteExecute(Sender: TObject);
begin
  if Assigned(ListView1.Selected) then
    ListView1.Items.Delete(ListView1.Selected.Index);
end;

procedure TMainForm.acNewExecute(Sender: TObject);
begin
  //check "There's more" section
  //about InputQuery deprecation
  InputQuery('TODO', 'Write your new TODO', '',
    procedure(const AResult: TModalResult;
      const AValue: string)
    var
      LValue: string;
    begin
      LValue := AValue;
      if (AResult = mrOk) and (LValue.Trim.Length > 0) then
        AddItem(LValue);
    end);
end;
```

7. Directly on the **ActionList1** component, create the **OnUpdate** event handler and fill it in with the following code. This code makes the **Delete** button invisible when no item is selected on the list:

```
procedure TMainForm.ActionList1Update(Action: TBasicAction;
                          var Handled: Boolean);
begin
  acDelete.Visible := Assigned(ListView1.Selected);
end;
```

8. Go to the main form declaration and in the `private` section, declare the following variables:

```
private
  FDataFileName: String;
  procedure LoadFromFile;
  procedure SaveToFile;
  procedure AddItem(const TODO: String);
```

9. Hit *Ctrl + Shift + C* and fill in the method bodies with the following code:

```
procedure TMainForm.LoadFromFile;
var
  LFileReader: TStreamReader;
begin
  ListView1.Items.Clear;
  if TFile.Exists(FDataFileName) then
  begin
    LFileReader := TFile.OpenText(FDataFileName);
    try
      while not LFileReader.EndOfStream do
      begin
        AddItem(LFileReader.ReadLine);
      end;
    finally
      LFileReader.Close;
    end;
  end;
end;

procedure TMainForm.SaveToFile;
var
  LItem: TListViewItem;
  LFileWriter: TStreamWriter;
begin
  LFileWriter := TFile.CreateText(FDataFileName);
  try
    for LItem in ListView1.Items do
    begin
      LFileWriter.WriteLine(LItem.Text);
    end;
  finally
    LFileWriter.Close;
  end;
end;

procedure TMainForm.AddItem(const TODO: String);
var
```

```
    LItem: TListViewItem;
begin
    LItem := ListView1.Items.Add;
    LItem.Text := TODO;
    ListView1.ItemIndex := LItem.Index;
end;
```

10. As you can see, the name of the file used to store the data is in the
    `FDataFileName` variable.

11. Create the **OnCreate** and **OnSaveState** event handlers for the form:

```
procedure TMainForm.FormCreate(Sender: TObject);
begin
    FDataFileName := TPath.Combine(TPath.GetDocumentsPath,
    'datafile.txt');
    LoadFromFile;
end;
procedure TMainForm.FormSaveState(Sender: TObject);
begin
    SaveToFile;
end;
```

12. The last thing to do is to connect the `acNew` and `acDelete` actions to two
    buttons. Drop two **TButton** components on the lower **TToolbar** named **Footer**,
    name them `btnDelete` and `btnNew`, and set the following properties:

```
object btnDelete: TButton
  Action = acDelete
  Align = alLeft
  StyleLookup = 'trashtoolbutton'
end
object btnNew: TButton
  Action = acNew
  Align = alRight
  StyleLookup = 'additembutton'
end
```

13. Run the app. For testing purposes, you can run the app using the 32-bit Windows
    target.

How it works...

When the app starts, it looks in its documents path for a file named `datafile.txt`. If it exists, it is loaded and all the lines become items in the `TListView`. Remember that Delphi allows you to write cross-platform applications, so you must be aware of the way Delphi allows you to normalize the differences between operating systems; otherwise, you risk thwarting the power of Delphi and FireMonkey. The `TPath` class is useful for keeping us ignorant about system default paths, path separators, and other stuff related to the file system. We want to put our data into the `documents` folder. However, in Android, the `document` folder is different from the iOS one (and if your code has to run in the desktop environment as well, the paths are also different). Therefore, using the `TPath` class, we can be completely ignorant about where the file is actually stored. We can know the path, but we do not want to explicitly define it; let `TPath` do its job. These are some well-known paths that `TPath` already knows. Whenever you need the specific path, ask `TPath`:

```
class function GetHomePath: string; static;
class function GetDocumentsPath: string; static;
class function GetSharedDocumentsPath: string; static;
class function GetLibraryPath: string; static;
class function GetCachePath: string; static;
class function GetPublicPath: string; static;
class function GetPicturesPath: string; static;
class function GetSharedPicturesPath: string; static;
class function GetCameraPath: string; static;
class function GetSharedCameraPath: string; static;
class function GetMusicPath: string; static;
class function GetSharedMusicPath: string; static;
class function GetMoviesPath: string; static;
class function GetSharedMoviesPath: string; static;
class function GetAlarmsPath: string; static;
class function GetSharedAlarmsPath: string; static;
class function GetDownloadsPath: string; static;
class function GetSharedDownloadsPath: string; static;
class function GetRingtonesPath: string; static;
class function GetSharedRingtonesPath: string; static;
```

Let's go back to our app. When the items are loaded into the listview, the `acNew` and `acDelete` actions allow the user to add and remove items from the list. When the form is about to go into the background, the `FormSaveState` event saves all the items, one item for a line into the `datafile.txt` file.

In a more complex situation, it is much better to have an in-memory representation of your data model that isn't bound to any visual control. Suppose you need to access the data in another form. How do we do that? If your data is bound to the GUI, you are bound to it too! The state of your app should not be stored only on the visual controls. However, for a simple situation like this recipe, it is not a big problem.

There's more...

The power of Delphi is a great advantage in mobile development. In many cases, some fundamental features of mobile apps can be tested as a desktop app as well. For instance, if you need to test the usual cycle:

- Get data from a remote web service
- Organize and save the data in local storage
- Retrieve the data and show it in the GUI

There are chances that you can test these features as a desktop application, simplifying and speeding up the deployment and the debug phase, and then when these fundamental parts work as expected, focusing on the mobile-related problems. If you have ever developed using other environments (apart from the scripted ones), you securely appreciate this possibility.

Another important note for Android developers is the modal dialogs. Android OS doesn't support modal dialog boxes. Instead of calling `ShowModal` (or `InputBox`, `InputQuery`, and so on), you should call `Show` (or one of the overloaded versions of `InputBox`, `InputQuery`, and so on) and have the form return and call your event. Embarcadero recommends that we not use modal dialog boxes on either of the mobile platforms because unexpected behavior can result. Avoiding usage of modal dialog boxes eliminates potential problems in debugging and supporting your mobile apps.

Moreover, while still used in this recipe code for backward compatibility, since Delphi 10.1 Berlin, `InputBox`, `InputQuery`, and `MessageDlg` are deprecated on FireMonkey. If your code has to only compile on Berlin or later, you should use the new `TDialogService` class, which contains the following methods:

- `ShowMessage`
- `InputQuery`
- `MessageDialog`

These three methods work in a similar way but show different dialogs. For instance, `TDialogService.MessageDialog` displays a dialog box with a custom message, dialog type, a set of buttons, and help context ID. `MessageDialog` can work synchronously or asynchronously depending on the preferred mode. `MessageDialog` internally calls `MessageDialogAsync` or `MessageDialogSync`. When `PreferredMode` is set to `Platform`, the following applies:

- On desktop platforms (Windows and OS X), `MessageDialog` behaves synchronously. The call finishes only when the user closes the dialog box.
- On mobile platforms (Android and iOS), `MessageDialog` behaves asynchronously. The call finishes instantaneously; it does not wait for the user to close the dialog box.

To force a specific behavior for the different platforms, set `PreferredMode` to `Sync` or `ASync` (however, `ASync` is not supported by Android).

Going back to our recipe, here is the `acNewExecute` method using a conditional compilation to use the `TDialogService` class if present. Remember, new projects should only use the `TDialogService` (defined in `FMX.DialogService.pas`):

```
procedure TMainForm.acNewExecute(Sender: TObject);
begin
{$IF CompilerVersion >= 24}
  // "Berlin" (or better) specific code
  TDialogService.InputQuery('TODO', ['Write your new TODO'], [''],
    procedure(const AResult: TModalResult;
      const AValues: array of string)
    var
      LValue: string;
    begin
      LValue := AValues[0];
      if (AResult = mrOk) and (LValue.Trim.Length > 0) then
        AddItem(LValue);
    end);
```

```
{$ELSE}
  //"Seattle" and previous versions up to XE7
  InputQuery('TODO', 'Write your new TODO', '',
    procedure(const AResult: TModalResult; const AValue: string)
    var
      LValue: string;
    begin
      LValue := AValue;
      if (AResult = mrOk) and (LValue.Trim.Length > 0) then
       AddItem(LValue);
    end);
{$ENDIF}
end;
```

Using SQLite databases to handle a to-do list

Usually, mobile apps read or write data using the network. In many cases, however, you need local storage to save your data. A local database can be useful for a number of things:

- To buffer information while the internet connection is not available
- To save information that will be realigned on the central server when back at the office
- To allow you a fast search on a relatively small set of data retrieved from a central database and stored on the device
- To store some structured data

In all these cases, you have to handle a database. This recipe will show you how to do it.

Getting ready

This recipe is about a to-do list. It is similar to the *Using TListView to show and search local data* recipe, but in this case, we'll use an SQL database and show data to the user using LiveBindings. Moreover, we'll see how to create output converters for LiveBindings.

How to do it...

When you need a mobile database, you have two choices in Delphi: SQLite (an open source embedded database) and InterBase ToGo.

Since version XE6, RAD Studio has included InterBase ToGo and IBLite editions for embedded application development. You can deploy your mobile applications to iOS or Android devices with an InterBase ToGo license (at a cost) or IBLite license (free).

If your app is a bit more complex and needs encryption, stored procedures, or a number of data types, you definitely have to go for InterBase ToGo. Otherwise, you can use SQLite. IBLite is the same engine as InterBase ToGo, but with limits. The biggest limit is the lack of encryption. However, an app that uses IBLite doesn't require code updates if you need to scale to InterBase ToGo; change the license and you will be okay.

This recipe is very simple in terms of database requirements, so we'll use SQLite. However, the same concepts are applicable to InterBase ToGo and IBLite.

Open the TODOList.dproj project. The main form has all the components that are required to access the database (in a real-world app, consider using a data module for this, just like the desktop applications). The app has been created using the **Header/Footer** mobile template. The first TabItem contains the to-do lists, while the second TabItem allows you to update an existing to-do list or create a new to-do list.

When the application starts, the **TFDConnection** components connect to the database. If the database file doesn't exist, the SQLite engine is configured to create a new database file from scratch. This feature is very useful and can be configured by setting the OpenMode parameter to CreateUTF8. The UTF8 encoding is almost always the best choice for international applications; in this case, it is the default setting for the **TFDConnection** components. Here's the relevant part of the **TFDConnection** parameters:

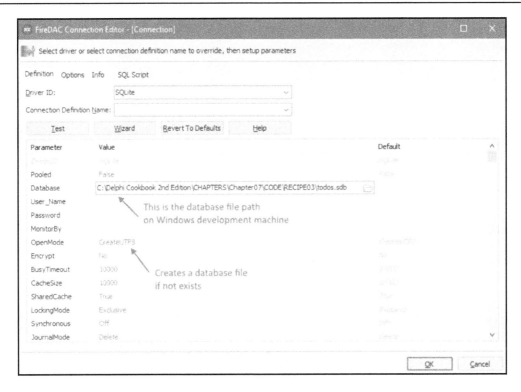

Figure 8.4: The connection parameters

Another problem to solve is related to the database path. In Windows, you can develop your mobile app using the 32-bit Windows target and a local path on your system; however, when the app runs on the device, you have to use another path. How do we solve this? In the connection's `BeforeConnect` event handler, consider the following code:

```
procedure TMainForm.ConnectionBeforeConnect(Sender: TObject);
begin
{$IF DEFINED(IOS) or DEFINED(ANDROID)}
  Connection.Params.Values['Database'] :=
        TPath.GetDocumentsPath + PathDelim + 'todos.sdb';
{$ENDIF}
end;
```

With this code, the database will be created in the proper iOS or Android `document` folder.

The next problem is related to the database structure. When and how do we create the table that we need? Let's check the `AfterConnect` event handler on the connection:

```
procedure TMainForm.ConnectionAfterConnect(Sender: TObject);
begin
  Connection.ExecSQL('CREATE TABLE IF NOT EXISTS TODOS ( ' +
    ' ID INTEGER PRIMARY KEY AUTOINCREMENT NOT NULL, ' +
    ' DESCRIPTION     CHAR(50) NOT NULL, ' +
    ' DONE            INTEGER  NOT NULL ' +
    ')');
  qryTODOs.Active := True;
end;
```

Just after the database is created, and at any subsequent run, the app tries to create the database table if it doesn't exist yet.

Then, open the dataset connected to the bind source to show the data that's present. The list view is configured with the following code:

```
ItemAppearance.ItemAppearance = 'ListItemRightDetail'
ItemAppearance.ItemHeight = 100
SearchVisible = True
```

The second tab contains a **TMemo** component, a **TSwitch** component, and two **TLabel** components. The `TBindSourceDB` data source connected to the `qryTODO` dataset is connected to the list and to the `detail` component placed on the second `TabItem` as well. This is shown in the following screenshot (integrated with some clarifying text):

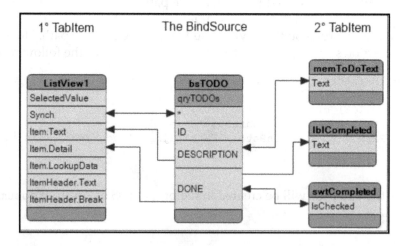

Figure 8.5: The LiveBinding designer showing the binding connections between the Bindsource, the list view, and the detail components

All the code used to handle the dataset is normal dataset-oriented code, just like the code used to manage datasets on a desktop application.

This recipe shows a nasty problem. SQLite doesn't have the Boolean field type, so the **DONE** field in the **TODO** table is of type integer, where 1 means `true` and 0 means `false`. However, we want to connect the **DONE** field to a `TSwitch.IsChecked` property of type Boolean. In this situation, when you try to change the switch value, you will get an error like the following:

Figure 8.6: The exception raised by Delphi when you try to connect an Integer field to a Boolean component property

How do we solve this? The LiveBinding engine has a powerful mechanism to convert data from one type to another. When the result of an expression is of type X and the property where that value needs to be written is of type Y, the engine looks for a valid output converter that is able to convert type X to type Y. The available output converters are shown on the `BindingList1.OutputConverters` property. As you can see, there isn't an `OutputConverter` from `String` to `Boolean` that acts as we need it to. To solve our problem, we have to register another `OutputConverter` object that's able to convert a Boolean value (`swtCompleted.IsChecked`) to a string value (because LiveBinding uses the `TField.SetText` method to set a value of a field). This output converter is registered in the `BoolToStringConverterU.pas` unit. The procedure used to register the new converter and make it visible to the LiveBinding engine is shown in the following code:

```
const
  sBoolToString = 'BoolToString';

procedure RegisterOutputConversions;
begin
  TValueRefConverterFactory.UnRegisterConversion(TypeInfo(Boolean),
    TypeInfo(String));
  TValueRefConverterFactory.RegisterConversion(TypeInfo(Boolean),
    TypeInfo(String), TConverterDescription.Create(
    procedure(const InValue: TValue; var OutValue: TValue)
```

```
begin
  if InValue.AsBoolean then
    OutValue := '1'
  else
    OutValue := '0';
  end, sBoolToString, sBoolToString, '', True, sBoolToString, nil));
end;
```

Now, the app works correctly. However, be careful, as now all the conversions from Boolean to string will be considered true when 1 and false when 0. This internal mechanism of LiveBindings needs to be clearly understood because it can cause a lot of headaches in non-trivial cases. The full code of this `OutputConverter` is available in the `BoolToStringConverterU.pas` unit.

On the second tab, there is a label that describes the meaning of the switch. When the switch is checked, the label says **The task is completed**; otherwise, it says **The task is not completed**. This feature is implement using LiveBinding expressions. Go to the LiveBinding designer and select the arrow that connects the **DONE** field to the `lblCompleted.Text` property. Now, hit *F11* to show the **Object Inspector** window and check the `CustomFormat` property. Here, logic is used by the label. The expression is reported as follows:

```
"The task is " + IfThen(value = 1, "completed","not completed")
```

This code is a relational expression that transforms a read value from a dataset field to a text value shown in a label. Normally, the value is read from the source component and written on the target property component. However, using the `CustomFormat` property, you can change this default behavior to get more complex and useful information. This expression is a good example of that.

There's more...

As you can see, mobile development is a mix of well-known things and new things. The LiveBindings framework is a big new thing, and you may be frightened by it. However, don't be afraid: all your needs are there. Here are some useful links to go deeper with the concepts shown in this recipe:

- Another approach to the Integer-As-Boolean problem can be found at http://www.malcolmgroves.com/blog/?p=1490
- Information on formatting fields using LiveBindings can be found at http://www.malcolmgroves.com/blog/?p=1226

- Documentation about output converters can be found
  at `http://docwiki.embarcadero.com/RADStudio/en/LiveBindings_Output_Conv`
  `erters`
- Some tutorials on LiveBindings in RAD Studio can be found
  at `http://docwiki.embarcadero.com/RADStudio/en/LiveBindings_in_RAD_Stud`
  `io`

Do not block the main thread!

Long requests to external systems such as storage, databases, hardware, and networks have always been difficult to handle from a user experience point of view. For programmers, it is easy to simply run the long request and, when finished (after seconds, minutes, or hours), inform the user that their data is there. However, we should think about user experience even more in the mobile world.

Getting ready

If your app runs a long-running request and the UI is frozen, the user might think that something is going wrong and start to tap here and there to try and unblock the app. After a few seconds, either the operating system itself will close the app, or the user will push the **Home** button to close the app and then, usually, uninstall it. Yes, user experience is one of the most important things on a mobile. Just as with a desktop, the user experience should be of primary importance, but what I want to emphasize is while on a desktop, you may have patient users because they are sitting in front of a PC (or a Mac), on a mobile, you will certainly have impatient users who want immediate feedback from your app. Mobile apps can be used on the move, so the user may be busy doing something else while they are using your app, so the app must be fast and should give feedback as soon as possible. If a long operation is running, the app should inform the user and the GUI should never freeze. In this recipe, we aren't looking at, not how to have 0 seconds latency, but how to inform the user that the app is actually working as expected, and the only thing the user has to do is wait!

How to do it...

The scenarios described in this recipe are very common, so this demo will have to face real timings and real problems. We'll do, as a long-running request, a REST call to an open web service that provides weather forecasts. I've used this app for a while and the forecasts even seem accurate! Cool!

The service is provided by `http://openweathermap.org` and we will issue the REST request at the following endpoint: `http://api.openweathermap.org/data/2.5/forecast`.

All the parameters required for the request will be defined at runtime by the app. To use this service, you need an API key. To get the API key, follow the instructions here: `http://openweathermap.org/appid`. After obtaining the API key, we can start to create the app:

1. Create a new mobile app by navigating to **File** | **New** | **Other...** | **Delphi Projects** | **Multi-Device Application**.
2. Choose the **Header/Footer** template and click on **OK**.
3. As soon as Delphi creates the project template, save all the files with the following names:
   - Save the project as `WeatherForecasts.dproj`
   - Save the form as `MainFormU.pas`
4. Drop a **TPanel** component just below the header toolbar and align it to **alTop**.
5. In the panel you have just dropped, drop two **TEdits** components and a **TButton** component and name them `EditCity`, `EditCountry`, and `btnGetForecasts`, respectively. Then, set the other properties, as shown in the following code:

```
object btnGetForecasts: TButton
  Align = Right
  Size.PlatformDefault = False
  StyleLookup = 'refreshtoolbutton'
end
object EditCity: TEdit
  Align = Client
  Margins.Left = 10.000000000000000000
  Margins.Top = 10.000000000000000000
  Margins.Bottom = 10.000000000000000000
  TextPrompt = 'City'
end
object EditCountry: TEdit
  Align = Right
  Margins.Top = 10.000000000000000000
  Margins.Bottom = 10.000000000000000000
  TextPrompt = 'Country'
End
```

6. Drop a **TAniIndicator** component into the header toolbar and align it to the **Right**. Set its **Margins** property to `10` for each side.

7. Drop a **TListView** component on the form's center and set the following properties (the relevant properties extracted from the `MainFormU.fmx` file):

```
object ListView1: TListView
  AllowSelection = False
  Align = Client
  ItemAppearanceObjects.ItemObjects.Text.WordWrap = True
  ItemAppearanceObjects.ItemObjects.Text.Height = 50
  ItemAppearanceObjects.ItemObjects.Accessory.Visible =
    False
  CanSwipeDelete = False
end
```

8. Drop a **TLabel** component into the footer toolbar, align it to **Client**, and name it `lblInfo`.

9. Drop the **TRESTClient** and **TRESTResponse** components and leave the default properties and names.

10. Your form, at design time, should look like the following:

Figure 8.7: The Weather forecasts form at design time

11. Now, let's write some code. In the `private` section of the form, declare a string instance field called `FOSLang`. This variable will contain the current operating system language so that we can request the proper localized text from the service.

12. Create the `FormCreate` event handler and fill it in with the following code:

```
procedure TMainForm.FormCreate(Sender: TObject);
var
  LLocaleService: IFMXLocaleService;
begin
  if
TPlatformServices.Current.SupportsPlatformService(IFMXLocaleService
) then
    begin
      LLocaleService := TPlatformServices.Current.GetPlatformService
        (IFMXLocaleService) as IFMXLocaleService;
      FOSLang := LLocaleService.GetCurrentLangID;
    end
  else
    FOSLang := 'US';

  EditCountry.Text := FOSLang;
  RESTClient1.BaseURL := 'http://api.openweathermap.org/data/2.5';
  RESTRequest1.Resource :=
'forecast?q={country}&mode=json&lang={lang}&units=metric&APPID={APP
ID}';
  RESTRequest1.Params.ParameterByName('APPID').Value := APPID;
  AniIndicator1.Visible := False;

end;
```

13. In the `implementation` section of `uses`, add the following units:

```
uses
  System.JSON, System.DateUtils, FMX.Platform;
```

14. Create an **OnClick** event handler for the **btnGetForecasts** button and fill it in with the following code:

```
procedure TMainForm.btnGetForecastsClick(Sender: TObject);
begin
  ListView1.Items.Clear;
  RESTRequest1.Params.ParameterByName('country').Value :=
    string.Join(',', [EditCity.Text, EditCountry.Text]);
  RESTRequest1.Params.ParameterByName('lang').Value := FOSLang;
  AniIndicator1.Visible := True;
  AniIndicator1.Enabled := True;
```

```
    btnGetForecasts.Enabled := False;
    RESTRequest1.ExecuteAsync(
      procedure
      var
        LForecastDateTime: TDateTime;
        LJValue: TJSONValue;
        LJObj, LMainForecast, LForecastItem, LJObjCity: TJSONObject;
        LJArrWeather, LJArrForecasts: TJSONArray;
        LTempMin, LTempMax: Double;
        LDay, LLastDay: string;
        LItem: TListViewItem;
        LWeatherDescription: string;
        LAppRespCode: string;

      begin
        try
          LJObj := RESTRequest1.Response.JSONValue as TJSONObject;

          // check for errors
          LAppRespCode := LJObj.GetValue('cod').Value;
          if LAppRespCode.Equals('404') then
          begin
            lblInfo.Text := 'City not found';
            Exit;
          end;
          if not LAppRespCode.Equals('200') then
          begin
            lblInfo.Text := 'Error ' + LAppRespCode;
            Exit;
          end;

          // parsing response...
          LJArrForecasts := LJObj.GetValue('list') as TJSONArray;
          for LJValue in LJArrForecasts do
          begin
            LForecastItem := LJValue as TJSONObject;
            LForecastDateTime :=
  UnixToDateTime((LForecastItem.GetValue('dt')
                as TJSONNumber).AsInt64);
            LMainForecast := LForecastItem.GetValue('main') as
  TJSONObject;
            LTempMin := (LMainForecast.GetValue('temp_min')
                as TJSONNumber).AsDouble;
            LTempMax := (LMainForecast.GetValue('temp_max')
                as TJSONNumber).AsDouble;
            LJArrWeather := LForecastItem.GetValue('weather') as
  TJSONArray;
            LWeatherDescription := TJSONObject(LJArrWeather.Items[0])
```

```
                .GetValue('description').Value;
            LDay := FormatDateTime('ddd d mmm yyyy',
    DateOf(LForecastDateTime));
            if LDay <> LLastDay then
            begin
              LItem := ListView1.Items.Add;
              LItem.Purpose := TListItemPurpose.Header;
              LItem.Text := LDay;
            end;
            LLastDay := LDay;
            LItem := ListView1.Items.Add;
            LItem.Text := FormatDateTime('HH', LForecastDateTime) + '
    ' +
              LWeatherDescription + Format(' (min %2.2f max %2.2f)',
              [LTempMin, LTempMax]);
          end;

          // display the city name at the bottom
          LJObjCity := LJObj.GetValue('city') as TJSONObject;
          lblInfo.Text := LJObjCity.GetValue('name').Value + ', ' +
            LJObjCity.GetValue('country').Value;

        finally
          // stop the waiting animation
          AniIndicator1.Visible := False;
          AniIndicator1.Enabled := False;
          btnGetForecasts.Enabled := True;
        end;

      end);
  end;
```

15. The parsing code is not simple, but now you should have all the information needed to correctly understand what's going on with this code.

16. Remember to replace APPID constant with your:

```
15. const
        // get your own APPID here http://openweathermap.org/appid
        APPID = 'your-app-id';
```

16. Hit *F9* and you should see the application running.

17. Insert a city name and a state code (such as Roma and IT or London and GB), and you will get the weather forecasts for the upcoming days, organized day by day.

How it works...

This recipe is simple from an architectural point of view. There are two parameters the user can enter. These parameters affect the request to the server that will respond with a JSON structure. Apart from the parsing code, interesting things happen when the request is sent to the server. If we had sent a normal synchronous request to the server, the UI would be blocked until the response arrives at the client. Using the `ExecuteAsynch` method executes the actual request on a background thread so that the main thread remains free to update the UI. When the request finishes the execution, then an anonymous method is called in the main thread context. The **TAniIndicator** component is started just before the request starts and is stopped after the parsing is finished. In this way, the user is aware that something is happening. Consider that any request to an external system could potentially last for hours. Be aware of this!

The code used to fill the list uses the grouping feature of the **TListView** component to show the forecasts organized day by day.

Another thing to note is that the web service can use a localized response for descriptive text. Therefore, in the `FormCreate` event, we use the `IFMXLocaleService` service to read the current system language. Later, we use that language code to inform the remote service about the preferred localization language.

Here's the app running in the mobile preview on an Italian PC:

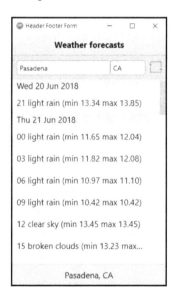

Figure 8.8: The Weather forecasts app running in the mobile preview

There's more...

Multithreading can be difficult, but the built-in features in the REST client library allow you to send HTTP requests in a background thread in a very simple manner. You can use it as much as you want. If you are not so confident with the REST client library, here's some documentation:

- *Delphi XE5 Mobile REST Client Demo* at
  `https://www.youtube.com/watch?v=OkRVbgF4VMI`
- *REST Client Library* at
  `http://docwiki.embarcadero.com/RADStudio/en/REST_Client_Library`

Another topic that should be deeply understood to correctly design and implement FireMonkey applications (and mobile apps are only a particular type of FireMonkey application) are the FireMonkey platform services. More information on platform services can be found
at `http://docwiki.embarcadero.com/RADStudio/en/FireMonkey_Platform_Services`.

A challenge today (especially in mobile development) is to decouple different parts/layers of your application while still allowing them to communicate efficiently. Here are some patterns and frameworks that can help you:

- Delphi Event Bus (explained in the *Communication made easy with Delphi Event Bus* recipe of `Chapter 5`, *The Thousand Faces of Multithreading*): `https://github.com/spinettaro/delphi-event-bus`
- MVW (model-view-whatever) patterns: `https://www.beyondjava.net/blog/model-view-whatever/`

Using a styled TListView to handle a long list of data

The **TListBox** control is very flexible. You can customize every aspect of each item in the list. However, it is not suitable if you want to handle a long list of data because flexibility comes at the cost of the system being slow when the number of data rows grow. Embarcadero specifies that you should use **TListView** to display a collection of items in a list that is optimized for LiveBindings and for fast and smooth scrolling.

Getting ready

In this recipe, we will use the *Do not block main thread!* recipe as a base to customize a list view using custom styles. In that recipe, we get a list of weather forecasts from a REST web service and then fill in the list view with that data using a standard style. In this recipe, that data will be nicely inserted into a custom list view with colors, alignment, and a summary footer. There is no design-time support with this approach because all the controls created in each item are created at runtime; however, this approach can be very useful if you want complete control over the look and feel of your list. To be clear, the recommended approach in this case is to write a custom style for the **TListView** component; put the component in a package, install it into the IDE, and then use it from the **Object Inspector** window. To have two samples of this approach, check the following projects that are provided as samples (the `Sample` folder on my machine is
`C:\Users\Public\Documents\Embarcadero\Studio\19.0\Samples`, where `19.0` is the version of the IDE).

Within the `Sample` folder, open `Object Pascal\Multi-Device Samples\User Interface\ListView\`.

In this folder, you have a number of projects and packages that show you how to use some advanced stuff related to the **TListView** components. To see the new style, you have to install the package and open the related demo project:

- The package to install the `RatingListItem` list item style is `SampleListViewRatingsAppearancePackage.dproj`
- The project that shows how to use the `RatingListItem` style is `SampleListViewRatingsAppearanceProject.dproj`
- The package to install the `MultiDetailItem` list item style is `SampleListViewMultiDetailAppearancePackage.dproj`
- The project that shows how to use the `MultiDetailItem` style is `SampleListViewMultiDetailAppearanceProject.dproj`

It is not too complex to create a custom list item style, and we'll see how to do this in the next recipe. In this recipe, we'll create the list item style element directly in the code. When you are satisfied with the result, you can create the proper package, as shown in the next recipe.

Since RAD Studio 10.1 Berlin, there is the ListView Items Editor, which helps if you want to define the aspect of your list view items at design time. The approach defined in this and the following chapters is more general and applicable to previous RAD Studio versions as well.

How to do it...

1. Copy the *Do not block main thread!* recipe into a new folder.
2. Open the project and save it as `WeatherForecastsEx.dproj`.
3. Change the code, as shown in the following steps.
4. In the `private` section of the form declaration, add the following methods:

```
private
  Lang: string;
  procedure AddFooter(AItems: TAppearanceListViewItems;
    const LMinInTheDay, LMaxInTheDay: Double);
  procedure AddHeader(AItems: TAppearanceListViewItems;
const ADay: String);
  procedure AddForecastItem(AItems:
    TAppearanceListViewItems;
    const AForecastDateTime: TDateTime;
    const AWeatherDescription: String;
    const ATempMin, ATempMax: Double);
```

5. Press *Ctrl* + *Shift* + *C* to create the method bodies, and then add the following code:

```
procedure TMainForm.AddHeader(AItems:
  TAppearanceListViewItems;
  const ADay: String);
var
  LItem: TListViewItem;
begin
  LItem := AItems.Add;
  LItem.Purpose := TListItemPurpose.Header;
  LItem.Objects.FindDrawable('HeaderLabel').Data := ADay;
end;

procedure TMainForm.AddForecastItem(AItems:
  TAppearanceListViewItems;
  const AForecastDateTime: TDateTime;
  const AWeatherDescription: String;
  const ATempMin, ATempMax: Double);
var
  LItem: TListViewItem;
begin
  LItem := AItems.Add;
  LItem.Objects.FindDrawable('WeatherDescription').Data :=
    FormatDateTime('HH', AForecastDateTime) + ' ' +
      AWeatherDescription;
  LItem.Objects.FindDrawable('MinTemp').Data :=
```

```
      FormatFloat('#0.00', ATempMin) + '°';
    LItem.Objects.FindDrawable('MaxTemp').Data :=
      FormatFloat('#0.00', ATempMax) + '°';
  end;

procedure TMainForm.AddFooter(AItems:
  TAppearanceListViewItems;
  const LMinInTheDay, LMaxInTheDay: Double);
var
  LItem: TListViewItem;
begin
  LItem := AItems.Add;
  LItem.Purpose := TListItemPurpose.Footer;
  LItem.Text := Format('min %2.2f°C max %2.2f°C',
    [LMinInTheDay, LMaxInTheDay]);
end;
```

6. Now, we have to use these methods of the form. In the `btnGetForecastsClick` method, substitute the code with the following:

```
procedure TMainForm.btnGetForecastsClick(Sender: TObject);
begin
  ListView1.Items.Clear;
  RESTRequest1.Params.ParameterByName('country').Value :=
    String.Join(',', [EditCity.Text, EditCountry.Text]);
  RESTRequest1.Params.ParameterByName('lang').Value := Lang;
  AniIndicator1.Visible := True;
  AniIndicator1.Enabled := True;
  btnGetForecasts.Enabled := False;
  RESTRequest1.ExecuteAsync(
    procedure
    var
      LForecastDateTime: TDateTime;
      LJValue: TJSONValue;
      LJObj, LMainForecast, LForecastItem, LJObjCity: TJSONObject;
      LJArrWeather, LJArrForecasts: TJSONArray;
      LTempMin, LTempMax: Double;
      LDay, LLastDay: string;
      LWeatherDescription: string;
      LAppRespCode: string;
      LMinInTheDay: Double;
      LMaxInTheDay: Double;

    begin
      try
        LJObj := RESTRequest1.Response.JSONValue as TJSONObject;
        // check for errors
        LAppRespCode := LJObj.GetValue('cod').Value;
```

```
    if LAppRespCode.Equals('404') then
    begin
      lblInfo.Text := 'City not found';
      Exit;
    end;
    if not LAppRespCode.Equals('200') then
    begin
      lblInfo.Text := 'Error ' + LAppRespCode;
      Exit;
    end;

    // parsing forecasts
    LMinInTheDay := 1000;
    LMaxInTheDay := -LMinInTheDay;
    LJArrForecasts := LJObj.GetValue('list') as TJSONArray;
    for LJValue in LJArrForecasts do
    begin
      LForecastItem := LJValue as TJSONObject;
      LForecastDateTime :=
UnixToDateTime((LForecastItem.GetValue('dt')
        as TJSONNumber).AsInt64);
      LMainForecast := LForecastItem.GetValue('main') as
TJSONObject;
      LTempMin := (LMainForecast.GetValue('temp_min')
        as TJSONNumber).AsDouble;
      LTempMax := (LMainForecast.GetValue('temp_max')
        as TJSONNumber).AsDouble;
      LJArrWeather := LForecastItem.GetValue('weather') as
TJSONArray;
      LWeatherDescription := TJSONObject(LJArrWeather.Items[0])
        .GetValue('description').Value;
      LDay := FormatDateTime('ddd d mmm yyyy',
DateOf(LForecastDateTime));
      if LDay <> LLastDay then
      begin
        if not LLastDay.IsEmpty then
        begin
          AddFooter(ListView1.Items, LMinInTheDay,
LMaxInTheDay);
        end;
        AddHeader(ListView1.Items, LDay);
        LMinInTheDay := 1000;
        LMaxInTheDay := -LMinInTheDay;
      end;
      LLastDay := LDay;
      LMinInTheDay := Min(LMinInTheDay, LTempMin);
      LMaxInTheDay := Max(LMaxInTheDay, LTempMax);
```

```
        AddForecastItem(ListView1.Items, LForecastDateTime,
          LWeatherDescription, LTempMin, LTempMax);
      end; // for in

      if not LLastDay.IsEmpty then
        AddFooter(ListView1.Items, LMinInTheDay, LMaxInTheDay);

      LJObjCity := LJObj.GetValue('city') as TJSONObject;
      lblInfo.Text := LJObjCity.GetValue('name').Value + ', ' +
        LJObjCity.GetValue('country').Value;

    finally
      AniIndicator1.Visible := False;
      AniIndicator1.Enabled := False;
      btnGetForecasts.Enabled := True;
    end;

  end);
end;
```

7. The main difference between the *Using SQLite databases to handle a to-do list* recipe
    and this recipe is the complete flexibility of data visualization. To get this
    flexibility, we added individual controls to each list item. We defined all the
    needed properties, width, alignment, colors, and so on. When the device switches
    to landscape orientation, some alignments need to be changed according to the
    larger horizontal space available. For this situation, a very handy list box
    UpdateObjects event handler is available. Create an UpdateObjects event
    handler on the list box and add the following code:

```
procedure TMainForm.ListView1UpdateObjects(
  const Sender: TObject;
  const AItem: TListViewItem);
var
  AQuarter: Double;
  lb: TListItemText;
begin
  case AItem.Purpose of
    TListItemPurpose.None:
      begin
        AQuarter := (AItem.Parent.Width -
          TListView(AItem.Parent).ItemSpaces.Left -
          TListView(AItem.Parent).ItemSpaces.Right) / 4;

        // AItem.Objects.Clear;
        AItem.Height := 24;

        lb := TListItemText.Create(AItem);
```

```
      lb.PlaceOffset.X := 0;
      lb.TextAlign := TTextAlign.Leading;
      lb.Name := 'WeatherDescription';

      lb := TListItemText.Create(AItem);
      lb.TextAlign := TTextAlign.Trailing;
      lb.TextColor := TAlphaColorRec.Blue;
      lb.Name := 'MinTemp';
      lb.PlaceOffset.X := AQuarter * 2;
      lb.Width := AQuarter;

      lb := TListItemText.Create(AItem);
      lb.TextAlign := TTextAlign.Trailing;
      lb.TextColor := TAlphaColorRec.Red;
      lb.Name := 'MaxTemp';
      lb.PlaceOffset.X := AQuarter * 3;
      lb.Width := AQuarter;
    end;

  TListItemPurpose.Header:
    begin
      AItem.Height := 48;
      lb := TListItemText.Create(AItem);
      lb.TextAlign := TTextAlign.Center;
      lb.Align := TListItemAlign.Center;
      lb.TextColor := TAlphaColorRec.Red;
      lb.PlaceOffset.Y := AItem.Height / 4;
      lb.Name := 'HeaderLabel';
    end;

  TListItemPurpose.Footer:
    begin
      AItem.Objects.TextObject
        .TextAlign := TTextAlign.Trailing;
    end;
  end;
```

8. With this adjustment, text inside the list item is always aligned correctly.

9. Run the app. For testing purposes, you can run the app using the 32-bit Windows target. Here's the app running on an Android phone:

Figure 8.9: The Weather forecasts ex app running in portrait and landscape modes on an Italian Android phone; note how the temperature columns are realigned between the two orientations

How it works...

After reading the JSON using the **TRESTClient** component, in the parsing code, we added controls to each item to represent the three columns we require. You cannot add every kind of control to the **TListViewItem** component; we can only add those that inherit from **TListItemDrawable**. However, you can inherit your own class from **TListViewItem** to implement all the advanced visualizations you require.

The relevant part about the customization happens in the `ListView1UpdateObjects` method:

```
procedure TMainForm.ListView1UpdateObjects(
  const Sender: TObject;
  const AItem: TListViewItem);
var
  AQuarter: Double;
  lb: TListItemText;
  lListView: TListView;
```

```
begin
  lListView := Sender as TListView;
  //different item purpose, different customization.
  case AItem.Purpose of
    TListItemPurpose.None:
      begin
        //calculate ¼ of the available horizontal space
        AQuarter := (lListView.Width – lListView.ItemSpaces.Left –
            lListView.ItemSpaces.Right) / 4;
        AItem.Height := 24;
        //1st column, the textual description
        //Check if the item is already created. This check
        //saves resource and make the app faster
        lb :=  TListItemText(AItem.Objects
              .FindDrawable('WeatherDescription'));
        if not Assigned(lb) then
        begin
          //if needed, create and initialize the component
          lb := TListItemText.Create(AItem);
          lb.PlaceOffset.X := 0;
          lb.TextAlign := TTextAlign.Leading;
          lb.Name := 'WeatherDescription';
        end;
        //offset must be always updated depending
        //on device orientation
        lb.PlaceOffset.X := 0;

        //2nd column, the min temperature
        lb := TListItemText(AItem.Objects
              .FindDrawable('MinTemp'));
        if not Assigned(lb) then
        begin
          lb := TListItemText.Create(AItem);
          lb.TextAlign := TTextAlign.Trailing;
          lb.TextColor := TAlphaColorRec.Blue;
          lb.Name := 'MinTemp';
        end;
        //offset and width must be updated
        lb.PlaceOffset.X := AQuarter * 2;
        lb.Width := AQuarter;

        //3rd column, max temperature
        lb := TListItemText(AItem.Objects
              .FindDrawable('MaxTemp'));
        if not Assigned(lb) then
        begin
          lb := TListItemText.Create(AItem);
          lb.TextAlign := TTextAlign.Trailing;
```

```
              lb.TextColor := TAlphaColorRec.Red;
              lb.Name := 'MaxTemp';
          end;
          lb.PlaceOffset.X := AQuarter * 3;
          lb.Width := AQuarter;
        end;

    TListItemPurpose.Header:
      begin
        AItem.Height := 48;
        //headers have only one itemtext
        lb := TListItemText(AItem.Objects
              .FindDrawable('HeaderLabel'));
        if not Assigned(lb) then
        begin
          lb := TListItemText.Create(AItem);
          lb.TextAlign := TTextAlign.Center;
          lb.Align := TListItemAlign.Center;
          lb.TextColor := TAlphaColorRec.Red;
          lb.Name := 'HeaderLabel';
        end;
        lb.PlaceOffset.Y := AItem.Height / 4;
      end;
    TListItemPurpose.Footer:
      begin
        //footer item doesn't have heavy
        //customization, only alignment
        AItem.Objects.TextObject.TextAlign := TTextAlign.Trailing;
      end;
    end;
  end;
```

First, we check that the objects have already been created. When you change the device's orientation, the objects are still there. So, if possible, it is better to reuse them than destroy and recreate them from scratch. Then, all the new items are created and added to the `ListViewItem` instance. Note that each control has a name. Each name is used to check their existence, and by the parsing code, to write the actual text into the `Data` property of the control. So, just to be clear, if in the item you add a **TListItemText** control named `MinTemp`, you can use `Item.Objects.FindDrawable('MinTemp').Data` to read and write (it depends on the actual object, but technically it is possible) generic data on the `MinTemp.Data` property. As you know, the `Data` property is handled by each control with a different meaning. In this specific case, all the `Data` properties represent the text written in the controls.

Then, the problem related to the orientation change is handled by the very useful `UpdateObjects` event on the list view. Here, we organize the horizontal space and split it into four columns, giving the first two columns to the weather description, the third to the minimum temperature, and the fourth to the maximum temperature. You can organize all the cool things you need in this event because it's called every time there is an update in the object's visualization.

There's more...

List views are tremendously helpful in mobile development and you must be familiar with them to implement good looking and efficient apps. Using the custom list item style in a package, you also get LiveBinding support, while the solution exposed in this recipe doesn't provide this support. Consider developing a custom list item style and packaging it in a package if you want design-time support. This recipe gives you the starting point for developing a custom style for the list view items. When you are satisfied with the result, create the proper package, as shown in the samples provided by Embarcadero.

Customizing the TListView

As already said in the *Using a styled TListView to handle a long list of data* recipe, the `TListView` is the best control for handling long lists of data. We already know how to change the default style using the `UpdateObjects` event. However, this approach lacks the Delphi RADness approach; no visual preview, no object inspector, no Visual LiveBindings, no live data. In this recipe we'll look at how to create a `TListView` style which can be installed in the Delphi IDE and used at design time in the object inspector and in the Visual LiveBindings designer.

Getting ready

`TListView` uses the **Appearance Class** to define how it looks at runtime. An **Appearance Class** is nothing more than a class derived from `TAppearanceObjects` (or one of its inherited classes). You can create and install a new customized appearance class and use it in your design by installing a new package. This package defines the classes that implement a custom appearance for list view items. You can customize the fields as necessary in order to implement specialized graphics or text.

How it works

Open the project group
Chapter08\CODE\RECIPE06\CustomAppearanceGroup.groupproj.

The group is composed of the following two projects:

- DelphiCookbookListViewAppearance.dproj: This is a package containing the appearance class that we defined for our purpose. This package needs to be installed and its path must be in the Delphi library path for all the necessary platforms.
- WeatherForecastCustomAppearance.dproj: The actual application. It uses the appearance class defined in the previous package and uses LiveBindings to show the weather forecasts.

Let's talk about the appearance class. Open the
DelphiCookbookListViewAppearanceU.pas file contained in the package. The main class in this unit is TDelphiCookbookItemAppearance. As you can see, this file is declared in the implementation section so that no one can directly use it. So, who can never use it? Check the unit initialization and finalization section:

```
initialization

TAppearancesRegistry.RegisterAppearance(TDelphiCookbookItemAppearance,
   TDelphiCookbookAppearanceNames.ListItem,
[TRegisterAppearanceOption.Item],
   sThisUnit);

finalization

TAppearancesRegistry.UnregisterAppearances
(TArray<TItemAppearanceObjectsClass>.Create(TDelphiCookbookItemAppearance))
;

end.
```

Using the initialization and finalization sections, the new appearance class is registered as soon as the package, and the units contained, are loaded into the IDE. But what does TDelphiCookbookItemAppearance look like?

Here's its declaration:

```
type
  TDelphiCookbookItemAppearance = class(TPresetItemObjects)
  public
    const DEFAULT_HEIGHT = 40;
  private
    FMinTemp: TTextObjectAppearance;
    FMaxTemp: TTextObjectAppearance;
    procedure SetMinTemp(const Value: TTextObjectAppearance);
    procedure SetMaxTemp(const Value: TTextObjectAppearance);
  protected
    function DefaultHeight: Integer; override;
    procedure UpdateSizes(const FinalSize: TSizeF); override;
    function GetGroupClass: TPresetItemObjects.TGroupClass; override;
  public
    constructor Create(const Owner: TControl); override;
    destructor Destroy; override;
  published
    property Accessory;
    property Text;
    property MinTemp: TTextObjectAppearance read FMinTemp write SetMinTemp;
    property MaxTemp: TTextObjectAppearance read FMaxTemp write SetMaxTemp;
  end;
```

As any appearance class, it inherits from TPresetItemObjects, which defines the standard behaviour for a generic TListViewItem. Then, we added two more text objects: MinTemp and MaxTemp. These objects are not real controls but a set of properties used internally to define actual graphical objects. It is possible to add a number of custom elements into an appearance class. See the following list for all the available elements:

- TTextObjectAppearance: A text label.

- TImageObjectAppearance: An image.

- TAccessoryObjectAppearance: A flexible graphical indicator that provides different predefined icons. It is used to graphically inform the user that more information is available if the item is tapped/clicked.

- TTextButtonObjectAppearance: This is a button with a predefined label. It is used in the standard item when **delete on swipe** is used.

- TGlyphButtonObjectAppearance: This is another kind of button, but has an icon instead of text.

The appearance class can define the default height for every item. In our case, the default height is 40. Two methods require particular attention: the constructor, where all the objects must be initialized, and the `UpdateSize`, which is called when an item is added to the list and every time you rotate the device or resize the list view.

Here's the constructor:

```
constructor TDelphiCookbookItemAppearance.Create(const Owner: TControl);
var
  LInitTextObject: TProc<TTextObjectAppearance>;
begin
  inherited;

  //The 2 text objects are initialized in the same way, so let's
  //create an anonymous method to initialize them
  LInitTextObject := procedure(pTextObject: TTextObjectAppearance)
    begin
      //initialization is quite standard for all objects. There
      //are however some fundamental steps...
      //notify the container when the control change
      pTextObject.OnChange := ItemPropertyChange;
      //setting the default properties
      pTextObject.DefaultValues.Align := TListItemAlign.Leading;
      pTextObject.DefaultValues.VertAlign := TListItemAlign.Center;
      pTextObject.DefaultValues.TextVertAlign := TTextAlign.Center;
      pTextObject.DefaultValues.TextAlign := TTextAlign.Trailing;
      pTextObject.DefaultValues.PlaceOffset.Y := 0;
      pTextObject.DefaultValues.PlaceOffset.X := 0;
      pTextObject.DefaultValues.Width := 80;
      pTextObject.DefaultValues.Visible := True;

      //reset the control to the default just defined
      pTextObject.RestoreDefaults;

      //the object is owned by the appearance object instance
      pTextObject.Owner := Self;
    end;

  //create and initialize the text label
  //for the min temperature
  FMinTemp := TTextObjectAppearance.Create;
  FMinTemp.Name := TDelphiCookbookAppearanceNames.MinTemp;
  //by default the text is blue
  FMinTemp.DefaultValues.TextColor := TAlphaColorRec.Blue;
  LInitTextObject(FMinTemp);

  //create and initialize the text label
```

```
//for the max temperature
FMaxTemp := TTextObjectAppearance.Create;
FMaxTemp.Name := TDelphiCookbookAppearanceNames.MaxTemp;
//by default this text will be red
FMaxTemp.DefaultValues.TextColor := TAlphaColorRec.Red;
LInitTextObject(FMaxTemp);

//Now we've to define LiveBindings members for this appearance
//class. Remember, the Visual LiveBinding designer
//will show the members defined here and will use these
//expressions to set the value of the member at runtime.

//define LiveBindings members related to mintemp
FMinTemp.DataMembers := TObjectAppearance.TDataMembers.Create
  (TObjectAppearance.TDataMember.Create(MIN_TEMP_MEMBER,
  // Displayed by LiveBindings
  Format('Data["%s"]',
    [TDelphiCookbookAppearanceNames.MinTemp])));
  // Expression to access value from TListViewItem


//define LiveBindings members related to maxtemp
FMaxTemp.DataMembers := TObjectAppearance.TDataMembers.Create
  (TObjectAppearance.TDataMember.Create(MAX_TEMP_MEMBER,
  // Displayed by LiveBindings
  Format('Data["%s"]',
    [TDelphiCookbookAppearanceNames.MaxTemp])));
  // Expression to access value from TListViewItem

// Define the appearance objects
AddObject(Text, True);
AddObject(MinTemp, True);
AddObject(MaxTemp, True);
end;
```

As you can see in the constructor, there is no information about size. We set alignments, colors, and so on, but no size. Where are sizes defined? In the `UpdateSizes` method, which is called whenever the `TListView` needs to know how big each component is. Moreover, we can set visibility and other details in this method. In this case, we want to hide the `MinTemp` text if the width of the control is not enough. This change is done automatically because `UpdateSizes` is called repeatedly:

```
procedure TDelphiCookbookItemAppearance.UpdateSizes(const FinalSize:
TSizeF);
var
  LColWidth: Extended;
  LFullWidth: Boolean;
```

```
begin
  BeginUpdate;
  try
    inherited;
    //we define a virtual layout of 12 columns based on
    //current listitem width
    LColWidth := FinalSize.Width / 12;
    //is the listitem wide enough to contein the full
    //set of information?
    LFullWidth := LColWidth * 4 >= MinTemp.Width;


    if LFullWidth then
    begin
      //mintemp is visible, the default text is large 6 virtual
      //columns and the other texts are 2 columns wide
      MinTemp.Visible := True;
      Text.InternalWidth := LColWidth * 6;
      MinTemp.PlaceOffset.X := LColWidth * 6;
      MinTemp.InternalWidth := LColWidth * 2;
      MaxTemp.PlaceOffset.X := LColWidth * 9;
      MaxTemp.InternalWidth := LColWidth * 2;
    end
    else
    begin
      //mintemp is not visible, the default text is large 8
      //virtual columns and the maxtemp texts is 4 columns wide
      MinTemp.Visible := False;
      Text.InternalWidth := LColWidth * 8;
      MaxTemp.PlaceOffset.X := LColWidth * 8;
      MaxTemp.InternalWidth := LColWidth * 4;
    end;
  finally
    EndUpdate;
  end;
end;
```

It is time to see this app running! Install the package (right-click on the project inside the **Project Manager** and click on **Install**). Just after the IDE confirms the correct installation, add the path of the file into the **Delphi Library**
Path: DelphiCookbookListViewAppearanceU.pas. From now on, for every TListView that you drop in a form, in the ItemAppearance.ItemAppearance property, there will be a new value selectable named DelphiCookbookWeatherAppearance. That is the appearance class registered by the package.

Now, select the `WeatherForecastCustomAppearance.dproj` project, select a suitable target (on a mobile, you can also see what has happened by rotating the device, while on Windows you can mimic this by resizing the window), and run the project. Here's how the app looks on an Android Phone:

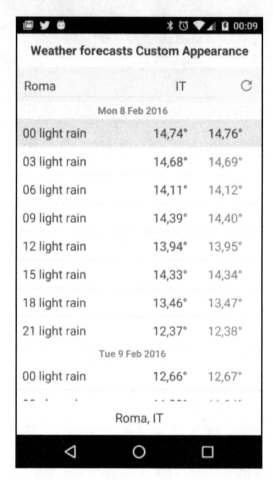

Figure 8.10: The app running on an Android phone. Note that the columns are right-aligned.

If you try to rotate the phone or resize the window, if you are doing your test on a Windows machine, you can see how the columns are correctly repositioned:

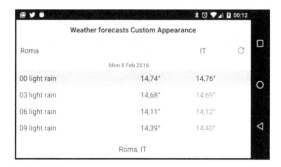

Figure 8.11: The app in landscape mode

If there isn't enough room available, the `mintemp` columns are hidden. Try to resize the application when it runs on a Windows machine. You will get a layout similar to the following:

Figure 8.12: The application running on a Windows machine. If there is not enough space when resizing the window, the mintemp column is automatically hidden.

Remember, from now on, the new appearance class is available for all the `TListView` at design time as well. You can create all the styles that you need, install them into the IDE as a package, and then use them in your application on mobiles and desktops.

The `ItemAppearance.ItemAppearance` property will show all the available styles, as shown in the following screenshot:

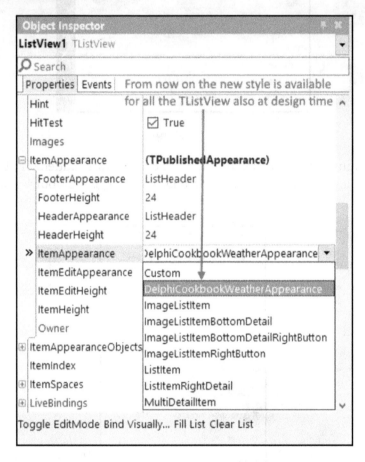

Figure 8.13: The object inspector is aware about our new style and lists it among the others

Moreover, each element defined in the appearance class as a published property is available in the object inspector as a single editable property under the `ItemAppearanceObjects.ItemObjects` property as well:

Figure 8.14: The single style objects are also available in the object inspector as singular properties

There's more...

Appearance class development can be very difficult. The style designed in this recipe is a read-only style and provides only the `ItemAppearance` variation. However, you can also create:

- `HeaderAppearance`
- `FooterAppearance`
- `ItemEditAppearance`

The good news is that the development process is the same, so after this recipe, you will be ready to create completely customized list views.

Taking a photo and location and sending it to a server continuously

In this recipe, we will talk about many things. We will see how to continuously get an image from the camera, how to get location information, and how to send binary data to a web server. Then, moving on to the server side, we will see how to read binary data from the client and how to generate content on the fly. All of these things will be used to implement a simple monitoring system.

Getting ready

This recipe is divided into client and server sides. The client side is a mobile app acting as a *special* camera that's able to get image and location data and then send it to a remote server. There is also a live preview on the main form so that you can see what you are sending to the server. The server simply gets the information and stores it in the file system. This recipe is quite complex, so I avoided an actual SQL (or NoSQL) database to store all the information and used the file system.

How to do it...

Launch two instances of Delphi and open one project in each of them (this will help in the debug phase). The server project is `MonitorServer.dproj`, while the client app is `MonitorMobile.dproj`.

Let's start with the client side.

The client side

On the main form, there are the **TCameraComponent** and **TLocationSensor** components; the **TButton** control on the top of the form is used to activate them. As soon as the camera has enough data to create a frame, the **TCameraComponent** calls its `SampleBufferReady` event handler, and now the process begins. Here's the code in the `SampleBufferReady` event handler:

```
procedure TMainForm.CameraComponent1SampleBufferReady(Sender: TObject;
  const ATime: TMediaTime);
var
  lFrame: TFrameInfo;
  lBitmapToSend: TBitmap;
```

```
begin
  CameraComponent1.SampleBufferToBitmap(FSnapshot, True);
  Image1.Bitmap.Assign(FSnapshot);

  if SecondsBetween(now, FLastSent) >= 4 then
  begin
    lBitmapToSend := GetResizedBitmap(FSnapshot);
    try
      lFrame := TFrameInfo.Create;
      lFrame.TimeStamp := now;
      lFrame.Lat := CurrLocation.Latitude;
      lFrame.Lon := CurrLocation.Longitude;
      lBitmapToSend.SaveToStream(lFrame.Stream);
      lFrame.Stream.Position := 0;
      FSenderThread.ImagesQueue.PushItem(lFrame);
    finally
      lBitmapToSend.Free;
    end;
    FLastSent := now;
  end;
end;
```

Information retrieved by the camera is converted into an actual bitmap using the handy
`SampleBufferToBitmap` method provided by **TCameraComponent** itself. Now, we have
an image. Where does the location information come from? The `TLocationSensor`
component has the `OnLocationChanged` event that is called whenever the actual location,
considering the different ways to get the location (such as GPS, Wi-Fi, and GPS combined
with Wi-Fi), actually changes. In the `LocationSensor1LocationChanged` procedure, we
save the new location in a form field, as shown in the following code:

```
procedure TMainForm.LocationSensor1LocationChanged(Sender: TObject;
  const OldLocation, NewLocation: TLocationCoord2D);
begin
  CurrLocation := NewLocation;
end;
```

Now, go back to the `CameraComponent1SampleBufferReady` event handler. The
information is used to fill an instance of `TFrameInfo` and then this instance is pushed to
the `TThreadedQueue<TFrameInfo>` thread property. The main thread pushes the
`TFrameInfo` instances into the queue, while the background thread reads the `TFrameInfo`
instances, creates a proper HTTP request, and then sends it to the server. The `TFrameInfo`
type contains all the information that's required by the server:

```
type
  TFrameInfo = class
  private
```

```
    { . . . some private declarations . . . }
  public
    constructor Create;
    property Stream: TStream read FStream;
    property Lat: Double read FLat write SetLat;
    property Lon: Double read FLon write SetLon;
    property TimeStamp: TDateTime read FTimeStamp write SetTimeStamp;
  end;
```

The complex stuff actually runs on the background thread. Let's see its `Execute` method:

```
procedure TImageSenderThread.Execute;
var
  lHTTPClient: THTTPClient;
  lFrameInfo: TFrameInfo;
  lEncodedParams: string;
begin
  inherited;
  FFilesToDelete := TList<String>.Create;
  lHTTPClient := THTTPClient.Create;
  lHTTPClient.ConnectionTimeout := 2000;
  lHTTPClient.ResponseTimeout := 1000;
  while not Terminated do
  begin
    try
      if FImagesQueue.PopItem(lFrameInfo) <> wrTimeout then
      begin
        //prepare the request parameters
        lEncodedParams := Format('ts=%s&lat=%s&lon=%s',
          [FormatDateTime('YYYY-MM-DD HH-NN-SS', lFrameInfo.TimeStamp),
          FormatFloat('##0.00000000', lFrameInfo.Lat, FFormatSettings),
          FormatFloat('##0.00000000', lFrameInfo.Lon, FFormatSettings)]);

        TNetEncoding.URL.EncodeQuery(lEncodedParams);
        //actually send the http request
        lHTTPClient.ContentType := 'image/png';
        lHTTPClient.Post(MONITORSERVERURL + '/photo?' +
          lEncodedParams, lFrameInfo.Stream);
      end;
    except
        //the best way to handle this exception, and keep this
        //code simple, is to send the next frame.
        //The same approach of the video
        //streaming protocols: "in case of error, send the next
        //frame" so, do nothing
    end;
  end;
end;
```

In the usual thread loop, we try to read the next TFrameInfo instance from the queue. If such an instance is present, we create an HTTP request using a simple HTTP POST method with the image file in the request body and the other information in the querystring parameter. In order to avoid unnecessary I/O operations, the file is not saved on the file system, but the stream itself is sent. In this code, you might note some suppressed exceptions (try..except with an empty except block). Usually, this is not good. However, in this case, if we lose a frame for some reason, the best way to fix the problem is to send the next one. So, in some places, the exceptions are suppressed because the next frame will solve the problem. Moreover, the threaded queue had a size of only two elements. If the main thread tried to append a third FrameInfo object in the queue, it is stopped for 10 milliseconds; if it still cannot append the data, that data is lost. This is one of the approaches available when you are dealing with queues: if the queue is full, new data is discarded until the queue consumes its current content. To save space and battery energy, the image is resized before being sent. The actual resizing is done by the GetResizedBitmap method just before the image taken by the camera is assigned to the stream property of the TFrameInfo instance:

```
function TMainForm.GetResizedBitmap(const Value: TBitmap;
  const MaxSize: UInt16 = 640): TBitmap;
var
  lProp: Extended;
  lLongerSide: Double;
begin
  Result := TBitmap.Create;
  Result.Assign(Value);
  lLongerSide := Max(Value.Width, Value.Height);
  if lLongerSide > MaxSize then
  begin
    lProp := MaxSize / lLongerSide;
    Result.Resize(Trunc(Value.Width * lProp), Trunc(Value.Height * lProp));
  end;
end;
```

The server side

The server side is a WebBroker project with only two actions configured, as shown in the following table:

Action name	PathInfo	HTTP method
DefaultHandler	/	mtGet
waPhoto	/photo	mtPost

The `waPhoto` action receives the client's request, reads the data, and saves them on the file system. This action saves two files:

- The actual image file as a `.png` image file
- Another file containing all the location information in JSON format

Here's the code for the `waPhoto` action:

```
procedure TwmMain.wmMainwaPhotoAction(Sender: TObject; Request:
TWebRequest;
Response: TWebResponse; var Handled: Boolean);
var
  lFileStream: TFileStream;
  lByteStream: TBytesStream;
  lFileName: string;
  lLat: Double;
  lLon: Double;
  lInfoObject: TJSONObject;

  procedure SaveInfoFile;
  begin
    lInfoObject := TJSONObject.Create;
    if TryStrToFloat(Request.QueryFields.Values['lat'], lLat,
FFormatSettings)
    then
       lInfoObject.AddPair('lat', TJSONNumber.Create(lLat));
    if TryStrToFloat(Request.QueryFields.Values['lon'], lLon,
FFormatSettings)
    then
       lInfoObject.AddPair('lon', TJSONNumber.Create(lLon));
    TFile.WriteAllText('images' + PathDelim + lFileName + '.info',
       lInfoObject.ToString);
  end;

  function QueryFieldsValidation: Boolean;
  begin
    Result := true;
    Result := Result and (not Request.QueryFields.Values['ts'].IsEmpty);
    Result := Result and (not Request.QueryFields.Values['lat'].IsEmpty);
    Result := Result and (not Request.QueryFields.Values['lon'].IsEmpty);
  end;

begin
  if not QueryFieldsValidation then
  begin
    Response.StatusCode := 400;
    Response.Content := 'Invalid query fields';
```

```
    Exit;
  end;
  if not SameText(Request.ContentType, 'image/png') then
  begin
    Response.StatusCode := 400;
    Response.Content := 'Invalid content type';
    Exit;
  end;

  TDirectory.CreateDirectory('images');
  lFileName := Request.QueryFields.Values['ts'] + '.png';
  lFileStream := TFileStream.Create('images' + PathDelim + lFileName,
fmCreate);
  try
    lByteStream := TBytesStream.Create(Request.RawContent);
    try
      lFileStream.CopyFrom(lByteStream, 0);
    finally
      lByteStream.Free;
    end;
  finally
    lFileStream.Free;
  end;
  SaveInfoFile;
  Response.StatusCode := 200;
  DeleteFiles;
end;
```

Now, data is saved in a couple of files with names similar to the following:

- `2016-04-27 23-14-53.png`: This is a plain `.png` image file
- `2016-04-27 23-14-53.png.info`: This is a JSON text file containing location information related to the previous file

Now, the `DefaultHandler` action is used to generate some HTML to let the remote user see the image and location information. Here's the code for this action:

```
procedure TwmMain.WebModule1DefaultHandlerAction(Sender: TObject;
Request: TWebRequest; Response: TWebResponse; var Handled: Boolean);
var
  lHTMLOut: TStringBuilder;
  lFileName, lJSONInfoString: string;
  lStart, lFileTimeStamp: TDateTime;
  lTimes: Integer;
  lJSONInfo: TJSONObject;
  lLat, lLon: Double;
const
```

```
      MONITORED_MINUTES = 5;
begin
  lHTMLOut := TStringBuilder.Create;
  try
    lHTMLOut.AppendLine('<!doctype html><html><head>');
    lHTMLOut.AppendLine('<style>');
    lHTMLOut.AppendLine
      (' body {font-family: Verdana; padding: 40px 10px 0px 50px; }');
    lHTMLOut.AppendLine(' pre {font-size: 200%;}');
    lHTMLOut.AppendLine('</style>');
    lHTMLOut.AppendLine('<meta http-equiv = "refresh" Content = "4">');
    lHTMLOut.AppendLine('</head><body>');
    lHTMLOut.AppendLine('<h1>Delphi Cookbook Mobile Monitor</h1>');
    lStart := Now;
    lTimes := 0;
    while true do
    begin
      lTimes := lTimes + 1;
      lFileTimeStamp := lStart - OneSecond * lTimes;
      lFileName := 'images' + PathDelim + FormatDateTime(DATEFORMAT,
        lFileTimeStamp) + '.png';
      if TFile.Exists(lFileName) then
      begin
        lHTMLOut.AppendFormat('<h3>Last update %s</h3>',
          [DateTimeToStr(lFileTimeStamp)]);
        lHTMLOut.AppendFormat('<img src="%s"><br>', [lFileName]);
        if TFile.Exists(lFileName + '.info') then
        begin
          try
            lJSONInfoString := TFile.ReadAllText(lFileName + '.info');
            lJSONInfo := TJSONObject.ParseJSONValue(lJSONInfoString)
              as TJSONObject;
            if Assigned(lJSONInfo) then
            begin
              lLat := (lJSONInfo.GetValue('lat') as TJSONNumber).AsDouble;
              lLon := (lJSONInfo.GetValue('lon') as TJSONNumber).AsDouble;
              lHTMLOut.AppendFormat('<pre>Lat: %3.8f Lon: %3.8f</pre>',
                [lLat, lLon]);
            end
            else
              lHTMLOut.Append('<pre>Invalid metadata information');
          except
            on E: Exception do
            begin
              lHTMLOut.AppendFormat('<pre>Invalid metadata information:
%s',
                [E.Message]);
            end;
```

```
          end;
        end
        else
        begin
          lHTMLOut.Append('<pre>No others info available</pre>');
        end;
        Break;
      end
      else if lTimes >= 60 * MONITORED_MINUTES then
      begin
        lHTMLOut.AppendFormat
          ('<h2>No image availables in the last %d minutes</h2>',
          [MONITORED_MINUTES]);
        Break;
      end;
    end;
    lHTMLOut.AppendLine('</body></html>');
    Response.Content := lHTMLOut.ToString;
  finally
    lHTMLOut.Free;
  end;
end;
```

This method creates some HTML on the fly and looks for the most recent snapshot saved on the server. When it finds an image, it inserts the image's file name into the HTML to let the browser request it. Then, it opens the `.info` JSON file, reads the location information, and inserts it into the HTML as well. Note that this monitoring app doesn't have a proper synchronization mechanism between file writing and file reading, so in many parts of the code, you see an empty `try except` block. For this recipe, it is enough. However, in more critical systems, a proper mechanism (such as critical sections, monitors, or mutex) is required to synchronize file access and avoid empty frames, especially with multiple clients.

To update the image displayed on the HTML page, there is a special meta tag in the HTML document header, as follows:

```
<meta http-equiv = "refresh" Content = "4">
```

With this line, the page is updated every four seconds (more information about the `http-equiv` meta tags can be found at `http://www.w3schools.com/Tags/att_meta_http_equiv.asp`).

To try the application, launch the server and navigate to the `http://localhost:8080` URL on your browser.

You should see a page like the following:

Figure 8.15: The monitoring system page when it hasn't found images for the last 5 minutes

Now, in the mobile project, open the `ImageSenderThreadU.pas` unit and locate `const` `MONITORSERVERURL`. Change the `const` value to point to your machine's IP. Note that the phone (or tablet) and your PC must be on the same Wi-Fi network. In my case, the constant is configured as follows:

```
const
    MONITORSERVERURL = 'http://192.168.1.100:8080';
```

Replace the IP with yours, and leave the protocol (`http`) and the port (`8080`) as is.

In a real-world app, put a small configuration section in the mobile app to let the user enter the actual URL where the server listens.

Run the mobile app, activate the camera using the button in the upper-right corner, and after a couple of seconds, you should see an image and the location information in the web page. The final web page should look like the following:

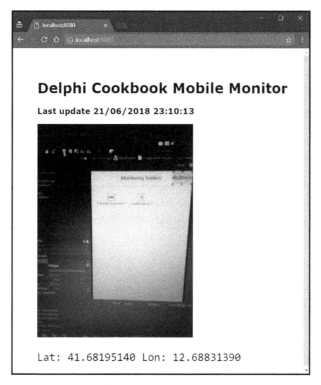

Figure 8.16: The monitoring system running while showing a sort of recursive image of itself

There's more...

This recipe is like training for a lot of concepts. If you want to go deeper into them, you can read the following articles and information:

- *Using Location Sensors* at:
  http://docwiki.embarcadero.com/RADStudio/en/Mobile_Tutorial:_Using_Loc
  ation_Sensors_(iOS_and_Android)
- *Uses Permissions* at: http://docwiki.embarcadero.com/RADStudio/en/Uses_
  Permissions
- *FMX.Media.TCameraComponent* at:
  http://docwiki.embarcadero.com/Libraries/en/FMX.Media.TCameraComponent

Talking with the backend

This recipe will introduce you to real-world business mobile apps and their related application servers. It is not a simple world. It is full of well-known and specific traps related to your infrastructure, your business logic, your application transactions, and so on. Just to be clear, you have to take care of your design and the way you implement it to a greater extent compared to a classic client/server application. Upon going deeper into mobile programming (and in general, in all asynchronous scenarios), you will see that things become harder than usual. In the mobile world, things can get messy really fast and your customers will complain even faster. Be warned!

This recipe is a mobile client for the **People Manager** application server developed in the *Implementing a RESTful interface using WebBroker* recipe in `Chapter 6`, *Putting Delphi on the Server*.

Getting ready

As already mentioned, this recipe is composed by the application server and the mobile client. The UI is not blocking so that all the REST requests are executed in a background thread using the built-in features of **TRESTClient**.

How to do it...

The app is based on the **Header/Footer with Navigation** mobile template. In the first `TTabItem` object, there is a list of people. In the second `TTabItem` object, there is the selected person's details. Data is read from the `REST` services that are exposed by the `PeopleManager.dproj` server.

The client implements a simple CRUD operation and uses a subset of the server's services. The service used and the relative URL are mentioned in the following table (you can implement search functionality as an exercise):

HTTP verb	URL	Description
GET	/people	This returns a JSON array containing one JSON object for each record present in the PEOPLE table. In each object, the property name is the name of the field, while the property values are the value of the fields.
POST	/people	This creates a new person in the PEOPLE table. This requires a request body containing the new person's data to create a JSON object. The content-type request must be application/json.
PUT	/people/{id}	This updates the person with id with the data passed in the request body. This requires a request body containing the person to update as a JSON object. The content-type request must be application/json.
DELETE	/people/{id}	This deletes the person with id.

The GET people/:id method is available from the server too, but the client doesn't use it because the GET /people method already returns an array with all the complete entities. In a real-world app, you may have lots of entities, or a lot of entity attributes or nested objects, so it makes sense to use the GET verb to get the full single entity representation.

Locally, the data is stored in a TDataSet component, a TFDMemTable component (yes, I love it) to be precise, and is loaded using the class helper declared in MVCFramework.DataSet.Utils.pas (contained in the DelphiMVCFramework project, and has also already been used in Chapter 6, *Putting Delphi on the Server*).

All the logic is implemented in a data module created before the main form is created (go to **Project** | **Options** | **Forms** to check the form creation order). Methods provided by the data module to the main form are as follows:

```
public
  procedure SavePerson(AOnSuccess: TProc;
            AOnError: TProc<Integer, String> = nil);
  procedure DeletePerson(AOnSuccess: TProc;
            AOnError: TProc<Integer, String> = nil);
  procedure LoadAll(AOnSuccess: TProc;
            AOnError: TProc<Integer, String> = nil);
  function CanSave: Boolean;
```

CanSave is used to enable or disable UI actions depending on the dsPeople dataset state. The LoadAll method is called from the FormShow event handler, and it requests data for the server and populates the in-memory dataset. Seeing that all remote requests are asynchronous, we need a callback to update the UI after the request is finished in order to show data in the case of success, or to show error messages in the case of errors. Here's the code for the data module's LoadAll method:

```
procedure TdmMain.LoadAll(AOnSuccess: TProc; AOnError: TProc<Integer,
String>);
begin
  if dsPeople.State in [dsInsert, dsEdit] then
    dsPeople.Cancel;    //cancel all unposted data
  dsPeople.Close;

  RESTRequest.ClearBody;
  RESTRequest.Resource := 'people';
  RESTRequest.Method := TRESTRequestMethod.rmGET;
  //execute remote request asynchronously
  //WARNING! The anonymous method passed as parameter to the
  //ExecuteAsynch is execute within the main thread, so there is
  //no need to synchronize UI access
  RESTRequest.ExecuteAsync(
    procedure
```

```
      begin
        if RESTRequest.Response.StatusCode = 200 then
        begin
          //load response jsonarray in the dataset
          dsPeople.Active := True;
          dsPeople.AppendFromJSONArrayString(
            RESTRequest.Response.JSONValue.ToString);
          dsPeople.First;
          if Assigned(AOnSuccess) then
          //call the 'success' user callback
          AOnSuccess();
        end
        else
        begin
          if Assigned(AOnError) then
              //call the 'error' user callback
              AOnError(RESTRequest.Response.StatusCode,
RESTRequest.Response.StatusText);
        end;
      end);
  end;
```

This method is declared in the data module. How do we call this method in the `acRefreshData` action within the main form? Here's the code:

```
  procedure TMainForm.acRefreshDataExecute(Sender: TObject);
  begin
    DoStartWait('Please wait while retrieving the people list');
    dmMain.LoadAll(
      procedure
      begin
        DoEndWait;
      end,
      procedure(StatusCode: Integer; StatusText: String)
      begin
        DoEndWait;
        ShowError(Format('Error [%d]: %s', [StatusCode, StatusText]));
      end);
  end;
```

Remember, a call to the `LoadAll` method is not blocking for the main thread. So, any code after a call to `LoadAll` is executed as soon as possible (as the OS decides) and not after the data is retrieved. This is the reason why we need the callbacks. The first anonymous method is our `success` callback, and it is executed when data is already in the dataset and the user can see it in the list view. The second anonymous method is our `error` callback, and it is executed if errors occur in the call. The other remote calls work in the same manner.

If you, for some reason, would like to use a different HTTP component to perform the REST HTTP calls, and this library doesn't support asynchronous client requests, you can always rely on the good old anonymous thread (or PPL since Delphi XE7). The following code is included in the `LoadAll` method, but it is commented to show an alternative way to perform a remote call without using the `ExecuteAsynch` method:

```
TThread.CreateAnonymousThread(
  procedure
  begin
    try
      //synch call, but executed in an anonymous thread
      RESTRequest.Execute;
      TThread.Synchronize(nil,
        procedure
        begin
          if RESTRequest.Response.StatusCode = 200 then
          begin
            dsPeople.Active := True;
            dsPeople.AppendFromJSONArrayString(
              RESTRequest.Response.JSONValue.ToString);
            if Assigned(AOnSuccess) then
              AOnSuccess();
          end
          else
            AOnError(RESTRequest.Response.StatusCode,
                     RESTRequest.Response.StatusText);
        end);
    except
      on E: Exception do
      begin
        if Assigned(AOnError) then
        begin
          ErrMsg := E.Message;
          TThread.Synchronize(nil,
            procedure
            begin
              //Passing 'Zero' to the callback means that some
              //non-protocol related exception has been raised
              AOnError(0, ErrMsg);
            end);
        end
      end;
    end;
  end).Start;
```

An important feature of well-designed mobile apps is the feedback to the user. Your user *must* know what your application is doing after their input; otherwise, he/she would probably stop it. Therefore, we need to show a **Please wait** screen. To do so, this app uses a TPopup component. This component has a property called IsOpen that is used to show it or hide it. Just before each request, we set an instance form variable to true and after the request, when the response is visible somewhere in the UI, we set that variable to false. Here's the code to handle the **Please wait** screen:

```
procedure TMainForm.DoEndWait;
begin
  BackgroundOperationRunning := False;
end;

procedure TMainForm.DoStartWait(AWaitMessage: String);
begin
  //this label is placed inside the "Please wait" screen
  lblMessage.Text := AWaitMessage;
  BackgroundOperationRunning := True;
end;
```

How do we actually show the TPopup component? The setter of the BackgroundOperationRunning is a good place to do so. Here's the code:

```
procedure TMainForm.SetBackgroundOperationRunning(const Value: Boolean);
begin
  FBackgroundOperationRunning := Value;
  acRefreshData.Enabled := not FBackgroundOperationRunning;
  AniIndicator1.Visible := FBackgroundOperationRunning;
  TabItem1.Enabled := not FBackgroundOperationRunning;
  ppMessage.IsOpen := FBackgroundOperationRunning;
end;
```

Data is linked to the UI using the LiveBindings engine. Here's the LiveBindings designer showing the links:

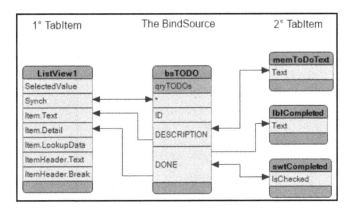

Figure 8.17: The LiveBindings designer showing the links between the dsPeople and the UI

Before lunching the app you have to replace `SERVERBASEURL` constant with your server ip address:

```
const
  SERVERBASEURL = 'http://192.168.1.242:8080';
```

After launching the app, you will get the following wait screen:

Figure 8.18: The wait screen

Then, when the data is retrieved, parsed, and loaded, you will get the following screen:

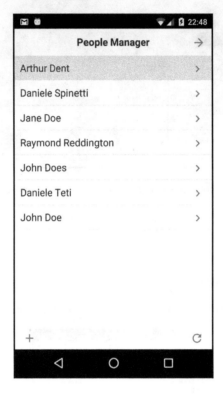

Figure 8.19: The list of people loaded in the listview

If you tap an item, you will get the editing screen:

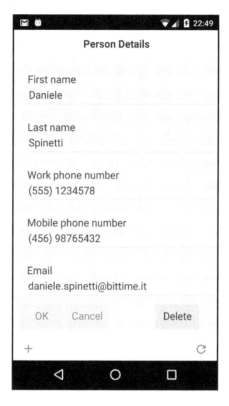

Figure 8.20: The editing screen showing the person's information

There's more...

A lot of topics have been covered in this recipe! Mobile apps can be really complex, as this simple example demonstrates. However, using the LiveBindings engine, the local storage offered by SQLite and IBLite, and the nice Delphi components to load data in-memory, you can create mobile apps easily enough. Here are some other demos about the technologies involved in developing these types of apps:

- *FireDAC IBLite with Delphi XE6*
  at `https://www.youtube.com/watch?v=jbRJCqNgNDc`
- *Delphi XE5 Mobile REST Client Demo*
  at `https://www.youtube.com/watch?v=OkRVbgF4VMI`

- *Delphi XE5 Mobile REST Client Demo Source*
  at `http://delphi.org/2013/09/delphi-xe5-mobile-rest-client-demo-source/`
- *The New REST Client Library, A Tool of Many Trades* at:
  `https://www.youtube.com/watch?v=nPXYLK4JZvM`

Making a phone call from your app

Many mobile devices, especially in the consumer market, are phones or devices that can make phone calls. In some cases, your mobile app may have the capability of making a call or just monitor incoming or outgoing calls.

Getting ready

In this recipe, we'll see how to make a call and how to monitor current calls as well. Also, in this case, the useful FireMonkey platform services framework comes in handy.

How to do it...

1. Create a new mobile app by navigating to **File** | **New** | **Multi-Device Application Delphi**.
2. Select the **Header/Footer** template and click on **OK**.
3. Drop the following components on the main form:
   - **TEdit** (`edtPhoneNumber`)
   - **TButton** (`btnCall`)
   - **TListBox** (`lbCalls`)
   - **TListBox** (`lbInfo`)
4. Arrange the components as shown in the following screenshot:

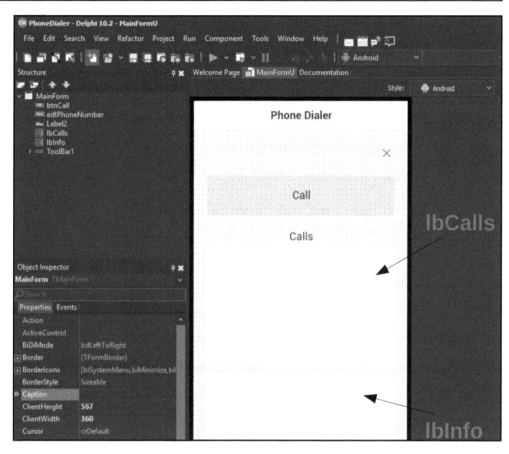

Figure 8.21: The form with all the controls arranged

5. Put in some labels to explain what the list boxes will contain, as shown in the preceding screenshot.

6. Now, create the `FormCreate` event handler and fill it in with this code:

```
procedure TMainForm.FormCreate(Sender: TObject);
begin
  lbInfo.Clear;
  if
TPlatformServices.Current.SupportsPlatformService(IFMXPhoneDialerSe
rvice,
    IInterface(FPhoneDialerService)) then
  begin
    FPhoneDialerService.OnCallStateChanged := CallStateChanged;
    lbInfo.ItemHeight := lbInfo.ClientHeight / 4;
    lbInfo.Items.Add('Carrier Name: ' +
```

```
        FPhoneDialerService.GetCarrier.
            GetCarrierName);
        lbInfo.Items.Add('ISO Country Code: ' +
            FPhoneDialerService.GetCarrier.GetIsoCountryCode);
        lbInfo.Items.Add('Network Code: ' +
    FPhoneDialerService.GetCarrier.
            GetMobileCountryCode);
        lbInfo.Items.Add('Mobile Network: ' +
    FPhoneDialerService.GetCarrier.
            GetMobileNetwork);
        btnCall.Enabled := True;
      end
      else
        lbInfo.Items.Add('No Phone Dialer Service');
    end;
```

7. In the form's `private` section, declare the following methods:

```
    private
        FPhoneDialerService: IFMXPhoneDialerService;
        procedure CallStateChanged(const ACallID: string; const AState:
    TCallState);
        function CallStateAsString(AState: TCallState): String;
```

8. Press *Ctrl + Shift + C* and fill in the methods we've just created with the following code:

```
    function TMainForm.CallStateAsString(AState: TCallState): String;
    begin
      case AState of
        TCallState.None:
          Result := 'None';
        TCallState.Connected:
          Result := 'Connected';
        TCallState.Incoming:
          Result := 'Incoming';
        TCallState.Dialing:
          Result := 'Dialing';
        TCallState.Disconnected:
          Result := 'Disconnected';
        else
          Result := '<unknown>';
      end;
    end;

    procedure TMainForm.CallStateChanged(const ACallID: string;
      const AState: TCallState);
    begin
```

```
lbCalls.Items.Add(Format('%-16s %s', [ACallID,
CallStateAsString(AState)]));
end;
```

9. Now, create the **OnClick** event for the `btnCall` method and fill it with this code:

```
procedure TMainForm.btnCallClick(Sender: TObject);
begin
  if not edtPhoneNumber.Text.IsEmpty then
    FPhoneDialerService.Call(edtPhoneNumber.Text)
  else
  begin
    ShowMessage('No number to call, please type a phone number.');
    edtPhoneNumber.SetFocus;
  end;
end;
```

10. Run the app on your phone. Note the `lbInfo` method showing all the information about your mobile network. Write a phone number in the editing area and click the **Call** button. Note what happens to the `lbCalls` method during the outgoing calls and during the incoming calls. This activity is shown in the following screenshot:

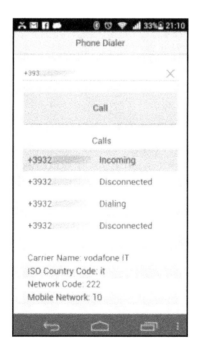

Figure 8.22: The Phone Dialer app running on a phone, after some in/out calls; note the events in the first list

How it works...

This recipe is very simple. All the work is done at the beginning when the `FormCreate` event handler asks the system whether it supports the `IFMXPhoneDialerService` interface. This interface has the following methods:

```
{ Interface of Phone Dialer }
IFMXPhoneDialerService = interface (IInterface)
['{61EE0E7A-7643-4966-873E-384CF798E694}']
// Make a call by specified number
function Call(const APhoneNumber: string): Boolean;
// Get current carrier
// The object the Carrier is deleted automatically with deleting
service
function GetCarrier: TCarrier;
// Get all curtrent calls. If the current calls aren't present, the
array will be empty
// The developer shall delete array cells after use
function GetCurrentCalls: TCalls;
// Getter, Setter and property for work with event of tracing of state
change of a call
function GetOnCallStateChanged: TOnCallStateChanged;
procedure SetOnCallStateChanged(const AEvent: TOnCallStateChanged);
property OnCallStateChanged: TOnCallStateChanged read
GetOnCallStateChanged write SetOnCallStateChanged;
end;
```

There's more...

Using the monitor functionality, you can implement a system to track the phone calls' duration and type (incoming or outgoing). Using this service, you can implement a list of contacts that are centralized on a server and allow your user to call those contacts without having the contact in the phone's address book. Another utilization is to monitor the allowed incoming calls, and if a special blocked number calls, you can send a notification to a remote server. There are endless possibilities – explore them yourself.

Tracking the application's life cycle

In the "safe" Microsoft Windows desktop application development land, our application has a life cycle, but it is not crucial to take care of it. Usually, you have a set of events to handle such as `FormCreate`, `FormClose` (at the form level), or `Application OnRestore` or `Application OnTerminate`. In some cases, you have to handle the state where the main application window is minimized, and this is still simple. In the mobile world, as usual, things are a bit more complex. The concept of a life cycle is evidence of this. Just to make things messier, the Android activity's life cycle is different from the iOS view life cycle. Remember, when an app is in the background, it can be completely destroyed.

Getting ready

But, hey! Why I should care about the life cycle? That's a very good point! There are a lot of things that you should, or must, do while your application is switching from one state to another.

Here are some examples:

- Handle the current input control's state. You can save or discard data, but you cannot send the **Do you want to save?** message to the user. If a user touches the **Home** button, you cannot stop them. For this specific situation, it could be useful to have the `OnSaveState` form event too, which is an abstraction of what we are talking about.
- Stop or restart CPU intensive work related to a calculation.
- Look for some previously saved data on the file system.
- Search Bluetooth devices or app tethering-enabled applications.
- Update a remote resource more frequently than when the app was running in the background. In the background, you may check a particular HTTP resource once an hour, while if the app is in the foreground, you can decide whether to check that resource once a minute.
- Append a system notification to remind the user of something just before terminating an app.
- Stop the audio output (if applicable for your app).
- Stop the GPS monitoring (if applicable for your app).
- Go into power saving mode, whatever it means for your app, and many more.

As you can see, the application's life cycle is very important. Let's see how we can hook up to it.

How to do it...

This recipe is not a standard recipe. We will not build a complete app, but a reference app. You can launch this app every time you want to know which event (app event or form event) is fired and when.

As the first thing to do, in the `FormCreate` event, we have to hook up to the FireMonkey messaging system and subscribe to the `TApplicationEventMessage` message type:

```
procedure TMainForm.FormCreate(Sender: TObject);
begin
  TPlatformServices.Current.SupportsPlatformService(IFMXLoggingService,
    IInterface(FLoggingService));

  FSubscrID := TMessageManager.DefaultManager.SubscribeToMessage
    (TApplicationEventMessage,
    procedure(const Sender: TObject; const Msg: TMessage)
    var
      AppEvent: TApplicationEventMessage;
    begin
      AppEvent := TApplicationEventMessage(Msg);
      case AppEvent.Value.Event of
        TApplicationEvent.FinishedLaunching:
          LogEvent('App Finished Launching');
        TApplicationEvent.BecameActive:
          LogEvent('App Became Active');
        TApplicationEvent.WillBecomeInactive:
          LogEvent('App Will Become Inactive');
        TApplicationEvent.EnteredBackground:
          LogEvent('App Entered Background');
        TApplicationEvent.WillBecomeForeground:
          LogEvent('App Will Become Foreground');
        TApplicationEvent.WillTerminate:
          LogEvent('App Will Terminate');
        TApplicationEvent.LowMemory:
          LogEvent('App Low Memory');
        TApplicationEvent.TimeChange:
          LogEvent('App Time Change');
        TApplicationEvent.OpenURL:
          LogEvent('App Open URL');
      end;
    end);

  LogEvent('Event FormCreate');
end;
```

With this code, every time the system raises a message regarding our app, we'll be informed. The `System.Messaging.pas` unit, added in the `implementation uses` clause, contains the classes needed to access the system's messaging system.

How does this messaging system work? Once you have an instance of `TMessageManager`, you can subscribe message handling methods to specific types of messages. Message handling methods may be methods of an object or anonymous methods. In our case, we have used an anonymous method. This messaging mechanism can also be used in your app or component as an independent messaging system. However, FireMonkey also uses it to send system messages using the default messaging manager instance. In this recipe, we're using it to subscribe to the system messages.

An instance of `TApplicationEvent`, the type on which we're doing the big case statement, represents the application-related messages and may have any of the following values:

Event	Description
`BecameActive`	This indicates that an application has gained focus.
`EnteredBackground`	This indicates that the application is running in the background because the user is no longer using it.
`FinishedLaunching`	This indicates that the application has been launched.
`LowMemory`	This event is a warning for the application that the device is running out of memory. In this case, the application should reduce memory usage, freeing structures and data that are not fundamental or that can be reloaded as per requirements at a later point.
`OpenURL`	This indicates that the application has received a request to open a URL (only for iOS).
`TimeChange`	This indicates that there has been a significant change in time (only for iOS). This event might happen, for example, when the day changes or when the device changes to or from daylight saving time.
`WillBecomeForeground`	This indicates that the user is now using the application, which was previously running in the background.
`WillBecomeInactive`	This indicates that the application is going to lose focus and become inactive.
`WillTerminate`	This indicates that the user or the operating system is quitting the application.

Remember to be a good FireMonkey citizen, when you subscribe to a system notification, you have to unsubscribe to it too. We do this in the `FormDestroy` event just after logging, at the end:

```
procedure TMainForm.FormDestroy(Sender: TObject);
begin
  LogEvent('Event FormDestroy');
  TMessageManager.DefaultManager.Unsubscribe(TApplicationEventMessage,
FSubscrID, True);
end;
```

The `LogEvent` method appends the message text to the list box and writes the same message to the system log as well using the reference to `IFMXLoggingService` retrieved in the `FormCreate` event handler. Moreover, whereas the form events could be many, there is a checkbox to exclude them from the logging. Here's the code for the `LogEvent` method:

```
procedure TMainForm.LogEvent(Msg: string);
begin
  if (not CheckBox1.IsChecked) and Msg.StartsWith('event', True) then
    Exit;
  Memo1.Lines.Add(Format('%s: %s', [TimeToStr(Now), Msg]));
  Memo1.GoToTextEnd; //memo goes to the last line
  if Assigned(FLoggingService) then
    FLoggingService.Log('LifeCycle: %s', [Msg]); //syslog
end;
```

This is the infrastructure code, but what events are we waiting for? In the main form, there are some test buttons that raise specific system and form events. Here's the app while it is logging form events and system messages:

Figure 8.23: The app while it is logging form events and system messages

In the main form, every interesting event that could be raised, whether the app state changes or not, is filled with code similar to the following one:

```
procedure TMainForm.FormActivate(Sender: TObject);
begin
  LogEvent('Event FormActivate');
end;
```

Now, everything is traced, including app state changes and form events. Now you can connect your device, if it's not already connected, and launch the proper tool to see the device logger (launch `Monitor.bat` for Android devices or see the device console for iOS devices). Start the app and play with the buttons.

Try to tap the **Open Form** button and then close the newly opened form by tapping on the **Close** button. As you can see in the list, only form events are called (`FormDeactivate` and `FormActivate`), and this is reasonable. Now, tap on the **ShowMessage** button and see what happens. Form events are not raised, but an app message arrives. Look! The app goes into the inactive state for a `ShowMessage` call! This is a case where this sort of testing tool is very handy. If you don't know exactly when an app switches its state from one to another, you cannot rely on this state change to do anything useful and reliable. But now, you have the right tool!

There's more...

Experimenting with the life cycle, you can find interesting utilizations that will make your user happy with your app.

Another interesting thing that I suggest you study is the messaging system based on a variation of the well-known and more general **Observer** design pattern of the GoF framework. Simply speaking, this messaging system is just something that triggers an event to which anyone can listen. Different libraries offer different implementations and for different purposes, but the basic idea is to provide a framework for issuing events and subscribing to them.

More information about the `System.Messaging.pas` unit can be found in the following articles:

- *Sending and Receiving Messages Using the RTL*
  at `http://docwiki.embarcadero.com/RADStudio/en/Sending_and_Receiving_Messages_Using_the_RTL`
- *List of FireMonkey Message Types*
  at `http://docwiki.embarcadero.com/RADStudio/en/List_of_FireMonkey_Message_Types`
- System.Messaging (Delphi)
  at `http://docwiki.embarcadero.com/CodeExamples/en/System.Messaging_(Delphi)`
- Where the messaging system began: the Observer design pattern
  at `http://en.wikipedia.org/wiki/Observer_pattern`

Building your own SMS sending service with the REST API

For some types of applications, it's necessary to identify the user uniquely during the registration phase: just think about services like WhatsApp or Telegram. To do this, a text message (SMS) is usually sent to a telephone number containing a code to be validated to complete the registration process. In this recipe, we will see how to build an SMS sending service to be used, for example, for the situation we have just described.

Getting ready

In this recipe, we will show you how to build your own SMS Sending Service: a RESTFul server part will provide REST APIs to queue and dequeue SMS messages; a mobile application (built only for Android OS) will query these REST APIs and if an SMS is to be sent, it will use its SMS manager to send it. A third VCL application is created to test and push SMS messages within the REST server.

The server application uses DMVC, a Delphi open source framework based on WebBroker that allows you to create powerful RESTful web services. You can find the project code here: `https://github.com/danieleteti/delphimvcframework`. You can find information about its configuration in Delphi, in `Chapter 6`, *Putting Delphi on the Server*.

REST server

Perform the following steps:

1. Navigate to **Delphi Project | Web Broker | Web Server Application**.
2. Now, the wizard will ask you what type of web server application you want to create. This demo will be built as a console application, so select the standalone console application and click **Next**.
3. The wizard proposes a TCP port where the service will listen. Click on the **Test** port; if the test port succeeds, use it, otherwise change the port until the test passes. In this recipe, port 8080 is used.
4. Click **Finish**.
5. Save all. Name the project `SMSServerManager.dproj` and the WebModule `WebModuleU.pas`.
6. We will start from the business object classes. This web service will manage SMS, so let's create a new unit and declare the following class:

```
TSMS = class
public
  property FROM: String;
  property DESTINATION: String;
  property TEXT: String;
end;
```

7. Hit *Ctrl* + *Shift* + *C* to autocomplete the declaration; save the file as `SmsBO.pas`. Note that in projects where you have a lot of different types of classes (business objects, controllers, data modules, and so on), it can be useful to organize the units into different folders. So, I saved `SmsBO.pas` in a folder named `BusinessObjects`. Feel free to do this as well.

8. It's time to create a **TDataModule**. Add a new **TDataModule**, name it `SmsModule`, and save it into the `Modules` folder as `SmsModuleU.pas`.

9. Fill `SmsModuleU.pas` with the following code:

```
unit SmsModuleU;

interface

uses
   System.SysUtils, System.Classes, SmsBO,
System.Generics.Collections;

type
   TSmsModule = class(TDataModule)
   public
      procedure Push(ASMS: TSMS);
      function Pop: TSMS;
   end;

implementation

{$R *.dfm}

var
   SMSQueue: TThreadedQueue<TSMS>;

   { TSmsModule }

procedure TSmsModule.Push(ASMS: TSMS);
begin
   SMSQueue.PushItem(ASMS);
end;

function TSmsModule.Pop: TSMS;
begin
   Result := SMSQueue.PopItem;
end;

initialization

SMSQueue := TThreadedQueue<TSMS>.Create(1000, 1000, 1000);
```

```
finalization

SMSQueue.Free;

end.
```

10. The preceding code represents the part that actually accesses the data. In this recipe, we'll use a `TThreadedQueue` to manage the SMS queue. A `TThreadedQueue` was used compared to an access to the DB data to make the demo simple and easy to understand. Besides, `TThreadedQueue` does its job very well.

11. Now, it is time to create a DelphiMVCFramework controller. This is the class with all the code to handle the HTTP requests and responses. Here, there should not be business logic code.

12. Create a new unit, name it `SMSControllerU.pas`, and save it into the `Controllers` folder.

13. Fill `SMSControllerU.pas` with the following code:

```
unit SmsControllerU;

interface

uses
   MVCFramework, SmsModuleU, MVCFramework.Commons;

type

   [MVCPath('/sms')]
   TSmsController = class(TMVCController)
   private
      FSmsModule: TSmsModule;
   protected
      procedure OnAfterAction(Context: TWebContext; const
AActionNAme: string); override;
      procedure OnBeforeAction(Context: TWebContext; const
AActionNAme: string;
         var Handled: Boolean); override;
   public

      [MVCPath('/push')]
      [MVCHTTPMethod([httpPOST])]
      [MVCConsumes('application/json')]
      procedure PushSMS;

      [MVCPath('/pop')]
      [MVCHTTPMethod([httpPOST])]
```

```delphi
    [MVCConsumes('application/json')]
    procedure PopSMS;
  end;

implementation

uses
  SmsBO;

{ TSmsController }

procedure TSmsController.PushSMS;
var
  LSMS: TSMS;
begin
  LSMS := Context.Request.BodyAs<TSMS>;
  FSmsModule.Push(LSMS);
  Render(201, 'SMS pushed');
end;

procedure TSmsController.PopSMS;
var
  LSMS: TSMS;
begin
  LSMS := FSmsModule.Pop();
  if Assigned(LSMS) then
    Render(LSMS)
  else
    Render(404, 'Queue empty');
end;

procedure TSmsController.OnAfterAction(Context: TWebContext;
  const AActionNAme: string);
begin
  inherited;
  FSmsModule.Free;
end;

procedure TSmsController.OnBeforeAction(Context: TWebContext;
  const AActionNAme: string; var Handled: Boolean);
begin
  inherited;
  FSmsModule := TSmsModule.Create(nil);
end;

end.
```

14. We're about to finish. Go back to `WebModuleU.pas` and create the `OnCreate` event handler. Here, we have to configure the DelphiMVCFramework starting point. It is really simple; just two lines of code:

```
procedure TwmMain.WebModuleCreate(Sender: TObject);
begin
  FMVC := TMVCEngine.Create(Self);
  FMVC.AddController(TSmsController);
end;
```

15. The `FMVC` variable must be declared in the private section of the class and you have to add the `SmsControllerU` unit in the implementation uses clause.

16. Now your project should compile. If not, check the dependencies between all the units.

Mobile sending application

Perform the following steps:

1. Navigate to **File | New| Multi-Device Application**.
2. Now the wizard will ask you what type of multi-device application you want to create. Select **Header/Footer** application and click **Ok**.
3. Select the folder where you want to save it (I recommend you save it in the same path of the server but in different folders).
4. Click **Select Folder**.
5. Rename your project in `SendSMS`. The form unit in `MainFormU` and the form class variables are `TSMSSendingForm` and `SMSSendingForm`, respectively.
6. Now, it is time to create a service to perform queries to the server APIs. Create a new unit, name it `SMS.RESTAPIs.ServiceU.pas`, and save it into the `Services` folder.
7. Fill `SMS.RESTAPIs.ServiceU.pas` with the following code:

```
unit SMS.RESTAPI.ServiceU;

interface

uses
  SmsBO;

type

  ISMSRESTAPIService = interface
```

```
    ['{CA2FCE5E-C368-4230-A2A2-D1734288E9DE}']

    function PopSMS: TSMS;
    procedure PushSMS(ASMS: TSMS);
  end;

function BuildSMSRESTApiService: ISMSRESTAPIService;

implementation

uses
  MVCFramework.Serializer.JSON, System.Net.HTTPClient, REST.Types,
  MVCFramework.Serializer.Intf, System.Classes, System.SysUtils;

const
  // change this server URL with your
  BASE_URL = 'http://192.168.1.242:8080/sms';

type

  TSMSRESTAPIService = class(TInterfacedObject, ISMSRESTAPIService)
  private
    procedure CheckResponse(AResp: IHTTPResponse);
    procedure InitializeRequest(AReq: IHTTPRequest);
  public
    function PopSMS: TSMS;
    procedure PushSMS(ASMS: TSMS);
  end;

function BuildSMSRESTApiService: ISMSRESTAPIService;
begin
  Result := TSMSRESTAPIService.Create;
end;

{ TSMSRESTAPIService }

procedure TSMSRESTAPIService.InitializeRequest(AReq: IHTTPRequest);
begin
  AReq.AddHeader('Content-Type', CONTENTTYPE_APPLICATION_JSON);
end;

procedure TSMSRESTAPIService.CheckResponse(AResp: IHTTPResponse);
begin
  if AResp.StatusCode >= 400 then
    raise Exception.Create(AResp.ContentAsString);
end;

function TSMSRESTAPIService.PopSMS: TSMS;
```

```
var
  lHTTP: THTTPClient;
  lReq: IHTTPRequest;
  lResp: IHTTPResponse;
  LSerializer: IMVCSerializer;
begin
  lHTTP := THTTPClient.Create;
  try
    lReq := lHTTP.GetRequest('POST', BASE_URL + '/pop');
    InitializeRequest(lReq);
    lResp := lHTTP.Execute(lReq);
    CheckResponse(lResp);
    if lResp.ContentAsString.IsEmpty then
      Exit(nil);
    LSerializer := TMVCJSONSerializer.Create;
    Result := TSMS.Create;
    LSerializer.DeserializeObject(lResp.ContentAsString(), Result);
  finally
    lHTTP.Free;
  end;
end;


procedure TSMSRESTAPIService.PushSMS(ASMS: TSMS);
var
  lHTTP: THTTPClient;
  lReq: IHTTPRequest;
  lResp: IHTTPResponse;
  LSerializer: IMVCSerializer;
  LSerializedObj: string;
  LSS: TStringStream;
begin
  lHTTP := THTTPClient.Create;
  try
    LSerializer := TMVCJSONSerializer.Create;
    LSerializedObj := LSerializer.SerializeObject(ASMS);
    LSS := TStringStream.Create(LSerializedObj);
    try
      lReq := lHTTP.GetRequest('POST', BASE_URL + '/push');
      lReq.SourceStream := LSS;
      InitializeRequest(lReq);
      lResp := lHTTP.Execute(lReq);
      CheckResponse(lResp);
    finally
      LSS.Free;
    end;
  finally
    lHTTP.Free;
  end;
```

```
    end;

    end.
```

8. Change the `BASE_URL` constant accordingly with your server ip.

9. Now, we have to create a service that will use a **façade** to make requests to the server and send the text message to the recipient. Create a new unit, name it `SMS.ServiceU.pas`, and save it into the `Services` folder.

10. Fill `SMS.ServiceU.pas` with the following code:

```
unit SMS.ServiceU;

interface

type
  ISMSService = interface
    ['{C44D7371-6E5B-427D-A017-2163DFC6DA34}']
    procedure SendSMS(const Number: string; Text: string);
    procedure SendNextSMS;
  end;

function GetSMSService: ISMSService;

implementation

uses
  Androidapi.Helpers, Androidapi.JNI.JavaTypes,
Androidapi.JNI.Telephony,
  SMS.RESTAPI.ServiceU, System.Rtti, SmsBO, System.Threading,
System.SysUtils,
  System.UITypes;

type
  TSMSService = class(TInterfacedObject, ISMSService)
  public
    procedure SendSMS(const Number: string; Text: string);
    procedure SendNextSMS;
  end;

  { TSMSService }

procedure TSMSService.SendNextSMS;
var
  LRESTApi: ISMSRESTAPIService;
begin
  // I use a TTask to perform the request in a background thread
  // an HTTP request must be never performed on main thread
```

```
TTask.Run(
  procedure
  var
    ASMS: TSMS;
  begin
    LRESTApi := BuildSMSRESTApiService;
    ASMS := LRESTApi.PopSMS;
    if not Assigned(ASMS) then
      exit;
    try
      SendSMS(ASMS.DESTINATION, ASMS.Text);
    finally
      ASMS.Free;
    end;

  end);
end;


procedure TSMSService.SendSMS(const Number: string; Text: string);
var
  LSmsManager: JSmsManager;
  LSmsTo, LSmsFrom, LText: JString;
begin
  // SMSManager is used to send SMS
  // very easy!
  LSmsManager := TJSmsManager.JavaClass.getDefault;
  // destination number
  LSmsTo := StringToJString(Number);
  LSmsFrom := nil;
  // sms text
  LText := StringToJString(Text);
  LSmsManager.sendTextMessage(LSmsTo, LSmsFrom, LText, nil, nil);
end;


function GetSMSService: ISMSService;
begin
  Result := TSMSService.Create;
end;

end.
```

11. It's time to create the interface for the mobile app. Put on the form two **TButton** (aligned on Top) a `TLabel` (aligned to the client), a **TTimer**, and a **TActionList**.

12. Make all the changes to make it graphically like the following screenshot:

Figure 8.24: Mobile app form designer

13. Create a new action using the `ActionList`, and name it `actSendSMS`.

14. Generate an **OnClick** event of the *Start Sending Service* button and an **OnExecute** event of the newly created action. Put this code within these two events:

```
procedure TSMSSendingForm.actSendSMSExecute(Sender: TObject);
begin
  GetSMSService.SendNextSMS;
end;

procedure TSMSSendingForm.btnStartSendingClick(Sender: TObject);
begin
  SendTimer.Enabled := not SendTimer.Enabled;
  if SendTimer.Enabled then
  begin
    Label2.Text := 'Sending Service Started...';
    Label2.FontColor := TAlphaColorRec.Red;
    btnStartSending.Text := 'Stop Sending Service';
    btnStartSending.FontColor := TAlphaColorRec.Red;
  end
  else
  begin
    Label2.Text := 'Sending Service Stopped...';
    Label2.FontColor := TAlphaColorRec.Blue;
    btnStartSending.Text := 'Start Sending Service';
```

```
      btnStartSending.FontColor := TAlphaColorRec.Blue;
   end;
end;
```

15. Now, set the `Action` property of *Send Next Sms* to `actSendSMS` and the **OnTimer** event of `SendTimer` to `actSendSMSExecute`.

16. One last thing: we have to enable **Send SMS** permission. To do this, go to **Project | Options | Uses Permissions** and check **Send SMS.**

How it works...

Our REST service handles an SMS queue that is stored in a `TThreadedQueue` called `SMSQueue`. It provides two REST APIs to push or pop an element to this queue:

HTTP VERB	URL	DESCRIPTION
POST	/sms/push	Pushes a new SMS in the queue. Requires a request body containing the new SMS to create it as a JSON object. The request content-type must be `application/json`.
POST	/sms/pop	Returns an item from the message queue. If there are no messages in the queue, a 404 error is returned.

It may seem strange that a request to retrieve information is made through the POST method instead of GET.

From `w3.org`:

Methods can also have the property of "idempotence" in that (aside from error or expiration issues) the side-effects of N > 0 identical requests is the same as for a single request. The methods GET, HEAD, PUT and DELETE share this property. Also, the methods OPTIONS and TRACE SHOULD NOT have side effects, and so are inherently idempotent.

In this specific case when you call `/sms/pop`, you have a side effect because the next call to `/sms/pop` is not equal to first. So in this case, for the property of "idempotence" POST is better than GET

Start the **REST Server**... It will start and listen to port `8080`.

In this recipe, there is also a third project, a VCL application to queue SMS messages inside the REST server. You can find it in the `Chapter 8\CODE\RECIPE11\TestConsole\TestConsole.dproj` recipe code. Open and run it. It will show up like in the following screenshot:

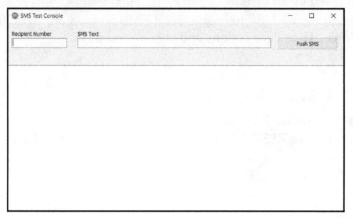

Figure 8.25: VCL SMS Test Console running

Use the console to queue the SMS on the server, and add as many as you want (be careful regarding the number of recipients entered. When the mobile application runs, it will be sent real SMS).

It's time to launch the mobile app and start the mechanism that's been created. Open the `Mobile SendSMS App` project, set **Target Platforms** on `Android`, and **run it**.

Once started, the app will work in two ways:

1. **One Shot**:
   - Performing a tap on **Send Next SMS**, the APP will query the server to receive a text message to send
2. **Continuous Mode**:
   - Performing a tap on **Start Sending Service**, the APP will start a timer that will query the server to receive a text message to send every second:

Figure 8.26: App running in continuous mode

To recap, do the following:

1. Start the server.
2. Run the mobile app in continuous mode.
3. Push, by **Test Console**, the SMS on the server queue.
4. An SMS will be sent to the recipient's number.

There's more...

The mobile app that we've created allows you to send SMS messages, but only if it is active, for example, if it is in the foreground (active and visible to the user). If you close the app or tap the **Home** button, it will stop sending messages. This is due to how the Activity life cycle of the Android OS functions. You could enhance this app by creating a service to allow it to operate, even when the app is in the background. You're curious, right? Well, in Chapter 9, *Delphi and IoT*, there is a recipe that's perfect for this!

However, I will leave you with some links so that you can better understand the concepts covered in this chapter:

- **Android activity life cycle:** `https://developer.android.com/guide/components/activities/activity-lifecycle`
- **Android services:** `https://developer.android.com/guide/components/services`

Using specific platform features

9

In this chapter, we will cover the following topics:

- Using Android SDK Java classes
- Using iOS Objective-C SDK classes
- Displaying PDF files in your app
- Sending Android intents
- Letting your phone talk—using the Android `TextToSpeech` engine
- Using Java classes in Android apps with Java2OP
- Doing it in the background, the right way—Android services

Introduction

There are situations where, if you need a particular Android or iOS feature, FireMonkey does not help you. FireMonkey does a very good job in supporting all the common mobile features, but not all the APIs have been already imported, polished, and wrapped in nice Object Pascal reusable classes or components. What can you do in these cases? The good news is that you can import classes from the underlying SDK (and NDK, in the case of Android) and wrap them just like Embarcadero did in the FireMonkey platform.

In this chapter, we will see some class import examples. Keep in mind that the code using imported classes is not cross-platform. That is, if you import an Android SDK class and your code uses it, you lose the possibility of compiling that specific code for iOS; however, you can, as usual, use some IFDEFs to statically select the Android-specific code from the iOS-specific code. Moreover, in the last recipe, we'll see how to use one of the most powerful Android features—services.

Using Android SDK Java classes

In this section, we'll talk about the mechanisms that the compiler offers to import classes from the Android SDK and NDK. This is not a standard recipe, but is more of a showcase showing the possibilities offered by the Delphi compiler and the process needed to fully use them when dealing with OS built-in libraries.

Getting ready

What we will do is import a well-known Android class used everywhere in the Android ecosystem—Toast. As the Android documentation says:

> *"A toast provides simple feedback about an operation in a small popup. It only fills the amount of space required for the message and the current activity remains visible and interactive."*

Toast is used to inform users about something in an unobtrusive way. They do not have an **OK** or **Close** button, because they automatically disappear after a defined timeout.

Well, how do we use a Toast in a Delphi app?

The first thing to do is to have a clear vision of the class methods and all the other types involved in their definition. Going to the official documentation on the `http://developer.android.com/reference/android/widget/Toast.html` website, you can get this information. Here are the most relevant as the Android Java SDK documentation explains them.

The following table shows Toast class constants:

Type	Constant
int	LENGTH_LONG
int	LENGTH_SHORT

The following table shows Toast public instance methods:

Type	Method
void	cancel()
int	getDuration()
int	getGravity()
float	getHorizontalMargin()

Type	Method
float	getVerticalMargin()
View	getView()
int	getXOffset()
int	getYOffset()
void	setDuration(int duration)
void	setGravity(int gravity, int xOffset, int yOffset)
void	setMargin(float horizontalMargin, float verticalMargin)
void	setText(int resId)
void	setText(CharSequence s)
void	setView(View view)
void	show()

The following table shows `Toast` public static methods (like the class methods in Delphi):

Type	Method
Toast	makeText(Context context, int resId, int duration)
Toast	makeText(Context context, CharSequence text, int duration)

This is the typical usage of the `Toast` class inside an Android activity:

```
Toast.makeText(getContext(), "Hello Toast World",
Toast.LENGTH_LONG).show();
```

Now, with this information, we can start to define our import Delphi class.

How to do it...

The Android Delphi compiler allows you to declare a specific class as a Generic Java Import of an SDK Java class. The class that does this magic is declared within the `Androidapi.JNIBridge.pas` unit, and is declared as follows:

```
TJavaGenericImport<C: IJavaClass; T: IJavaInstance>
```

The `TJavaGenericImport` is a generic class that we can use to make the declaration of imported Java object factories easier. Using this class, we split the class methods and instance methods into two interfaces. This class blends the two interfaces into one factory that can produce instances of Java objects, or provide a reference to an instance representing the Java class. Moreover, Android Java SDK uses Java String objects, while Delphi uses strings. If you need to pass a string to a method imported from the SDK that expects a `JString` (the type used by the Delphi compiler to match the Java string object), you have to use the `StringToJString` function defined in `Androidapi.Helpers.pas` to convert it.

So, the next step to use the `Toast` class is to define two interfaces. The first one declares all the class methods (static in Java) with the same signature as the Java ones. The second one declares all the instance methods with the same Java signature as well. So, how do we map Java types to the Delphi types? In the Delphi RTL, there are a lot of samples of imported Java classes, and this is a small summary of what you can understand from the already imported classes and from the `api-version.xml` file present in the Android SDK, which contains the declaration of all the SDK classes (using RAD Studio 10.2 Tokyo, the path is `<Public Documents>\Embarcadero\Studio\19.0\CatalogRepository\AndroidSDK-24.3.3\ platform-tools\api\api-version.xml`):

Java Type	Delphi Type
boolean	Boolean
byte	ShortInt
char	WideChar
double	Double
float	Single
int	Integer
long	Int64
short	SmallInt
void	If used as a return type, use procedure instead of function
java/lang/CharSequence	JCharSequence
android/view/View	JView
java/lang/String	JString

All methods must be declared with the cdecl calling convention to be compatible with the Java calling convention. Moreover, the interface declaring the interface methods must be decorated with the JavaSignature RTTI attribute, which defines the full Java package where the mapped class is declared in the SDK. It may seem complex, but the resultant code is not. Here is the final import declaration for the Toast class:

```
type

  [JavaSignature('android/widget/Toast')]
  JToast = interface(JObject)
    ['{AC116FB8-FE4D-47E8-BEC9-96E919A01CC7}']
    procedure cancel; cdecl;
    function getDuration: Integer; cdecl;
    function getGravity: Integer; cdecl;
    function getHorizontalMargin: Single; cdecl;
    function getVerticalMargin: Single; cdecl;
    function getView: JView; cdecl;
    function getXOffset: Integer; cdecl;
    function getYOffset: Integer; cdecl;
    procedure setDuration(duration: Integer); cdecl;
    procedure setGravity(gravity: Integer; xOffset: Integer;
      yOffset: Integer); cdecl;
    procedure setMargin(horizontalMargin: Single;
      verticalMargin: Single); cdecl;
    procedure setText(resId: Integer); cdecl; overload;
    procedure setText(s: JCharSequence); cdecl; overload;
    procedure setView(view: JView); cdecl;
    procedure show; cdecl;
  end;

  JToastClass = interface(JObjectClass)
    ['{127EA3ED-B569-4DBF-9BCA-FE1491FC615E}']
    function init(context: JContext): JToast; cdecl;
    function makeText(context: JContext; resId: Integer; duration: Integer)
      : JToast; cdecl; overload;
    function makeText(context: JContext; text: JCharSequence; duration:
Integer)
      : JToast; cdecl; overload;
  end;
```

BE CAREFUL! Java is a case sensitive language, while Delphi is not, so when you use classes in Delphi with the JavaSignature attribute it's like you're using Java itself, so the case is of fundamental importance. If you come across the error "Invoke error: method not found", this could be the reason.

Now, with these two interfaces, we can declare our `TJToast` class inheriting from the `TJavaGenericImport` using the following code:

```
TJToast = class(TJavaGenericImport<JToastClass, JToast>)
const
  LENGTH_LONG = 1;
  LENGTH_SHORT = 0;
end;
```

As you can see, the body of the class is almost empty because all the methods will be used by an internally created object returning an interface reference. `LENGTH_LONG` and `LENGTH_SHORT` are simple constants in Java, so I added them as `const` in the `TJToast` declaration. `TJToast` can be used as follows, using the same methods documented for the Android Java SDK:

```
procedure TMainForm.Button3Click(Sender: TObject);
var
  Toast: JToast;
begin
  Toast := TJToast.JavaClass.makeText(TAndroidHelper.Context,
    StrToJCharSequence('Hello World'), TJToast.LENGTH_SHORT);
  Toast.show();
end;
```

If you try to run this code with a previous version of Delphi Tokyo, you will get the following exception:

`Java.lang.RuntimeException: Can't create handler inside thread that has not called Looper.prepare()`

Among the changes made in Delphi Tokyo, we have the following:

Unification of Delphi and Java threads on Android `CallInUIThread` has been deprecated. All code is now running in the Java UI thread, removing the need for thread synchronization; however, in the recipe I left the code valid for the previous versions: **Old Working Raw Call**.

The ProcessMessages method used to call the CheckSynchronize method, which called callbacks from the Java-native thread. There used to be two threads: the Delphi UI thread and the Java native thread. After the refactoring work in 10.2, there is only one thread, the Java native thread, which receives all notifications from Android and forwards them to the Delphi event handlers. While it is blocked via the while cycle, no event handler can be executed, preventing an exit from the cycle. Callbacks and events in general are called if there is no code holding the app main thread.

– Embarcadero Docwiki

Now the code works, but the utilization pattern is not very Delphi-like. Indeed, we're using Java classes and methods, but using the Delphi syntax; however, we can write some helper code to make the `Toast` utilization more similar to the Delphi RTL and the Delphi programmer mindset:

```
interface

{$SCOPEDENUMS ON}

type
  TToastDuration = (Short = 0, Long = 1);
  TToastPosition = (default = 0, Top = 48, Bottom = 80, Center = 17,
    VerticalCenter = 16, HorizontalCenter = 1);

procedure ShowToast(const AText: string;
  const ADuration: TToastDuration = TToastDuration.Short;
  const APosition: TToastPosition = TToastPosition.Default);

implementation

uses
  FMX.Helpers.Android, AndroidAPI.Helpers;

procedure ShowToast(const AText: string;
  const ADuration: TToastDuration = TToastDuration.Short;
  const APosition: TToastPosition = TToastPosition.Default);
var
  Toast: JToast;
begin
  Toast := TJToast.JavaClass.makeText(TAndroidHelper.context,
    StrToJCharSequence(AText), Integer(ADuration));
  if APosition <> TToastPosition.Default then
    Toast.setGravity(Integer(APosition), 0, 0);
  Toast.show();
end;
```

In this version, we've also used the `setGravity` method to define the `Toast` position on the screen. To do this, we have used an enumerated type mapped to the same integer values defined in the `android.view.Gravity` class. Also, check the call to the `TAndroidHelper.Context` to get the activity context needed by the method.

Now, we can use the `Toast` class using a very Delphi-style function. Here are some sample calls:

```
ShowToast('Hello Toast World');
ShowToast('Hello Toast World', TToastDuration.Long,
    TToastPosition.Center);
ShowToast('Hello Toast World', TToastDuration.Short);
```

As a suggestion, try to make your imports as intuitive as possible for your Delphi users, even if you are the only user, because all the rest of your code is written using the Delphi libraries. Stay as homogeneous as possible; it is a good principle whatever language you use. Encapsulate the imported classes in proper Delphi code structures (including classes, records, functions, and whatever is appropriate) and the style of your code will benefit from it, being much more coherent in itself.

In the recipe code, there is the full app showing the different kinds of `Toast` utilization.

There's more...

Complex classes require more work to be imported, but there are tools that can help in this hard work. Some tools available for the Delphi versions before XE8 are **Java2Pas**: `http://www.softwareunion.lu/en/downloads/`. For users of a newer version, Embarcadero released `Java2OP`, which does a very good job. One of the next recipes will talk about it.

These tools do a good job and help in the boring methods declaration phase; however, you cannot simply import a class and use it in your Delphi code. In many cases, you have to do additional work to arrange a good class structure in your units to get around units' circular references; however, if you are interested and want to know more, I suggest checking the good presentation by Brian Long at the *CodeRage 8* conference, where he talks about accessing Android and iOS API (`http://blog.blong.com/2013/10/my-coderage-session-files.html`).

Since Delphi XE7, it is possible to use your own or third-party Java libraries in RAD Studio applications in a simpler way. Check out this link for more information: `http://docwiki.embarcadero.com/RADStudio/en/Using_a_Custom_Set_of_Java_Libraries_In_Your_RAD_Studio_Android_Apps`.

As FireMonkey and the mobile soul of Delphi matures, third-party mobile components have started to become available on the market. Even if you are not interested in native widgets, you can study the code from the project *D.P.F Delphi Android Native Components*, which can help you understand how this kind of interfacing job works (`http://sourceforge.net/projects/dpfdelphiandroid/`).

Moreover, you can also use native NDK `.so` files. To get an idea of how to do it, check the `Androidapi.Log.pas` unit, where the function used by the `IFMXLoggingService` service on Android is declared. As you will see in the following code, there is a declaration very similar to the declaration usually used for the Windows DLL:

```
function __android_log_write(Priority: android_LogPriority;
   const Tag, Text: MarshaledAString): Integer; cdecl;
   external AndroidLogLib name '__android_log_write';
```

As time goes by, Embarcadero will add more and more wrappers for the Android SDK classes and functionalities, but until then, if you need to use specific SDK classes or third-party Java classes (that will require a bit of work to be packaged in the generated APK), you can rely on the compiler support and the RTL class `TJavaGenericImport` to declare and use it.

Using iOS Objective-C SDK classes

Just as we saw regarding Android in the previous recipe, Delphi is able to access the iOS SDK as well. In this section, we'll talk about the mechanisms that the compiler offers to import classes from the iOS SDK. This is not a standard recipe, but more of a showcase showing the possibilities offered by the Delphi compiler and the process needed to fully use them when dealing with OS built-in libraries. The mechanism is similar to the Android ones, but there are some notable differences.

Getting ready

In Objective-C, all classes have `NSObject` as a common ancestor. The iOS SDK is composed of some frameworks. An iOS framework is a number of classes specialized for a single purpose. For example, UIKit is the framework containing all the basic classes related to the UI, the iAd framework contains all the stuff related to the advertising, and MapKit wraps up all the mapping-related classes.

Note that Objective-C uses the `NSString` objects, while Delphi uses strings. If you need to pass a string to an iOS API that expects an `NSString`, you can use the `StrToNSStr` function defined in `Macapi.Helpers.pas` to convert it.

Let's say we need to use the `UIDevice` class from the iOS SDK (the process is applicable for every class in the SDK). As the Apple documentation says:

> *"The UIDevice class provides a singleton instance representing the current device. From this instance you can obtain information about the device such as assigned name, device model, and operating system name and version."*

How it works...

The iOS Delphi compiler allows you to declare a specific class as a Generic Objective-C import of an SDK class. The class that does this magic is declared within the `Macapi.ObjectiveC.pas` unit and is declared as follows:

```
TOCGenericImport<C: IObjectiveCClass; T: IObjectiveCInstance>
```

The `TOCGenericImport` is a generic class that we can use to make the declaration of imported `ObjectiveC` object factories easier. Using this class, we split the class methods and instance methods into two interfaces. This class blends the two interfaces into one factory that can produce instances of `ObjectiveC` objects, or provide a reference to an instance representing the `ObjectiveC` class. But how do we define the methods in the two interfaces?

Reading the iOS documentation for the `UIDevice` class, you read method and property signatures. Let's translate some of the most significant ones.

The first property we want to translate is `model`. The `model` property returns the model of the device; which can be iPhone or iPod touch, or other values identifying the device model. The property is read-only.

This is the complete signature:

```
@property(nonatomic, readonly, retain) NSString *model
```

In Object, Pascal is translated as follows:

```
function model: NSString; cdecl;
```

As you can see, a read-only property is mapped to a function with the name of the property as the function name, and with the `ObjectiveC` property type as the `ObjectPascal` return value. But what about R/W (read/write) properties?

The next property we want to translate is `proximityMonitoringEnabled`, an R/W property of the type Boolean indicating whether proximity monitoring is enabled or not.

This is the complete signature:

```
@property(nonatomic,getter=isProximityMonitoringEnabled)
BOOL proximityMonitoringEnabled
```

In Object, Pascal is translated as follows:

```
procedure setProximityMonitoringEnabled(proximityMonitoringEnabled:
Boolean); cdecl;
function isProximityMonitoringEnabled: Boolean; cdecl;
```

An R/W property is mapped to a procedure (the setter) and a function (the getter). The procedure name starts with the set followed by the `ObjectiveC` property name (`proximityMonitoringEnabled` becomes `setProximityMonitoringEnabled`), and accepts a parameter of the same type of the property. The function name is defined by the property signature; in this case, it is `isProximityMonitoringEnabled`, returning a value of the same type of the property type. If the property signature does not impose the getter name, the translation is similar to the following code.

 BE CAREFUL! Objective-C is a case sensitive language, while Delphi is not, so when you use classes in Delphi from Objective-C you're using Objective-C itself, so the case is of fundamental importance.

In Objective-C, it is as follows:

```
@property(nonatomic, retain) NSString *accessibilityLabel
```

In Delphi, it is as follows:

```
function accessibilityLabel: NSString; cdecl;
procedure setAccessibilityLabel(accessibilityLabel: NSString); cdecl;
```

The `UIDevice` import will look like the following code, where only some methods were imported:

```
UIDeviceClass = interface(NSObjectClass)
    ['{A2DCE998-BF3A-4AB0-9B8D-4182B341C9DF}']
    function currentDevice: Pointer; cdecl;
```

```
    end;

UIDevice = interface(NSObject)
    ['{70BB371D-314A-4BA9-912E-2EF72EB0F558}']
    function batteryLevel: Single; cdecl;
    function batteryState: UIDeviceBatteryState; cdecl;
    function isBatteryMonitoringEnabled: Boolean; cdecl;
    function isMultitaskingSupported: Boolean; cdecl;
    function isProximityMonitoringEnabled: Boolean; cdecl;
    function localizedModel: NSString; cdecl;
    function model: NSString; cdecl;
    function name: NSString; cdecl;
    function orientation: UIDeviceOrientation; cdecl;
    procedure playInputClick; cdecl;
    function proximityState: Boolean; cdecl;
    function systemName: NSString; cdecl;
    function systemVersion: NSString; cdecl;
    function uniqueIdentifier: NSString; cdecl;
  end;

  TUIDevice = class(TOCGenericImport<UIDeviceClass, UIDevice>)
  end;
```

Now, the `UIDevice` class is usable as the following; however, the suggested use is as a singleton using the `currentDevice` property. Here, it is used as a normal instance just for example purposes:

```
var
  device: UIDevice;
begin
  device := TUIDevice.Create;
  ShowMessage(NSStrToStr(device.model));
end;
```

Note that the class methods defined in the `UIDevice` class can also be used by Delphi. You don't need to create an instance as with normal class methods, but the returning pointer must be wrapped in the proper class type:

```
var
  device: UIDevice;
  model: string;
begin
  //wraps the pointer to the proper type using the Wrap method
  device := TUIDevice.Wrap(TUIDevice.OCClass.currentDevice);
  model := NSStrToStr(device.model);
  ShowMessage(model);
end;
```

There's more...

The topic about the `ObjectiveC` class imports is huge, and a deep explanation of it is beyond the scope of this book; however, if you are interested and want to know more, I suggest you check out the presentation by Brian Long at the *CodeRage 8 conference*, where he talks about accessing iOS and Android API (`http://blog.blong.com/2013/10/my-coderage-session-files.html`).

As FireMonkey and the mobile soul of Delphi matures, third-party mobile components start to become available on the market. Even if you are not interested in native widgets, you can study the code in the project *D.P.F Delphi iOS Native Components* (`http://sourceforge.net/projects/dpfdelphiios/`).

Displaying PDF files in your app

In the mobile world, often you need to show your user PDF files. Maybe these PDF files are used as reports (usually generated by some reporting tool on the remote server), as a statement about something that the user should do, which is in the form of a small book products catalog. So, how do we display a PDF deployed within the app, or downloaded from some remote server and stored locally? How do we do it on Android and iOS? This is the topic of this recipe.

Getting ready

Let's say we have to create an app that contains some PDF files. In this case, we don't download the files, we simply deliver them within the app. Later, we'll see how to download them from the network.

To deploy additional files within our app, we use the **Deployment Manager** accessible by navigating to **Project** | **Deployment**. To find out how to use it, check the Embarcadero documentation (`http://docwiki.embarcadero.com/RADStudio/en/Deployment_Manager`).

The additional file will be deployed in the private documents folder. Under Android, the private documents folder is identified as `./assets/internal`, while on iOS it is identified as `.\Startup\Documents`. Using the Deployment Manager, put a PDF file in these folders for each platform so it will be included in the generated app package.

How it works...

All that's required to show the PDF is encapsulated in a single unit called
`xPlat.OpenPDF.pas`. The main form contains a button that, once clicked, calls the function
`OpenPDF`, passing the name of the file to be shown as follows:

```
procedure TMainForm.btnOpenPDFClick(Sender: TObject);
begin
  OpenPDF('samplefile.pdf');
end;
```

Let's analyze the `OpenPDF` function in the `xPlat.OpenPDF.pas` unit. Here's the full code:

```
unit xPlat.OpenPDF;

interface

procedure OpenPDF(const APDFFileName: string);

implementation

uses
  System.SysUtils, IdURI, FMX.Forms, System.Classes,
  System.IOUtils, FMX.WebBrowser, FMX.Types, FMX.StdCtrls
{$IF defined(ANDROID)}
    , Androidapi.JNI.GraphicsContentViewText
    , FMX.Helpers.Android
    , Androidapi.Helpers
    , AndroidSDK.Toast
    , Androidapi.JNI.Net
    , Androidapi.JNI.JavaTypes
{$ENDIF}
{$IF defined(IOS)}
    , iOSapi.Foundation
    , Macapi.Helpers
    , FMX.Helpers.iOS
    , FMX.Dialogs
{$ENDIF}
    ;

{$IF defined(ANDROID)}

procedure OpenPDF(const APDFFileName: string);
var
  Intent: JIntent;
  FilePath, SharedFilePath: string;
begin
```

```
  FilePath := TPath.Combine(TPath.GetDocumentsPath, APDFFileName);
  SharedFilePath := TPath.Combine(TPath.GetSharedDocumentsPath,
APDFFileName);
  if TFile.Exists(SharedFilePath) then
    TFile.Delete(SharedFilePath);
  TFile.Copy(FilePath, SharedFilePath);
  Intent := TJIntent.Create;
  Intent.setAction(TJIntent.JavaClass.ACTION_VIEW);
  Intent.setDataAndType(StrToJURI('file://' + SharedFilePath),
StringToJString('application/pdf'));
  try
    SharedActivity.startActivity(Intent);
  except
    on E: Exception do
      ShowToast('Cannot open PDF' + sLineBreak + Format('[%s] %s',
[E.ClassName, E.Message]),
        TToastDuration.Long);
  end;
end;
{$ENDIF}


{$IF defined(IOS)}
type
  TCloseParentFormHelper = class
  public
    procedure OnClickClose(Sender: TObject);
  end;


procedure TCloseParentFormHelper.OnClickClose(Sender: TObject);
begin
  TForm(TComponent(Sender).Owner).Close();
end;


procedure OpenPDF(const APDFFileName: string);
var
  NSU: NSUrl;
  OK: Boolean;
  frm: TForm;
  WebBrowser: TWebBrowser;
  btn: TButton;
  evnt: TCloseParentFormHelper;
begin
  frm := TForm.CreateNew(nil);
  btn := TButton.Create(frm);
  btn.Align := TAlignLayout.Top;
  btn.Text := 'Close';
  btn.Parent := frm;
  evnt := TCloseParentFormHelper.Create;
```

```
    btn.OnClick := evnt.OnClickClose;
    WebBrowser := TWebBrowser.Create(frm);
    WebBrowser.Parent := frm;
    WebBrowser.Align := TAlignLayout.Client;
    WebBrowser.Navigate('file://' + APDFFileName);
    frm.ShowModal();
  end;
{$ENDIF}

  end.
```

Showing the PDF file on Android

To display the PDF on Android, we use an Android-specific mechanism called **intents** (check the specific recipe to know more about Android intents). The file is actually shown by an external app already installed on the device; if such an app is not present, the PDF cannot be shown and a message is shown to the user. You can install Adobe PDF Reader or another app that is able to display PDFs, which is intent-compatible with the one from Adobe. In accordance with Android I/O security, to let another app read the PDF file bounded in our assets/internal folder, we have to copy the file from the private documents folder, which is private to the app and not accessible from other apps, to the shared documents folder (readable from all the other apps installed on the device).

Just after the copying the file, we create an intent and configure it to launch an app able to show that PDF. The configuration is simple enough, as shown in the following code:

```
//create the Intent directly from the Android SDK
Intent := TJIntent.Create;
//We need to show the PDF, so ACTION_VIEW is ok
Intent.setAction(TJIntent.JavaClass.ACTION_VIEW);
//Where is the file? Which mime type?
Intent.setDataAndType(
StrToJURI('file://' + SharedFilePath),
StringToJString('application/pdf'));
try
   //ask to the OS to find a proper app to handle the intent
   SharedActivity.startActivity(Intent);
except
   //TODO: there aren't apps able to show the PDF. Inform the user!
end;
```

Showing the PDF file on iOS

On iOS, there aren't any intents, but we can use another mechanism to show our PDF. The iOS WebView component is able to show PDFs, so we create a form on the fly containing a WebView and a button to close the form. The `OpenPDF` iOS implementation does not use iOS-specific mechanisms, apart from the WebView capabilities.

After having created the form at runtime (remember that if you don't have an `fmx` file associated with the `TForm` instance, you cannot use `TForm.Create()` to create the form but `TForm.CreateNew()`. The code is reproduced here, with some comments:

```
//create the form without an fmx
frm := TForm.CreateNew(nil);
//create the button used to close the form.
//On iOS there is not a "back" button as in Android
btn := TButton.Create(frm);
btn.Align := TAlignLayout.Top;
btn.Text := 'Close';
btn.Parent := frm;
evnt := TCloseParentFormHelper.Create;
//set the Button OnClick event handler
btn.OnClick := evnt.OnClickClose;
//create the TWebBroser component wich wraps the iOS WebView
WebBrowser := TWebBrowser.Create(frm);
WebBrowser.Parent := frm;
WebBrowser.Align := TAlignLayout.Client;
//point the webbroser to the local file under the private folder
WebBrowser.Navigate('file://' + APDFFileName);
frm.ShowModal();
```

There's more...

This code does its job; however, Android and iOS users do not have the same user experience. On Android, you can use whatever app you have installed on the device to show the PDF, so you could also change the file with annotations, highlights, and drawings directly on the file. Note that the file is readable also from apps other than yours. This can be a problem in some situations. On iOS, conversely, you cannot modify the PDF with annotations and do not have full control of the file, and the file remains private to your app. These facts must be carefully analyzed, and you have to be aware of the pros and cons of every choice you make. If you want to provide a uniform set of functionalities, additional work and third-party components and libraries are needed.

One particular mention is useful for the TMS iCL component suite (`http://www.tmssoftware.com/site/tmsicl.asp`). It is specific to FireMonkey on iOS, so it doesn't compile on Android, but it does contain a component called `TTMSFMXNativePDFLib`, which is able to create new PDF documents, open existing PDF documents, and do many other things.

 Using Google Docs Viewer:
If your PDF is located on a public URL, you can also use the PDF visualizer included in Google Docs. Point a WebView to the following URL and your PDF will show up:
`'https://docs.google.com/gview?embedded=true&url=' + PDFURL;`

Downloading the PDF file from the server

Let's say we have an application server that generates reports from some database data and saves them as PDF files.

We can download these files by simply using the `TidHTTP` component and storing them locally using code similar to the following:

```
var
  FileStream: TStream;
  FilePath: String;
begin
  FilePath := TPath.Combine(TPath.GetSharedDocumentsPath, 'myreport.pdf');
  FileStream := TFileStream.Create(FilePath, fmCreate);
  IdHttp1.Get('http://www.myserver.com/reports/myreport.pdf', FileStream);
end;
```

Sending Android Intents

One of the most useful things about Android development is the intents dispatching mechanism. As Google says, *An intent is an abstract description of an operation to be performed*, and just to be clearer, continues saying, *An Intent provides a facility for performing late runtime binding between the code in different applications. Its most significant use is in the launching of activities, where it can be thought of as the glue between activities. It is basically a passive data structure holding an abstract description of an action to be performed* (`http://developer.android.com/reference/android/content/Intent.html`).

Intents are widely used in Android, and if you want to fully integrate your Delphi app with the Android OS, you will probably have to deal with Intents. Delphi uses intents internally to deal with some fundamental Android services (`TShareSheetAction`, `TTakePhotoFromCameraAction`, and so on). In this recipe, we'll see with many examples how to directly use intents in our app.

Getting ready

The primary and mandatory pieces of information in an intent are:

- `action`: This is the general action to be performed, such as `ACTION_VIEW`, `ACTION_EDIT`, `ACTION_MAIN`, and so on
- `data`: This is the data to operate on, such as a person's record in a contacts database, expressed as a URI

There are two kinds of intent:

- **Explicit intent**: The app defines the target component directly in the intent
- **Implicit intent**: The app asks the Android system to evaluate registered components based on the intent data and other optional information

Using Java and the Android SDK, you can send an implicit intent with the following code:

```
Intent myIntent = new Intent(Intent.ACTION_VIEW,
Uri.parse("http://www.danielespinetti.it"));
startActivity(myIntent);
```

This code asks the Android system to view a web page. If the OS finds an activity able to handle this kind of information based on action and data, then that activity will be started and the intent data passed to it.

Intents are available also to Delphi users. The previous Java code can be translated in Delphi as follows:

```
var
  Intent: JIntent;
begin
  Intent := TJIntent.Create;
  Intent.setAction(TJIntent.JavaClass.ACTION_VIEW);
  Intent.setData(StringToJString('http://www.danielespinetti.it'));
  TAndroidHelper.Activity.startActivity(Intent);
end;
```

As you can see, the code is very similar to the Java version. Note that this code cannot be compiled on any platform but Android, so if you want to add this code in a cross-platform app (for Android, iOS, or also Windows, and macOS X), you have to surround it with some IFDEFs.

There are many components able to respond to intents; the Android documentations are very good on this topic. In this recipe, we'll open a web page, start Google Maps by pointing to a specific address, open an email client, open the Twitter app, and ask for speech-to-text recognition.

How it works...

In the main form, there are six buttons, a listbox, and some labels. Here's how the form is rendered at runtime (after using it to recognize the phrase `This is a book`):

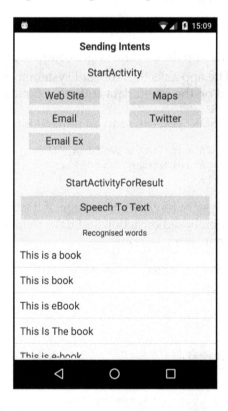

Figure 9.1: The app with the five buttons that will send intents

Let's open the `SendingAndroidIntents.dproj` project and study it. The first four buttons, as you can see while reading the events handler, call a form method called `LaunchViewIntent` after passing a URI:

```
procedure TMainForm.btnMapsClick(Sender: TObject);
begin
  //launch Google Maps (or similar app)
  LaunchViewIntent('geo://0,0?q=Piazza del Colosseo 1,00184 Roma');
end;

procedure TMainForm.btnEmailClick(Sender: TObject);
begin
  //launch an email client with an empty email
  LaunchViewIntent('mailto:d.spinetti@bittime.it', false);
end;

procedure TMainForm.btnTwitterClick(Sender: TObject);
begin
  //launch twitter client (if installed)
  LaunchViewIntent('http://twitter.com/spinettaro');
end;
```

The procedure `LaunchViewIntent` is defined as follows:

```
procedure TMainForm.LaunchViewIntent(AURI: string; AEncodeURL: boolean);
var
  Intent: JIntent;
  URI: JString;
begin
  if AEncodeURL then
    AURI := TIdURI.URLEncode(AURI);
  Intent := TJIntent.Create;
  Intent.setAction(TJIntent.JavaClass.ACTION_VIEW);
  URI := StringToJString(AURI);
  Intent.setData(TJnet_Uri.JavaClass.parse(URI));
  TAndroidHelper.Context.startActivity(Intent);
end;
```

This method executes all the steps needed to create an intent for the purpose of displaying something; indeed, the action `ACTION_VIEW` means, *I want to view something*, and asks the OS to show the information described in the `data` property, and other intent properties if present.

Firstly, we check whether the URI needs to be encoded; if so, we use the `TidURI.URLEncode` method from the `INDY` library to do the encoding. Then, an intent is created and configured with `ACTION_VIEW` as the action and the passed URI as the data. Having the intent configured, the last thing to do is to ask the OS what that intent is for. In this case, we want to start an activity to do the work defined in the intent. The Android context used by the FireMonkey framework is accessible using the `TAndroidHelper.Context` static method. So the last line uses `TAndroidHelper.Context.startActivity(Intent)` to actually send the intent. This kind of intent is the most simple.

More complex intent – sending a full flagged email

The fifth button, **Email Ex**, sends an email just like the **Email** button, but is more powerful because the prepared email will also have the subject, the body, and the CC and BCC fields correctly filled. Let's see how this is possible.

In this case, the simple `ACTION_VIEW` with some data is not enough. Here's the code used to send a more complex email:

```
procedure TMainForm.btnEmailExClick(Sender: TObject);
var
  Intent: JIntent;
  URI: JString;
  AddressesTo: TJavaObjectArray<JString>;
  AddressesCC, AddressesBCC: TJavaObjectArray<JString>;
begin
  Intent := TJIntent.Create;
  Intent.setAction(TJIntent.JavaClass.ACTION_SENDTO);
  Intent.setData(TJnet_Uri.JavaClass.parse(StringToJString('mailto:')));
  AddressesTo := TJavaObjectArray<JString>.Create(2);
  AddressesTo.Items[0] := StringToJString('d.spinetti@bittime.it');
  AddressesTo.Items[1] := StringToJString('john.doe@nowhere.com');
  AddressesCC := TJavaObjectArray<JString>.Create(1);
  AddressesCC.Items[0] := StringToJString('jane.doe@nowhere.com');
  AddressesBCC := TJavaObjectArray<JString>.Create(1);
  AddressesBCC.Items[0] := StringToJString('backup@mywebsite.com');
  Intent.putExtra(TJIntent.JavaClass.EXTRA_EMAIL, AddressesTo);
  Intent.putExtra(TJIntent.JavaClass.EXTRA_CC, AddressesCC);
  Intent.putExtra(TJIntent.JavaClass.EXTRA_BCC, AddressesBCC);
  Intent.putExtra(TJIntent.JavaClass.EXTRA_SUBJECT,
StringToJString('Greetings from Italy'));
  Intent.putExtra(TJIntent.JavaClass.EXTRA_TEXT,
    StringToJString('I''m learning how to use Android Intents!'+
    sLineBreak + 'They are very powerful!' +
    sLineBreak + sLineBreak + 'See you...'));
```

```
    SharedActivity.startActivity(Intent);
  end;
```

As you can see, we set more properties in the intent than in the previous example. Also, a `TJavaObjectArray<JString>` is used to pass a Delphi wrapper of a Java array to the intent. Also note how generics can be used while talking to the Android SDK.

Tapping this button, you will get a fully prepared email as in the following screenshot. Note how the subject, the CC, and the BCC tabs have been filled using information sent by the intent:

Figure 9.2: Gmail ready to send the email prepared by our app

Starting an activity for a result – the SpeechToText engine

Sometimes, you want to get a result back from an activity when it ends its job. For example, you may start an activity that lets the user pick a photo in a photo gallery and when it ends, it returns the selected image or a person in a list of contacts and when it ends, it returns the person that was selected.

To do this, we call the `TAndroidHelper.Activity.startActivityForResult` method. The result will come back through a FireMonkey message readable using the global `TMessageManager` instance.

The `startActivityForResult` property gets two parameters: the first one is the intent itself, while the second is an integer value identifying the request code. This request code will be passed to the message handler when the activity ends. Yes, since the `startActivityForResult` property is not blocking, when the launched activity ends you have to know from which request it has been launched.

When an activity exits, some data should be returned back to its parent. It must always supply a result code, which can be the standard result `RESULT_CANCELED`, `RESULT_OK`, or any custom values starting with `RESULT_FIRST_USER` (all these values are defined in the Android documentation here: `http://developer.android.com/reference/android/app/Activity.html`). In addition, it can optionally return an Intent containing any additional data it wants. All of this information appears back on the parent message handler along with the integer identifier it originally supplied.

The last button launches the `SpeechToText` engine activity, asks the user to say something, then ends and sends the possible recognized phrases to the parent activity:

```
procedure TMainForm.btnSTTClick(Sender: TObject);
var
  Intent: JIntent;
  ReqCode: Integer;
const
  STT_REQUEST = 1001;
  ACTION_RECOGNIZE_SPEECH = 'android.speech.action.RECOGNIZE_SPEECH';
  EXTRA_LANGUAGE_MODEL = 'android.speech.extra.LANGUAGE_MODEL';
  EXTRA_RESULTS = 'android.speech.extra.RESULTS';
begin

  //assign a code to this request
  ReqCode := STT_REQUEST;
  //create and configure the intent (check android SDK docs)
  Intent := TJIntent.Create;
```

```
  Intent.setAction(StringToJString(ACTION_RECOGNIZE_SPEECH));
  Intent.putExtra(StringToJString(EXTRA_LANGUAGE_MODEL),
StringToJString('en-US'));
 //when the launched activity ends, this handler will be called.
 //Here we've to read the data sent back from the launched activity
 TMessageManager.DefaultManager.SubscribeToMessage(
   TMessageResultNotification,
   procedure(const Sender: TObject; const Message: TMessage)
   var
     M: TMessageResultNotification;
     i: Integer;
     Words: JArrayList;
     TheWord: string;
   begin
     M := TMessageResultNotification(message);
     //is this request the right one?
     if M.RequestCode = ReqCode then
     begin
       //The request returned OK?
       if (M.ResultCode = TJActivity.JavaClass.RESULT_OK) then
       begin
         Words :=
M.Value.getStringArrayListExtra(StringToJString(EXTRA_RESULTS));
         ListBox1.Clear;
         //if there are recognized words, fill the listbox
         if Words.size > 0 then
         begin
           ListBox1.BeginUpdate;
           try
             for i := 0 to Words.size - 1 do
             begin
               TheWord := JStringToString(JString(Words.get(i)));
               ListBox1.Items.Add(TheWord);
             end;
           finally
             ListBox1.EndUpdate;
           end;
         end
         else
           ShowToast('Some problems occurred');
       end
       else
         ShowToast('Nothing to recognise');
     end;
   end);
```

```
//start the activity for result passing the specific ReqCode
  TAndroidHelper.Activity.startActivityForResult(Intent, ReqCode);
end;
```

The code is not trivial, but the main parts are clearly identifiable. Firstly, we configure the intent to launch the speech recognizer. Then, before launching the intent, we subscribe to system messages of the TMessageResultNotification type. This kind of message is sent by FireMonkey when an Android activity has been launched with startActivityForResult. Inside the message handler, we check whether the message is from our launched activity (so we check the ReqCode), and whether the activity returned with no errors (so we check the RESULT_OK). If everything is OK, we can read the information contained in the returned intent. This time, the intent is used to send back information from the launched activity to the parent app.

The speech-to-text engine activity is listening and will display as shown in the following screenshot:

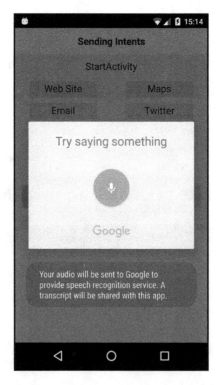

Figure 9.3: Play with the application and discover how the different kinds of intents work.

There's more...

Intents are fundamental parts of the Android ecosystem. FireMonkey uses them in the components and in the RTL, and a developer who wants to deeply integrate their app with the Android OS must know how they work and the possibilities that they open up. Every app installed on your device can be elegantly integrated in your app without too much effort. Good advice is to, study all the common intents available and usable in your Android device. You will learn useful things and will get in touch with many practical uses of intents.

All the available common intents are explained in this article: `http://developer.android.com/guide/components/intents-common.html`.

Letting your phone talk – using the Android TextToSpeech engine

In this recipe, we'll do some very fun stuff. On your Android phone, run an app with a listening UDP server on it. When another application—in our case, a VCL application—sends a UDP broadcast with some text, the Android app will pronounce the text using the android TTS engine.

Getting ready

The first thing to do is to import the TTS classes from the android SDK in our Delphi project. That is not a simple task. However, luckily, someone already did the job. Indeed, Jeff Overcash, the maintainer of the **InterBase Express** (**IBX**) components, wrote the *Android Text To Speech JNI Translation*. His translation with a simple demo app is available on **CodeCentral** (`http://cc.embarcadero.com/item/29594`).

In this recipe, we'll use the imported classes to let our android device read the text sent via a UDP broadcast. Note that the message will be read by each device that receives it. Thus, if you have two, three, or four phones, you will listen to the message read by all the phones almost simultaneously.

How it works...

Open the project group containing the mobile app and the VCL application.

In the mobile application, we have an empty form with a label on it aligned to the client. At the form startup (one second after form creation), we configure the TTS engine with the following code:

```
procedure TMainForm.Timer1Timer(Sender: TObject);
begin
  Timer1.Enabled := False;
  FTTS := TJTextToSpeech.JavaClass.init(TAndroidHelper.Context,
FTTSListener);
end;
```

The `FTTSListener` instance is a `TJavaLocal` descendant implementing the required `JTextToSpeech_OnInitListener` interface. The TTS system is getting initialized and, when done, the listener `onInit` method is called (check the unit `TTSListenerU.pas`). However, if the TTS engine is correctly initialized, we have to configure it by setting the language to be used when it talks. So, in the listener constructor, I've added an anonymous method that will be called by the listener to configure the engine after the initialization. The code is written inside the `FormCreate` event handler, as shown in the following code:

```
constructor TMainForm.Create(AOwner: TComponent);
begin
  inherited;
  FTTSListener := TttsOnInitListener.Create(
    procedure(AInitOK: boolean)
    var
      Res: Integer;
    begin
      if AInitOK then
      begin
        Res := FTTS.setLanguage(TJLocale.JavaClass.ENGLISH);
        if (Res = TJTextToSpeech.JavaClass.LANG_MISSING_DATA) or
          (Res = TJTextToSpeech.JavaClass.LANG_NOT_SUPPORTED) then
          Label1.Text := 'Selected language is not supported'
        else
        begin
          Label1.Text := 'READY To SPEAK!';
          IdUDPServer1.Active := True;
        end;
      end
      else
        Label1.Text := 'Initialization Failed!';
```

```
        end);
    end;
```

If the configuration goes well, the `TidUDPServer`, configured to listen on all interfaces on port `9999`, is activated. In the `idUDPServer1.OnUDPRead` event handler, there is a hook between the data sent over the network and the TTS engine:

```
procedure TMainForm.IdUDPServer1UDPRead(
AThread: TIdUDPListenerThread;
const AData: TIdBytes; ABinding: TIdSocketHandle);
var
  bytes: TBytes;
begin
  bytes := TBytes(AData);
  Speak(TEncoding.ASCII.GetString(bytes));
end;

procedure TMainForm.Speak(const AText: string);
begin
  FTTS.Speak(
    StringToJString(AText),
    TJTextToSpeech.JavaClass.QUEUE_FLUSH, nil);
end;
```

The `Speak` method is the entry point to the TTS engine. The mobile app is completed. Now, let's talk about the VCL application that has to send the UDP packets.

Open the `VCLTTSClient` project, and you will see a form like the following figure:

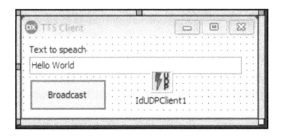

Figure 9.4: The simple VCL form that will send the UDP messages to the mobile app

This application is even simpler than the mobile one. Shortly, when the user clicks on the button, the event handler sends the text entered in the edit to all the available broadcast addresses (considering its subnet as a Class C network). In other words, if the PC where the application is running has a single IP address, let's say `192.168.1.50`, the UDP packet is sent to the broadcast address `192.168.1.255`, and so on for each Ethernet interface configured on the Windows machine.

To get all the IP addresses, I've used a handy class named `TIdStackLocalAddressList` that comes with INDY. Moreover, this is just a demo. So if you want to be sure about the broadcast addresses, you should do additional work, although this is a network-specific topic out of the scope of this book.

To replace the last address part (for example, `.50` must become `.255`), I've used a simple regular expression to replace the last octet. Note that this code actually works only for IPv4 addresses:

```
procedure TMainForm.btnSendClick(Sender: TObject);
var
  CurrIP, BrdcstIP: string;
  i: Integer;
begin
  for i := 0 to FAddressesList.Count - 1 do
  begin
    if FAddressesList.Addresses[i].IPVersion = Id_IPv4 then
    begin
      CurrIP := FAddressesList.Addresses[i].IPAddress;
      BrdcstIP := FToIPv4Broadcast.Replace(CurrIP, '.255');
      IdUDPClient1.Broadcast(Edit1.Text, 9999, BrdcstIP);
    end;
  end;
end;

procedure TMainForm.FormCreate(Sender: TObject);
begin
  FAddressesList := TIdStackLocalAddressList.Create;
  GStack.GetLocalAddressList(FAddressesList);
  FToIPv4Broadcast := TRegEx.Create('\.\d{1,3}$');
end;
```

That's it! Run the mobile app on your android phone and verify that it is currently connected to the same Wi-Fi where the PC is connected. Then, run the VCL application on your PC, write something on the edit, and hit the button. You android device should start to talk.

There's more...

The ability to set up TCP or UDP servers on our mobile devices opens a great range of possibilities. However, you should open ports on your phone conscientiously.

Thanks to Jeff Overcash, the TTS wrapper had greatly simplified the work required to let the android phone talk. If you want to go deeper into using the TTS engine, you should read the following Android documentation:

- Java documentation about the main class used in this recipe (`http://developer.android.com/reference/android/speech/tts/TextToSpeech.html`)
- The Java package where the classes have been imported from (`http://developer.android.com/reference/android/speech/tts/package-summary.html`)
- An introduction to the TTS engine on Android (`http://android-developers.blogspot.it/2009/09/introduction-to-text-to-speech-in.html`)

Using Java classes in Android apps with Java2OP

In this recipe, we will talk about the `Java2OP` command-line utility. `Java2OP.exe` (**Java To Object Pascal**) is a command-line tool that you can use to generate Delphi-native bridge files from Java libraries (JAR or class files). Once having generated the Object Pascal files describing the Java needed classes, you can then use them to provide your Delphi application's access to those Java libraries on Android.

`Java2OP.exe` is included in the most recent Delphi versions and is available at the following path: `<RAD Studio>\bin\converters\java2op\java2op.exe`.

Getting ready

This recipe is an upgraded version of the *Using Android SDK Java classes* recipe. Indeed, while in the first recipe, we created the Object Pascal files by hand, reading the Android documentation, here, we'll use the `Java2OP.exe` utility to automatically generate them. This is the process that you will use 90% of the time (if you have a modern version of Delphi). However, the first recipe is still relevant to correctly understand the process, and eventually changes something in the generated files if needed.

How to do it...

We start with a copy of the recipe *Using Android SDK Java classes*. Our objective is to recreate the same app using a generated interface file instead of the file created manually. Copy the project with all the files and put it in another folder, then open it in the RAD Studio IDE and remove the `AndroidSDK.Toast.pas` file from the project. Also, remove all references to the file in the other units. Change the project name from `ToastDemo` to `ToastDemoWithJava2OP`.

Now, open the command prompt and go to `<RAD Studio>\bin\converters\java2op\` (for example, `C:\Program Files (x86)\Embarcadero\Studio\19.0\bin\converters\java2op\`).

`Java2OP` can generate Object Pascal interface files from compiled Java classes and from JAR files. However, the output files do not include members that are part of the built-in RAD Studio Java libraries for Android unless you explicitly specify those members using the `-classes` parameter. Moreover, if any of the classes that you specified depend on members from the built-in RAD Studio Java libraries for Android, the resulting native bridge file does not re-declare those members, but it includes the RAD Studio units that already declare those members.

Information extrapolated from `http://docwiki.embarcadero.com/RADStudio/en/Java2OP.exe,_the_Native_Bridge_File_Generator_for_Android`.

While in the `java2op` folder, run the following command line:

```
Java2OP.exe -classes android.widget.Toast -unit AndroidSDK.Java2OP.Toast
```

Now you should see a new file called `AndroidSDK.Java2OP.Toast.pas` containing the generated Object Pascal interfaces. We asked the `Java2OP` folder to generate the interface file for the `android.widget.Toast` class, and to generate it in a unit called `AndroidSDK.Java2OP.Toast.pas`.

Copy this file into the project folder and add it to the project. You can also change the PATH environment variable to be able to call `Java2OP.exe` directly inside the project folder. Next, go to the main form and add this new file in the `uses` clause. Try to recompile. You should see some compiler errors. In the original recipe, we created some utility functions to simplify the utilization of the imported class. Now, we've got to do the same, but we should not change the generated file. Instead, we'll create another unit that uses the generated one and declares the utility functions. Let's create the new unit and name it `AndroidSDK.Toast.Utils.pas`. As you can see, we are using simple conventions for unit names. If the unit has been generated by the `Java2OP`, the name respects the following format:

```
AndroidSDK.Java2OP.<OriginalClassName>.pas
```

If the unit that contains the handwritten code is relative to the class utilization, respect the following format:

```
AndroidSDK.<OriginalClassName>.Utils.pas
```

However, any convention is good as far as it makes it simple to understand whether the file has been generated by `Java2OP` or has been handwritten.

In the new file, we have to declare some types and functions to make the raw Java interfaces more usable and more Delphi-like. The code is functionally the same as the first handwritten version. Here are the file contents:

```
unit AndroidSDK.Toast.Utils;

interface

{$SCOPEDENUMS ON}

type
  TToastDuration = (Short = 0, Long = 1);
  TToastPosition = (default = 0, Top = 48,
    Bottom = 80, Center = 17,
    VerticalCenter = 16, HorizontalCenter = 1);

procedure ShowToast(const AText: string;
  const ADuration: TToastDuration = TToastDuration.Short;
  const APosition: TToastPosition = TToastPosition.Default);
  .
implementation

uses
  FMX.Helpers.Android,
  AndroidAPI.Helpers,
```

```
    AndroidSDK.Java2OP.Toast;

  procedure ShowToast(const AText: string;
    const ADuration: TToastDuration = TToastDuration.Short;
    const APosition: TToastPosition = TToastPosition.Default);
  var
    Toast: JToast;
  begin
    Toast := TJToast.JavaClass.makeText(TAndroidHelper.Context,
      StrToJCharSequence(AText), Integer(ADuration));
    if APosition <> TToastPosition.Default then
      Toast.setGravity(Integer(APosition), 0, 0);
    Toast.show();
  end;

  end.
```

Now, if you try to recompile, you should get some compiler errors because
`TJToast.LENGTH_SHORT` is no defined. This is because the `Java2OP` generated file has this
identifier defined as `TJToast.JavaClass.LENGTH_SHORT`. Change it, as now the project
should compile and run as with the previous one, but using an auto-generated class
interface from the Java android classes.

There's more...

`Java2OP` is quite powerful. It can generate interfaces from classes, JAR files, or plain Java
source files. At the command prompt, type `Java2OP` without parameters to get a list of the
available switches.

Also, you can check the official documentation to properly understand the possibilities:
`http://docwiki.embarcadero.com/RADStudio/en/Java2OP.exe,_the_Native_Bridge_`
`File_Generator_for_Android`.

Just as a side note, in some Delphi installations Java2OP doesn't generate the correct
interfaces. If you have experienced this behavior, try to download the following version
directly from code central: `http://cc.embarcadero.com/item/30007`.

Doing it in the background, the right way – Android services

In this recipe, we'll be introduced to the fantastic world of Android services! As you probably know, Android was multitasking from the very first version. Multitasking is not a simple thing for an operating system running on limited hardware. Let's think about the memory that could be allocated for days, or weeks, to some specific processes with the user that runs new apps over and over again. At some point, the memory will run out and the OS will have to decide whether to prevent a new app from starting or to eliminate some old processes that the user hasn't used for a while. Obviously, the second option is the best. To allow new apps to run, the OS needs to free some memory still allocated to other apps. At this point, there is another question: *Which apps can be removed from the memory?*

Let's leave this question unanswered for a moment and talk about the Android OS components. Android is a complex OS composed of a lot of different components, but the principals are activities and services. As Google says (`http://developer.android.com/guide/components/activities.html`):

> *"An Activity is an application component that provides a screen with which users can interact in order to do something, such as dial the phone, take a photo, send an email, or view a map. Each activity is given a window in which to draw its user interface. The window typically fills the screen, but may be smaller than the screen and float on top of other windows."*

Let's say that, in Delphi terminology, an activity is defined as a form from the user's point of view (but technically speaking, for the Android OS, a Delphi app is composed of only one activity, which will contain all the created forms). What we've created so far using Delphi are owner-drawn forms that are hosted in a special native activity provided by Embarcadero and able to render FireMonkey graphics. Using the designer, we define the graphical aspect of the FireMonkey-based activity. What are services? Again, Google says (`http://developer.android.com/guide/components/services.html`):

> *"A Service is an application component that can perform long-running operations in the background and does not provide a user interface. Another application component can start a service and it will continue to run in the background even if the user switches to another application. Additionally, a component can bind to a service to interact with it and even perform interprocess communication (IPC). For example, a service might handle network transactions, play music, perform file I/O, or interact with a content provider, all from the background."*

As they have experienced it, Android developers very well know that there is a clear criteria and priority to decide which application's components can be killed to recover the memory. The simplified criteria is as follows:

- Activities that have more chance to be destroyed than services
- Background activities that have more chance to be destroyed than foreground activities

After this long introduction, let's talk about the recipe we'll be analyzing in this section. It is a simple app provided to a virtual newspaper reporter who has to collect people's answers to some questions. The virtual newspaper reporter walks around main city streets and asks people some questions, then he has to send this data to the central editorial staff for statistical analysis in real time. Obviously, this is a demo, and in a real-world app, it could be better to have a local database acting as the queue storage for the information to be sent. But for now, we want to keep things simple enough just to show you how to interact with the Android service. If you need a more robust solution, you can always integrate a database using the concept we have already seen in the recipe about databases on mobile.

Getting ready

Our app will have an activity to get the information and a service to push that information to the remote server. Using a service, we also know that if the reporter stops for lunch and his phone's memory is running low, Android will not kill the process where the service is running too fast.

As we mentioned, an Android service is an application without a user interface that performs background tasks.

There are essentially two types of services:

- **Started service**: This kind of service is started by an Android application component. The service can run in the background indefinitely, even if the application is closed. This type of service usually performs a single task and automatically stops after finishing.
- **Bound service**: This kind of service only runs while it is bound to an Android application. There is an interaction between the application and the service, and it remains active until the application unbinds. More than one application can bind to the same service.

Keep in mind that your service can work both ways. It can be started to run indefinitely, and also allow binding.

Remember that a **service** is simply a component that can run in the background even when the user is not interacting with your application. Thus, you should create a service only if you really need it. In other words, if you need to perform work outside your main thread while the user is interacting with your application, then you should probably use a background thread using *TThread*, or the *PPL*, and not a service. For example, if you want to play some music but only while your app is in the foreground, you might create a thread and do it. In this case, a service is not necessary.

We chose to use the started service for our needs. Android supports different kinds of services, and Delphi allows us to create the following types:

- Local service
- Local intent service
- Remote service
- Remote intent service

A local service is private to an application, while a remote service is public and other applications can access it. The intent variations refer to the way the intent is handled by the service, and whether it is in its main thread or not (more info about Intent Services can be found here: `http://developer.android.com/reference/android/app/IntentService.html`). However, remember that the `onStartCommand` service event is always called on the main application thread in any service.

In Delphi, we have to create the service and the app as two separated projects in the same project group. You can follow the detailed tutorial in the Embarcadero DocWiki to create an application and a local service (`http://docwiki.embarcadero.com/RADStudio/en/Creating_Android_Services`).

How it works...

Let's open the `Chapter09\CODE\RECIPE07\SurveyGroup.groupproj` group project and analyze each part.

The group project is composed of three projects:

- The application itself
- The local service that the app uses to send data to the server
- The REST web service that collects the information sent by the service

The app UI is quite simple and is shown in the following screenshot:

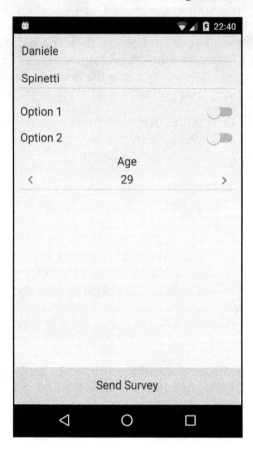

Figure 9.5: The UI or SurveyApp with some data

When the user writes the data and clicks on **Send Survey**, the inserted data is packaged in a JSON object and then sent to the local service. Here's the code under the button and the method that actually sends data to the service:

```
procedure TMainForm.actSendSurveyExecute(Sender: TObject);
var
  LJSurvey: TJSONObject;
begin
  LJSurvey := TJSONObject.Create;
  LJSurvey.AddPair('first_name', EditFirstName.Text);
  LJSurvey.AddPair('last_name', EditLastName.Text);
  LJSurvey.AddPair('option1', TJSONBool.Create(SwitchOption1.IsChecked));
  LJSurvey.AddPair('option2', TJSONBool.Create(SwitchOption2.IsChecked));
```

```
  LJSurvey.AddPair('age', TJSONNumber.Create(SpinBoxAge.Value));
  SendSurveyInterview(LJSurvey);
end;

procedure TMainForm.SendSurveyInterview(ASurveyData: TJSONObject);
var
  LIntent: JIntent;
begin
  LIntent := TJIntent.Create;
  LIntent.putExtra(StringToJString('data'),
StringToJString(ASurveyData.ToJSON));
  StartService('MySurveyService', LIntent);
end;

procedure TMainForm.StartService(const AServiceName: string;
  const AIntent: JIntent);
var
  LService: string;
begin
  LService := AServiceName;
  if not LService.StartsWith('com.embarcadero.services.') then
    LService := 'com.embarcadero.services.' + LService;
  AIntent.setClassName(TAndroidHelper.Context.getPackageName(),
    TAndroidHelper.StringToJString(LService));
  TAndroidHelper.Context.StartService(AIntent);
end;
```

So far, we've packaged data and sent that data to the service. But we would like to know when the data was actually sent to the WebService log, but only if the app is in the foreground. To do this, the service will broadcast an intent when the WebService log replies. To do that, we have to register and configure a BroadcastReceiver. We must carefully handle this registration code, because we have to keep the registration only for the necessary time, not longer. For instance, we cannot register the broadcast receiver and then forget about it; we have to unregister it in the Form-Close event, so that no notification looks for a receiver no longer alive:

```
procedure TMainForm.FormCreate(Sender: TObject);
var
  Filter: JIntentFilter;
begin
  //TMyReceiver is a custom     class which will
  //handle the notification
  FMyListener := TMyReceiver.Create;

  //configure a new broadcastreceiver using our own listener
  FBroadcastReceiver := TJFMXBroadcastReceiver.JavaClass.init(FMyListener);
```

```
    //to which intents are we interested? Let's configure
    //an intent filter to inform the Android OS
    //about our interests
    Filter := TJIntentFilter.JavaClass.init;
    //success response action
    Filter.addAction(StringToJString(TSurveyConstants.SURVEY_RESPONSE));
    //fail response action
 Filter.addAction(StringToJString(TSurveyConstants.SURVEY_RESPONSE_ERROR));

    //actually register the receiver
    TAndroidHelper.Context.registerReceiver(FBroadcastReceiver, Filter);
 end;

 procedure TMainForm.FormClose(Sender: TObject; var Action: TCloseAction);
 begin
    //unregister the receiver
 TAndroidHelper.Context.getApplicationContext.unregisterReceiver(FBroadcastR
 eceiver);
 end;
```

The action constants used in these event handlers are defined in a separate shared unit named ConstantsU.pas, as follows:

```
 type
   TSurveyConstants = class sealed
   public
     const
       SURVEY_RESPONSE = 'SURVEY_RESPONSE';
     const
       SURVEY_RESPONSE_ERROR = 'SURVEY_RESPONSE_ERROR';
   end;
```

What about the TMyReceiver class, which is the actual handler for our response data coming from the service? Here it is in the following code, as declared in the BroadcastReceiverU unit:

```
 unit BroadcastReceiverU;

 interface

 uses
   Androidapi.JNIBridge,
   Androidapi.JNI.Embarcadero,
   Androidapi.JNI.GraphicsContentViewText;

 typqe
   TMyReceiver = class(TJavaLocal, JFMXBroadcastReceiverListener)
   public
```

```
    procedure onReceive(context: JContext; intent: JIntent); cdecl;
  end;

implementation

uses
  Androidapi.Helpers, Androidapi.JNI.JavaTypes,
  Androidapi.JNI.Widget, UtilsU, ConstantsU, LogU;

procedure TMyReceiver.onReceive(context: JContext; intent: JIntent);
var
  LText: string;
begin
  //if the service reply with an error, show it to the user
  //with specific info
  if JStringToString(intent.getAction) =
TSurveyConstants.SURVEY_RESPONSE_ERROR then
  begin
    LText :=
JStringToString(intent.getStringExtra(StringToJString('error')));
    LText := 'Error: ' + sLineBreak + LText;
  end
  else
  begin
    //this is success response
    LText :=
JStringToString(intent.getStringExtra(StringToJString('result')));
    LText := 'Just Arrived: ' + sLineBreak + LText;
  end;

 //show a toast and log it as debug
  ShowToast(LText);
  LogInfo('Broadcast Received = %s -> %s',
[JStringToString(intent.getAction), LText], 'survey');
end;

end.
```

In this code, there are some utility units written to help the development. We will speak about them in the following sections.

Now that we know how the app works, we can talk about the service.

We chose an Android local service for our needs. This is the most typical choice when the Android application interacts directly with the Android service, both running on the same process.

The wizard creates a `Data Module` with the `TAndroidService` class from the `System.Android.Service` unit, with the necessary events:

```
unit MainServiceU;

interface

uses
  System.SysUtils,
  System.Classes,
  System.Android.Service,
  AndroidApi.JNI.GraphicsContentViewText,
  AndroidApi.JNI.Os;

type
  TAndroidServiceDM = class(TAndroidService)
    function AndroidServiceStartCommand(const Sender: TObject;
      const Intent: JIntent; Flags, StartId: Integer): Integer;
  private
    procedure BroadcastResponse(Value: String);
    procedure BroadcastException(const E: Exception);
  public
  end;

const
  BASE_URL = 'http://192.168.1.106:8080'; //put your IP

var
 AndroidServiceDM: TAndroidServiceDM;

implementation

{%CLASSGROUP 'FMX.Controls.TControl'}

{$R *.dfm}

uses
  AndroidApi.JNI.App, AndroidApi.Helpers, System.JSON,
  System.Net.HTTPClient, AndroidApi.JNI.JavaTypes,
  LogU, ConstantsU;
```

```
function TAndroidServiceDM.AndroidServiceStartCommand(
  const Sender: TObject;
  const Intent: JIntent; Flags, StartId: Integer): Integer;
var
  LJSONString: string;
Begin
  //Using START_STICKY in case of restarting the intent is not
  //redelivered, so we have to check its existence
  if Assigned(Intent) and Intent.hasExtra(StringToJString('data')) then
  begin
    LJSONString :=
JStringToString(Intent.getStringExtra(StringToJString('data')));
    TThread.CreateAnonymousThread(
      procedure
      var
        LHTTP: THTTPClient;
        LData: TStringStream;
        LResp: IHTTPResponse;
      begin
        LHTTP := THTTPClient.Create;
        LData := TStringStream.Create(LJSONString, TEncoding.UTF8);
        LData.Position := 0;
        LogInfo('Sending data to %s/surveys', [BASE_URL], 'surveyservice');
        try
          LResp := LHTTP.Post(BASE_URL + '/surveys', LData);
          BroadcastResponse(LResp.ContentAsString);
        except
          on E: Exception do
          begin
            BroadcastException(E);
          end;
        end;
        JavaService.stopSelfResult(StartId);
        LogWarning('Stopped StartId=%d', [StartId], 'surveyservice');
      end).Start;
  end
  else
  begin
    LogWarning('Service started, but no intent delivered', [],
'surveyservice');
  end;

  // We want this service to continue running until it is
  // explicitly stopped and restarted if killed,
  // so return START_STICKY.
  Result := TJService.JavaClass.START_STICKY;
end;
```

```
procedure TAndroidServiceDM.BroadcastException(const E: Exception);
var
  LJIntent: JIntent;
begin
  LJIntent := TJIntent.Create;
LJIntent.setAction(StringToJString(TSurveyConstants.SURVEY_RESPONSE_ERROR))
;
  LJIntent.putExtra(StringToJString('error'), StringToJString(E.ClassName +
': ' + E.Message));
  TAndroidHelper.Context.sendBroadcast(LJIntent);
end;

procedure TAndroidServiceDM.BroadcastResponse(Value: String);
var
  LJIntent: JIntent;
begin
  LJIntent := TJIntent.Create;
  LJIntent.setAction(StringToJString(TSurveyConstants.SURVEY_RESPONSE));
  LJIntent.putExtra(StringToJString('result'), StringToJString(Value));
  TAndroidHelper.Context.sendBroadcast(LJIntent);
end;

end.
```

All this simple service is contained inside the OnStartCommand event handler.

This event is called by the system every time a client explicitly starts the service by calling startService(), providing the arguments it supplied and a unique integer token representing the start request. Here is an explanation of their parameters.

The Intent parameter is the same passed to startService() by the activity. If the service restarts after its process is terminated, the value of the Intent could be nill. The Flags parameter contains extra content about this start request. The startId parameter is a unique integer provided by the Android OS to represent this specific request to start. It can be used with stopSelfResult(). The StartCommand event handler runs in the main thread, so we cannot perform any long operations here just as in the form code. Indeed, we first create a new background thread to do the actual work. Inside this thread, we make the actual HTTP request to the remote service and package the response for the listener.

When the response has been packaged in a proper JSON object, the thread calls `JavaService.stopSelfResult(StartId)`. What is that? Well, any Android service has its own life cycle. A started service's life cycle is handled by the service itself. The service continues to run after the `OnStartCommand()` event completes, and the reason the system could stop or destroy it is to recover memory. There are two different options to stop a service: in the first one, the service itself calls the `stopSelf()` function, while the other one is that another component calls the `stopService()` method. A started service can handle its stop independence of the activity that starts it. There are at least two ways in which a service can stop itself:

Method to stop the service	Meaning
`JavaService.stopSelf()`	Stop the service if it was previously started.
`JavaService.stopSelfResult(startId)`	Stop the service if the last time it was started was `startId`.

When the `stopSelf()` method is used to stop the service, the system destroys the service as soon as possible. However, it is necessary to go into detail of some aspects. For example, the service just mentioned handles concurrently multiple requests to `onStartCommand()`. Then it could be wrong to stop the service when you're processing a start request because you might have received a new start request since then (stopping at the end of the first request would terminate the second one). To avoid this problem, you can use `stopSelfResult(startId)` to ensure that your request to stop the service is always based on the most recent start request. That is, when you call `stopSelfResult(startId)`, you pass the `startId` of the start request (the `startId` delivered to the `OnStartCommand` event handler) to which your stop request corresponds. Then, if the service received a new start request before you were able to call `JavaService.stopSelfResult(startId)`, then the `startId` will not match and the service will not stop.

As a last note, let's talk about the value returned by the `OnStartCommand` event. This value is really important to understand the service life cycle. If `OnStartCommand` returns START_STICKY, the system will try to recreate your service after it is killed. If `OnStartCommand` returns START_NOT_STICKY, the system will not try to recreate your service after it is killed. As a last option, among the most popular, there is START_REDELIVER_INTENT, which is particularly handy. If `OnStartCommand` returns START_REDELIVER_INTENT if this service's process is killed while it is started, then it will be scheduled for a restart, and the last delivered intent will be redelivered to it again via a standard intent as it would be the first time.

But what about the data sent itself? Where is the serialized JSON object send? The third project in the group is called `SurveyWebService.dproj`, and is a `DelphiMVCFramework` RESTful service, which has one controller called `TSurveyCollector` and one action called `CreateSurveyResponse`. This project has been created from a nice wizard included in the `DelphiMVCFramework`. The `WebModule` is quite standard, while the controller is the following:

```
unit SurveysCollectorCtrlU;

interface

uses
  MVCFramework;

type
  [MVCPath('/surveys')]
  TSurveyCollector = class(TMVCController)
  public
    [MVCPath]
    [MVCHTTPMethod([httpPOST])]
    procedure CreateSurveyResponse;
  end;

implementation

uses
  System.SysUtils, System.IOUtils, System.JSON,
  MVCFramework.Commons, MVCFramework.Logger;

procedure TSurveyCollector.CreateSurveyResponse;
begin
  Log('Request data: ' + ctx.Request.Body);
  Log('Wait a bit...');
  Sleep(5000); //just to mimic a long operation
  if Context.Request.ThereIsRequestBody then
  begin
    //if there is a body request just send OK
    Render(HTTP_STATUS.OK, TJSONObject.Create(TJSONPair.Create('result',
'ok')))
  end
  else
  begin
      Render(HTTP_STATUS.BadRequest,
TJSONObject.Create(TJSONPair.Create('result', 'ko')));
  end;
```

```
   Log('Response sent');
  end;
  end.
```

As you can see, the server acts like a potentially complete backend system, but it is really only a fake server, which doesn't even do anything with the sent data and is just a way to show how it is possible to communicate with a remote server from an Android service.

So, do you want to run this distributed project? Let's start:

1. Run the `SurveyWebService.dproj` without debugging.
2. Retrieve your IP Address (`ipconfig` in the command prompt).
3. In the file `MainServiceU.pas`, change the constant `BASE_URL` according to your IP:

```
const
  BASE_URL = 'http://YOUR_IP_ADDRESS_HERE:8080';
```

4. Now, build the project `libMySurveyService.so`.
5. Select the project `MySurveyApp.dproj`, expand the node **Target Platform**, select **Android**, and then right-click on the **Android** platform and select **Add Android Service**.
6. From the resultant open dialog, select **Search file automatically from project base path** and then the folder where the service is – you can recognize the folder because you should see two folders named `Android` and `JavaClasses`). Click on **Select** and then follow the wizard to the end, where the final summary is shown. Check the files and click on **Finish**. This strange step is required to let the app know its service.
7. Now you can run the `MySurveyApp.dproj` project.
8. Write some data in the UI widget and tap **Send Survey**.
9. After some seconds, a `Toast` should inform you that the data has been correctly sent, and the `WebService` log should say the same thing (check the folder where the `WebService` executable is).

There's more...

Lot of things in this recipe! Delphi 10 Seattle is the first version of Delphi, that allows creating Android services. There is still some room for improvement, but this integration is already very powerful.

Before you start to write real-world apps using the Android service, I strongly encourage you to study the following documentation:

- Creating Android services: `http://docwiki.embarcadero.com/RADStudio/en/Creating_Android_Services`
- Differences and features of each kind of Android service (in Delphi): `http://docwiki.embarcadero.com/RADStudio/en/Android_Service`
- Android service introduction by Google: `http://developer.android.com/reference/android/app/Service.html`

As promised, I will talk about the other units involved in the project. At the time of writing, the Android system logger is provided by the unit `FMX.Types`. However, the unit is not compatible with services because it looks for an activity context. Waiting for an update, I wrote the `LogU.pas`, which completely exposes the power of the Android logger and works also in Delphi 10 Seattle services. The unit `UtilsU.pas` is simpler because it's only a small wrapper around the `TJToast` class. It defines a handy `ShowToast` procedure, which is simpler to call compared to the raw `TJToast` creation. Obviously, this unit can be used outside this project too.

10
Delphi and IoT

In this chapter, we will cover the following recipes:

- How to blink an LED using Arduino
- How to drive multiple relays with Arduino and Delphi
- Reading data from Arduino
- How to blink an LED using Raspberry Pi
- How to drive multiple relays with Raspberry Pi and Delphi
- Reading data from Raspberry Pi

Introduction

Internet of Things (**IoT**) is a buzzword today, so what is IoT? IoT is a network of physical devices (such as smartphones and vehicles) that connect and exchange data. In practical terms, IoT identifies the ability of physical devices to detect and collect data from the world around us, and then share this data through the internet. Once the data is shared, it can be processed and used for different purposes.

Imagine what it means to have access to any programmable electronic device with connectivity: from the car, to the lights in your home, or to your thermostat. Wherever we are, we can query them, we can control them, and we can command them.

In this chapter, we will see how to program and control the standard references for micro-controllers and single-board computers in the world—**Arduino** and **Raspberry Pi**, respectively.

Prerequisites

Before you start controlling the world around you, there are some prerequisites to respect.

Arduino

Arduino UNO is the board used in the following recipes because, as the Arduino documentation states, *The UNO is the best board to get started with electronics and coding—The UNO is the most used and documented board of the whole Arduino family.*

The recipes regarding Arduino expect that you have basic knowledge of the following:

- What Arduino is
- Which IDE to use
- What is Sketch is, how to write one, and how to upload one

If you do not have this knowledge, you need to acquire it before you continue because the aforementioned topics are not covered in this chapter. If you need it, here is the link to Arduino's getting started page for more information: `https://www.arduino.cc/en/Guide/HomePage`.

Raspberry Pi

Raspberry Pi 3 Model B is used in the follow recipes. The recipe's regarding Raspberry Pi expect that you have basic knowledge of it. In particular, in the following list, I provide the necessary points to continue with the recipes on Raspberry Pi. Each point is followed by a link with necessary documentation in case you do not already have this knowledge:

- What is Raspberry Pi? (`https://www.Raspberry Pi.org/`)
- Getting started (`https://projects.Raspberry Pi.org/en/projects/raspberry-pi-getting-started`).
- Be up and running with Raspbian OS (`https://www.Raspberry Pi.org/learning/software-guide/`).
- Heart of any Raspberry Pi help—documentation, troubleshooting, videos, hardware, and so on (`https://www.Raspberry Pi.org/help/`).
- Raspberry Pi GPIO (`https://www.Raspberry Pi.org/documentation/usage/gpio/`).

The **Raspberry Pi OS** (**Raspbian**) version used in these recipes is **Jessie**, but with due consideration with what is used in this specific recipe, it is very likely that it will work with later versions of the OS.

Arduino versus Raspberry Pi

I would like to list some features in order to clarify the differences between Arduino and Raspberry, as follows:

Arduino	Raspberry Pi
Is not a computer, at least not the way most of us think about computers	Is a **single-board computer** (**SBC**)
It is a micro-controller—a very simple processor designed to process information in a very simple way	It's just like a regular computer, except is very small and has easily accessed input and output pins

Both can work in symbiosis when it comes to the low-level flexibility of Arduino and Raspberry's most powerful computing power and operating system.

An example might be a case where one or more Arduino system deals with collecting data from sensors and sending it to the Raspberry Pi for archiving, managing, and viewing on the web.

How to blink an LED using Arduino

If you are a programmer, you know that the goal of any first program consists of writing enough code to show the sentence **Hello World!** on a screen. The blinking LED is the **Hello World!** of physical computing.

In this recipe, you will learn how write a Delphi application and the relative Arduino Sketch to send commands to your Arduino to make the Arduino's built-in LED blinking.

Getting ready

In this recipe, we have to communicate to the serial port to tell Arduino to turn the blinking LED light ON or OFF. To do this, we need a library/components for Serial Communications purposes—**ComPort** (`https://sourceforge.net/projects/comport/`). To install the ComPort library, perform the following steps:

1. Download it from this link: `https://sourceforge.net/projects/comport/`
2. Extract it in your preferred `Libs` folder (for example, `D:\Dev\Delphi\Libs`)
3. Open the design time project, `DsgnCPortDXE.dpk`, and install it
4. Add the `source` folder of this `Libs` to the `Library path` (*F6* and then type `Library Path`)

How to do it...

Perform the following steps:

1. Create a new **VCL Forms Application: File | New | VCL Forms Application**.
2. Click on the **Save All** icon and use these names: `MainFormU.pas` for the form unit file and `ArduinoBlinkALed` for the project's name.
3. On the form surface, put one **TComPort**, two **TButton**, one **TLabel**, and one **TToggleSwitch**. Make all the necessary changes to make sure that your form looks like the following screenshot:

Figure 10.1: MainForm at design time

4. Implement the **OnClick** event of `btnSetup` and put this code under the event:

```
procedure TMainForm.btnSetupClick(Sender: TObject);
begin
  ComPort1.ShowSetupDialog;
end;
```

5. Implement the **OnClick** event of `btnConnection` and put this code under the event:

```
procedure TMainForm.btnConnectionClick(Sender: TObject);
begin
  // If the port is connected.
  if ComPort1.Connected then
    ComPort1.Close // Close the port.
  else
    ComPort1.Open; // Open the port.
  UpdateComponentsState;
end;
```

6. Now, we have to implement the **OnClick** event of `blinkSwitch` to write on ComPort and send the command to Arduino:

```
procedure TMainForm.blinkSwitchClick(Sender: TObject);
begin
  case blinkSwitch.State of
    tssOff:
      ComPort1.WriteStr('BLINK_OFF');
    tssOn:
      ComPort1.WriteStr('BLINK_ON');
  end;
end;
```

7. Declare and implement the `UpdateComponentsState` procedure to make graphic changes depending on whether the ComPort is connected or not:

```
procedure TMainForm.UpdateComponentsState;
begin
  blinkSwitch.Enabled := ComPort1.Connected;
  Label1.Enabled := ComPort1.Connected;
  btnConnection.Caption := ifthen(ComPort1.Connected, 'Close',
'Open');
end;
```

8. Now, we have to make some considerations when the form is shown and when it is destroyed. Therefore, implement the **FormShow** and **FormDestroy** events:

```
procedure TMainForm.FormShow(Sender: TObject);
begin
  UpdateComponentsState;
end;

procedure TMainForm.FormDestroy(Sender: TObject);
begin
  if ComPort1.Connected then
    ComPort1.Close;
end;
```

9. The Delphi code is completed. It is necessary to write the Arduino Sketch to make it react correctly to the commands sent by the Delphi application. The following code is for Arduino Sketch. The code is easy to interpret and several comments have been added to make it easier to read:

```
/*
  Blink Built-in LED
  Activate built-in LED blinking when a specific command arrive.
*/

String command;
char receivedChar;
boolean blinkMode;

// the setup function runs once when you press reset or power the
board
void setup() {
  // initialize digital pin LED_BUILTIN as an output.
  pinMode(LED_BUILTIN, OUTPUT);
  // initialize Serial to max data rate
  Serial.begin(115200); // Port serial 115200 baud.
}

// read from serial and concat it to command
void readFromSerial() {
  while (Serial.available() > 0) {
    receivedChar = Serial.read();
    command.concat(receivedChar);
    delay(10);
  }
}

// do the led blink
```

```
void doBlink() {
  digitalWrite(LED_BUILTIN, HIGH); // turn the LED on (HIGH is the
voltage level)
  delay(1000); // wait for a second
  digitalWrite(LED_BUILTIN, LOW); // turn the LED off by making the
voltage LOW
  delay(1000); // wait for a second
}

// the loop function runs over and over again forever
void loop() {

  readFromSerial();

  if (command.equals("BLINK_ON") == true) {
    blinkMode = true;
  } else if (command.equals("BLINK_OFF") == true) {
    blinkMode = false;
  }

  if (blinkMode) {
    doBlink();
  }

  // Clean the command to receive the next
  command = "";
}
```

10. Verify and upload the Sketch on Arduino.
11. Start the Delphi application—**Run** | **Run** or hit *F9*:

Figure 10.2: Delphi Application at start-up

12. Click on the `ComPort Setup` button to set up the `Port` and `Baud Rate`. When the dialog is shown, set at `Port` the ComPort where Arduino is connected and the `Baud Rate` to 1,15,200.

> You can see in **Control Panel** | **System** | **Device Manager** under the **Ports (COM & LPT)** section, the ComPort where Arduino is connected.

13. Click **Open**.
14. If the connection is successful, the **Arduino LED Blinking** label and the switch will become enabled.
15. Click on the switch to make the LED blink on Arduino.

How it works...

When you click on `blinkSwitch`, one of two commands is sent to Arduino: `BLINK_ON` or `BLINK_OFF`, depending on the status of the switch. The fundamental part, the Delphi side, resides in the following function (this allows you to send the command):

```
procedure TMainForm.blinkSwitchClick(Sender: TObject);
begin
  case blinkSwitch.State of
    tssOff:
      ComPort1.WriteStr('BLINK_OFF');
    tssOn:
      ComPort1.WriteStr('BLINK_ON');
  end;
end;
```

The preceding code is used for the `WriteStr` function of `TComPort` to send the command to Arduino.

Regarding the Arduino Sketch (point *9* of *How to do it...*), it is appropriate to say a few words:

- In the `setup` function, the built-in LED is defined as an output pin and the serial port is initialized

- In the main loop, all the logic takes place:
  1. Reading from the serial is performed (`readFromSerial()`).
  2. If the `BLINK_ON` command is read from the serial, then the Boolean variable `blinkMode` is set to true. Unless the `BLINK_OFF` command is sent, the blinking described in the `doBlink()` function is performed `for each` loop.
  3. If the `BLINK_OFF` command is read from the serial, then the Boolean variable `blinkMode` is set to false and no operation on the built-in LED is performed.

There's more...

The built-in LED has been chosen so that no parts related to the electronic circuitry have to be added. This means that we have a minimal impact as the first demo. Obviously, the initialization of the LED and other features can easily be applied to the other pins and therefore to any other components. The base mechanism has already been shown. A recommendation I want to leave as we are talking about electronic circuits, however, is this:

> *"The voltage level output of a digital output pin is the same voltage that is powering the chip on the board. Most arduino boards use a +5vdc for the Vcc value, so output pins go from 0 to just short of +5vdc; however if the chip is powered from a 3.3vdc Vcc value, then the output pin will go up to just short of 3.3vdc. Output current limits are also related to what voltage the chip is running at, but absolute maximum when running at +5vdc is 40ma, with 20ma a recommended continuous working current rating limit."*

> *– Arduino official forum*

The baud rate indicates the data transmission speed. Arduino supports this rate for communicating with the computer serial: 300, 600, 1,200, 2,400, 4,800, 9,600, 14,400, 19,200, 28,800, 38,400, 57,600, or 1,15,200. In this recipe, we used the maximum rate.

See also

Finally, I will leave you with some really useful links:

- Arduino official website: `https://www.arduino.cc`
- Arduino official forum (for questions and troubleshooting): `https://forum.arduino.cc/`

How to drive multiple relays with Arduino and Delphi

A **relay** is a component that's able to open or close a circuit depending on the presence or absence of voltage at the input. In practice, it is a switch operated by an electromagnet. Let's understand this basic concept, learn how to open and close a circuit and make it programmable, and understand well how this is the starting point to open up more complex and interesting scenarios such as the remote control of devices, home automation, and more.

In this recipe, we will see how to open and close a circuit using a Relay Module and how to make its use programmable through a Delphi application.

Getting ready

The purpose of this recipe is to turn two different light bulbs ON or OFF through the use of a **2 Relay Module** and make it controllable through an application written in Delphi.

In this recipe, we are going to use the following:

- Arduino UNO
- Two Relay Modules (`http://wiki.sunfounder.cc/index.php?title=2_Channel_5V_Relay_Module`)
- Two light bulbs and relative supply power (the important thing is that the light bulbs have their own power supply)
- Jumper wires

In this recipe, we need a library/components for serial communications purposes—ComPort. Check the *Getting ready* section of the previous recipe to understand how to configure it and how it works.

How to do it...

Before moving to the software side, you must properly configure the electronic components:

1. Perform all the necessary steps to build the circuit, as shown in the following diagram:

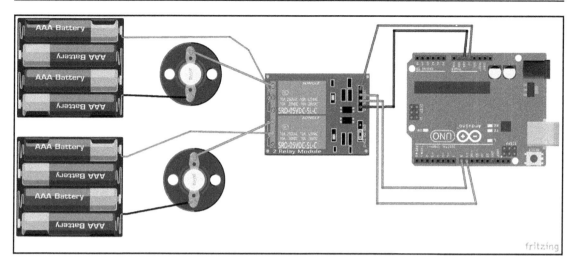

Figure 10.3: Circuit sketch

2. Compile and upload the following Sketch to Arduino:

```
int relay_pin_1 = 8;
int relay_pin_2 = 9;
String command;
char receivedChar;

// read from serial and concat it to command
void readFromSerial() {
  while (Serial.available() > 0) {
    receivedChar = Serial.read();
    command.concat(receivedChar);
    delay(10);
  }
}


void setup() {
  // put your setup code here, to run once:
  pinMode(relay_pin_1, OUTPUT);
  pinMode(relay_pin_2, OUTPUT);

  digitalWrite(relay_pin_1, HIGH);
  digitalWrite(relay_pin_2, HIGH);

  // initialize Serial to a data rate
  Serial.begin(9600); // Port serial 9600 baud.
}
```

```
void doCommand() {
    // turn light bulb 1 on
    if (command.equals("L1_ON") == true) {
        digitalWrite(relay_pin_1, LOW);
        return;
    }
    // turn light bulb 1 off
    if (command.equals("L1_OFF") == true) {
        digitalWrite(relay_pin_1, HIGH);
        return;
    }

    // turn light bulb 2 on
    if (command.equals("L2_ON") == true) {
        digitalWrite(relay_pin_2, LOW);
        return;
    }
    // turn light bulb 2 off
    if (command.equals("L2_OFF") == true) {
        digitalWrite(relay_pin_2, HIGH);
        return;
    }

}

void loop() {
    // put your main code here, to run repeatedly:
    readFromSerial();
    doCommand();
    delay(200); // wait for a second

    // Clean the command to receive the next
    command = "";

}
```

Now, it's time to develop the Delphi application. Perform the following steps:

1. Create a new **VCL Forms Application: File | New | VCL Forms Application**.
2. Click on the **Save All** icon and use these names: `MainFormU.pas` for the form unit file and `ArduinoMultipleRelays` for the project name.

3. On the form's surface, put one **TComPort**, two **TButton**, two **TLabel**, and two **TToggleSwitch**. Make all the necessary changes to make sure that your form looks like the following screenshot:

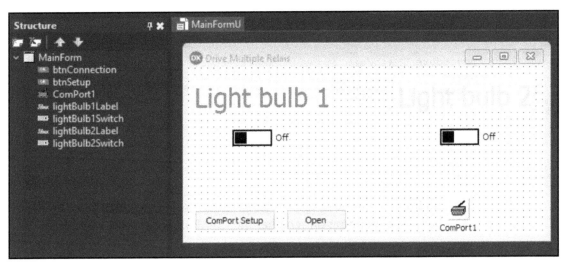

Figure 10.4: Arduino Multiple relays form at design time

4. The `btnSetup` button's **OnClick** event and the `btnConnection` button's **OnClick** event are the same as in the previous recipe:

```
procedure TMainForm.btnSetupClick(Sender: TObject);
begin
  ComPort1.ShowSetupDialog;
end;
procedure TMainForm.btnConnectionClick(Sender: TObject);
begin
  // If the port is connected.
  if ComPort1.Connected then
    ComPort1.Close // Close the port.
  else
    ComPort1.Open; // Open the port.
  UpdateComponentsState;
end;
```

5. Now, we have to implement the **OnClick** events of `lightBulb1Switch` and `lightBulb2Switch` to write on ComPort and send the command to Arduino:

```
procedure TMainForm.lightBulb1SwitchClick(Sender: TObject);
begin
```

```
      case lightBulb1Switch.State of
        tssOff:
          ComPort1.WriteStr('L1_OFF');
        tssOn:
          ComPort1.WriteStr('L1_ON');
      end;
    end;

    procedure TMainForm.lightBulb2SwitchClick(Sender: TObject);
    begin
      case lightBulb2Switch.State of
        tssOff:
          ComPort1.WriteStr('L2_OFF');
        tssOn:
          ComPort1.WriteStr('L2_ON');
      end;
    end;
```

6. Declare and implement the `UpdateComponentsState` procedure to make graphic changes, depending on whether the ComPort is connected or not:

```
    procedure TMainForm.UpdateComponentsState;
    begin
      lightBulb1Switch.Enabled := ComPort1.Connected;
      lightBulb2Switch.Enabled := ComPort1.Connected;
      lightBulb1Label.Enabled := ComPort1.Connected;
      lightBulb2Label.Enabled := ComPort1.Connected;
      btnConnection.Caption := ifthen(ComPort1.Connected, 'Close',
    'Open');
    end;
```

7. Now, we have to make some considerations when the form is shown and when it is destroyed. So, implement the **FormShow** and **FormDestroy** events like so:

```
    procedure TMainForm.FormShow(Sender: TObject);
    begin
      UpdateComponentsState;
    end;

    procedure TMainForm.FormDestroy(Sender: TObject);
    begin
      if ComPort1.Connected then
        ComPort1.Close;
    end;
```

8. Verify and upload the Sketch on Arduino.
9. Start the Delphi application: **Run | Run** or hit *F9:*

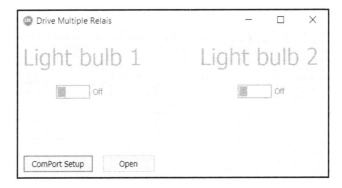

Figure 10.5: Delphi Application at start-up

10. Click on the `ComPort Setup` button to set up the `Port` and `Baud Rate`. When the dialog is shown, set the ComPort at `Port`, where Arduino is connected and the `Baud Rate` is set to 9,600.

> You can see in **Control Panel | System | Device Manager** under the **Ports (COM & LPT)** section the ComPort where Arduino is connected.

11. Click **Open**.
12. If the connection is successful, the labels and the switches will become enabled.
13. Click on the `lightBulb1Switch` or `lightBulb2Switch` switch to turn the specific light bulb ON or OFF on your Arduino.

How it works...

Basically, the relay is nothing more than a switch that can be operated through a digital signal (high or low). There are two types of configuration: **NO** and **NC**. They stand for normally open and normally closed. This means that to pass a current in the first configuration, a low signal must be sent, while for the second configuration, it is the opposite.

From this explanation, it is very easy to understand the Arduino Sketch:

1. In the `setup` function, the different pins are configured, but I want to dwell on two initializations:
   - `digitalWrite(relay_pin_1, HIGH);` and `digitalWrite(relay_pin_2, HIGH);`
   - We are connecting to the relay in NO (normally open configuration), so the preceding lines ensure that the relay is initially OFF (in this case, it's HIGH)

2. In the `loop()`, we are just switching the state from `HIGH` to `LOW` for both of the relays to make sure that the light bulbs light up. This happens in the `doCommand()` function, which, thanks to the serial reading performed by the `readFromSerial()` function, understands which is the input command and turns this into a command for the relays module:

String Command	Action	Code
L1_ON	Turn ON first light bulb	`digitalWrite(relay_pin_1, LOW);`
L1_OFF	Turn OFF first light bulb	`digitalWrite(relay_pin_1, HIGH);`
L2_ON	Turn ON second light bulb	`digitalWrite(relay_pin_2, LOW);`
L2_OFF	Turn OFF second light bulb	`digitalWrite(relay_pin_2, HIGH);`

3. Also, the Delphi code is very easy to understand. When you click on `lightBulb1Switch` or `lightBulb2Switch`, a command is sent to Arduino depending on which switch and its status in order to be consistent with the preceding table.

There's more...

In each demo the configuration for the ComPort is set but you have to know that there is a mechanism of storing and loading settings:

`Application` can easily store and load serial port settings using `StoreSettings` and `LoadSettings` methods. Settings can be stored into configuration file or registry. `StoredProps` property determines which properties need to be stored.

```
Example (Registry)
begin
  // store settings to registry
  ComPort1.StoreSettings(stRegistry,
'\HKEY_LOCAL_MACHINE\Software\ComPortTest');
  // load settings
```

```
  ComPort1.LoadSettings(stRegistry,
  '\HKEY_LOCAL_MACHINE\Software\ComPortTest');
end;

Example (Configuration file)
begin
  // store settings to configuration file
  ComPort1.StoreSettings(stIniFile, 'c:\ComPortTest.ini');
  // load settings
  ComPort1.LoadSettings(stIniFile, 'c:\ComPortTest.ini');
end;
```

See also

Here are some links to deepen your knowledge on the topics that we have covered in this recipe:

- The Relay Module used in this recipe: `http://wiki.sunfounder.cc/index.php?title=2_Channel_5V_Relay_Module`
- A tool to prototype electric circuits: `http://fritzing.org/home/`

Reading data from Arduino

In the previous recipes, we saw how to communicate something to Arduino. In this recipe, we are going to close the recipes about Arduino with the opposite operation—how to read the data that Arduino controls.

Getting ready

Communication with Arduino with the serial port provides mainly three functions:

- `Serial.read()`
- `Serial.write()`
- `Serial.print()`

The first one has already been discussed in the previous recipes, as it is used to communicate commands from Delphi to Arduino. Let's move on to the next two:

- `Serial.write()`: Writes binary data to the serial port. This data is sent as a byte or a series of bytes in order to send the characters representing the digits of a number to use the `print()` function instead (`https://www.arduino.cc/reference/en/language/functions/communication/serial/write/`).
- `Serial.print()`: Prints data to the serial port as human-readable ASCII text. This command can take many forms. Numbers are printed using an ASCII character for each digit. Floats are similarly printed as ASCII digits, defaulting to two decimal places. Bytes are sent as single characters. Characters and strings are sent as it is (`https://www.arduino.cc/reference/en/language/functions/communication/serial/print/`).

In this recipe, we need a library/components for Serial Communications purposes—ComPort. Check the *Getting ready* section of the *How to blink an LED using Arduino* recipe to understand how to configure it and how it works.

How to do it...

We will start from the previous recipe, making small changes so that you can read the status of the light bulbs and show this through the graphic components in the Delphi application.

This time, we're going to turn the light bulbs ON and OFF cyclically. So, this is the new Sketch:

```
int relay_pin_1 = 8;
int relay_pin_2 = 9;


void setup() {
  // put your setup code here, to run once:
  pinMode(relay_pin_1, OUTPUT);
  pinMode(relay_pin_2, OUTPUT);

  digitalWrite(relay_pin_1, HIGH);
  digitalWrite(relay_pin_2, HIGH);

  // initialize Serial to a data rate
  Serial.begin(115200); // Port serial 9600 baud.
}
```

```
void loop() {
  // put your main code here, to run repeatedly:
  delay(2000);
  // turn light bulb 1 on
  digitalWrite(relay_pin_1, LOW);
  Serial.print("L1_ON");
  delay(5000);

 // turn light bulb 1 off
  digitalWrite(relay_pin_1, HIGH);
  Serial.print("L1_OFF");
  delay(5000);

 // turn light bulb 2 on
  digitalWrite(relay_pin_2, LOW);
  Serial.print("L2_ON");
  delay(5000);

 // turn light bulb 2 off
  digitalWrite(relay_pin_2, HIGH);
  Serial.print("L2_OFF");
  delay(5000);

 }
```

Perform the following steps to develop the Delphi application:

1. Create a new **VCL Forms Application**: **File** | **New** | **VCL Forms Application**.
2. Click on the **Save All** icon and use these names: `MainFormU`.pas for the form unit file and `ReadDataFromArduino` for the project's name.

3. On the form surface, put one **TComPort**, two **TButton**, two **TLabel**, and two **TToggleSwitch**. Make all the necessary changes to make sure that your form looks like the following screenshot (or you can copy the previous recipe):

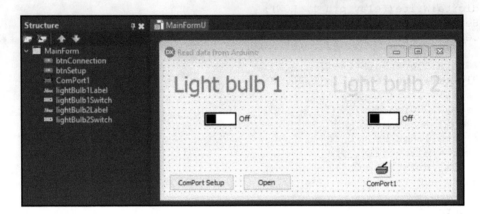

Figure 10.6: Reading data from the Arduino form at design time

4. We have to implement the `ComPort1RxChar` event to read messages from Arduino:

```
procedure TMainForm.ComPort1RxChar(Sender: TObject; Count:
Integer);
var
  Str: String;
begin
  // Receives messages from Arduino.
  ComPort1.ReadStr(Str, Count);

  if Str.ToUpper = 'L1_ON' then
    lightBulb1Switch.State := TToggleSwitchState.tssOn
  else if Str.ToUpper = 'L1_OFF' then
    lightBulb1Switch.State := TToggleSwitchState.tssOff
  else if Str.ToUpper = 'L2_ON' then
    lightBulb2Switch.State := TToggleSwitchState.tssOn
  else if Str.ToUpper = 'L2_OFF' then
    lightBulb2Switch.State := TToggleSwitchState.tssOff;

end;
```

5. The `btnSetup` button's **OnClick** event and the `btnConnection` button's **OnClick** event are the same as we saw in the previous recipe:

```
procedure TMainForm.btnSetupClick(Sender: TObject);
begin
  ComPort1.ShowSetupDialog;
end;
procedure TMainForm.btnConnectionClick(Sender: TObject);
begin
  // If the port is connected.
  if ComPort1.Connected then
    ComPort1.Close // Close the port.
  else
    ComPort1.Open; // Open the port.
  UpdateComponentsState;
end;
```

6. Declare and implement the `UpdateComponentsState` procedure to make graphic changes depending on whether the ComPort is connected or not:

```
procedure TMainForm.UpdateComponentsState;
begin
  lightBulb1Switch.Enabled := ComPort1.Connected;
  lightBulb2Switch.Enabled := ComPort1.Connected;
  lightBulb1Label.Enabled := ComPort1.Connected;
  lightBulb2Label.Enabled := ComPort1.Connected;
  btnConnection.Caption := ifthen(ComPort1.Connected, 'Close',
'Open');
end;
```

7. Now, we have to make some considerations when the form is shown and when it is destroyed. So, implement the **FormShow** and **FormDestroy** events:

```
procedure TMainForm.FormShow(Sender: TObject);
begin
  UpdateComponentsState;
end;

procedure TMainForm.FormDestroy(Sender: TObject);
begin
  if ComPort1.Connected then
    ComPort1.Close;
end;
```

8. Verify and upload the Sketch to Arduino.

9. Start the Delphi application: **Run | Run** or hit *F9*.

10. Click on the **ComPort Setup** button to set up the `Port` and `Baud Rate`. When the dialog is shown, set the ComPort at `Port`, where Arduino is connected and the `Baud Rate` is 1,15,200.

> You can see in **Control Panel | System | Device Manager** under the **Ports (COM & LPT)** section the ComPort where Arduino is connected.

11. Click **Open**.
12. If the connection is successful, the labels and the switches will become enabled.
13. Wait for the light bulbs to turn ON and OFF in the circuit and you will see that the switches change accordingly.

How it works...

The operation of this recipe is really simple. In the Arduino Sketch, the LEDs are switched ON and OFF with the same cyclicity. When a change occurs, this is communicated through the use of the `Serial.print()` method. This change is read by the Delphi application as the `ComPort1RxChar` event has been implemented. Within this method, the logic that understands how to update the graphical interface is inserted:

Command	Action
L1_ON	`lightBulb1Switch` state to ON
L1_OFF	`lightBulb1Switch` state to OFF
L2_ON	`lightBulb2Switch` state to ON
L2_OFF	`lightBulb2Switch` state to OFF

How to blink an LED using Raspberry Pi

As we have already seen for Arduino, the blinking LED is the **Hello World!** of physical computing. So, regarding Raspberry Pi, we are also going to start with this recipe. Let's go!

Getting ready

Delphi does not support direct deployment on Raspberry Pi with Rasbian OS, so for communication with the board, it was necessary to develop a TCP/IP server to send commands to the RPI so that they are interpreted and transformed into other ones (for example, the lighting of the LED). The server is developed using FreePascal, a language very close to Delphi. In order to use and make the most of the developed server, it is necessary to install FreePascal and Lazarus IDE. Here is the link: `http://wiki.freepascal.org/Lazarus_on_Raspberry_Pi`, which helps you install a precompiled, stable version of FPC and Lazarus on the Raspberry Pi.

The **Synapse library** was used to build the TCP/IP server (Synapse offers serial port and TCP/IP connectivity). It differs from other libraries in that you are only required to add some *Synapse Pascal* source code files to your code; no need for installing packages. The only exception is that you will need an external **Crypto** library if you want to use encryption such as SSL/TLS/SSH. See the documentation on the official site (`http://www.ararat.cz/synapse/doku.php/start`) for more details.

So, be careful! To use the TCP/IP server developed for these recipes on the RPI, it is also necessary to install the Synapse library following the steps provided in the previous link.

The fundamental part regarding this recipe is the part related to the use of GPIO Pins. I believe you already know that RPI presents 40 GPIO Pins:

- 26 are GPIO pins
- Others are power or ground pins
- Two are ID EEPROM
- Pins are 3.3V, not 5V like on Arduino
- They are connected directly to the Broadcom chip
- Sending 5V to a pin may kill the Pi

- The maximum permitted current draw from a 3.3V pin is 50mA:

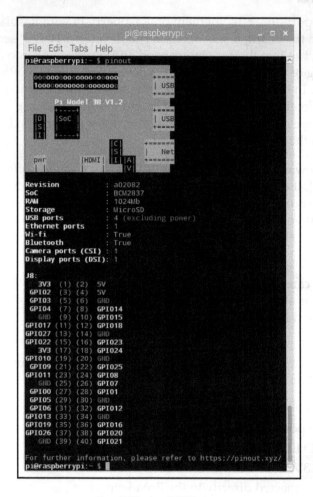

Figure 10.7 Raspberry Pi GPIO schema

There are different ways to access and use GPIOs. The method used in this recipe is to use the file system to generate particular files:

- The I/O lines are managed by the kernel and mapped in the **filesystem** (**FS**).
- In reality, nothing is really written on disk; it is a buffer/pipe that can be accessed using the interface of the FS, and that only exists in RAM.

- The buffers (we will call them files for convenience) that control the pins are found in `/sys/class/gpio/` and can be accessed from any programming language and also from the system shell.
- To perform operations on these files, you must have root privileges.
- There are two important files: `export` and `unexport`. Writing a number in export will expose the pin, and you can then access and use it.
- Instead, writing a number in `unexport` will release the pin.
- For example, if we want to use the pin `gpio1`, it will be enough to write 1 in export, and it is possible to do it as if export were a normal file:

    echo 1 > /sys/class/gpio/export

- After exporting a pin, you will create a folder in `/sys/class/gpio`, which is called the exported pin. In our case, we will create the folder, `/sys/class/gpio/gpio1`.
- Within this folder are the files to check the pin. The most important are direction and value.
- Within direction, we can read the direction associated with the pin, which can be in or out. If we write a value inside this file, we will set the pin as input or output.
- So, to set the pin `gpio1` as out, we can use the redirect, like we did previously:

    echo "out" > /sys/class/gpio/gpio1/direction

- In the value file, we can read or write the pin value, simply using 0 or 1. To turn on the pin, use the following code:

    echo 1 > /sys/class/gpio/gpio1/value

- To turn off the pin, use the following code:

    echo 0 > /sys/class/gpio/gpio1/value

Do not worry about this part. I wrote the TRaspberry Pi class (you can find it in the Commons folder of Raspberry Pi of this recipe, rpi.pas) through a high level interface that allows you to manage all the previously described cases. Here's the code of the interface section:

```
unit rpi;

{$mode objfpc}{$H+}

interface
```

```
uses
  Classes, SysUtils;

type
  TGPIODirection = (dirINPUT, dirOUTPUT);

  { TRaspberry Pi }

  TRaspberry Pi = class
  private
    FExportedPINs: array of integer;
    FSetupDelay: word;
    procedure SetSetupDelay(AValue: word);
  protected
    procedure UnExport; virtual; overload;
    procedure ExportPIN(const PIN: integer);
  public
    procedure UnExport(inputs: array of integer); overload;
    procedure SetPIN(const PIN: integer; const Status: boolean);
    function GetPIN(const PIN: integer): boolean;
    procedure SetPINDirection(const PIN: integer;
      const Direction: TGPIODirection);
    property SetupDelay: word read FSetupDelay write SetSetupDelay;
    constructor Create; virtual;
    destructor Destroy; override;
  end;
```

Now, let's take a closer look at the previous commands. We'll start with the `ExportPIN`
procedure:

```
procedure TRaspberry Pi.ExportPIN(const PIN: integer);
var
  lFile: integer;
begin
  lFile := fpopen('/sys/class/gpio/export', O_WrOnly);
  try
    if fpwrite(lFile, PChar(IntToStr(PIN)), 2) = -1 then
      raise Exception.CreateFmt('Cannot export PIN%d', [PIN]);
  finally
    fpclose(lFile);
  end;
end;
```

The following is how to set the pin's direction:

```
procedure TRaspberry Pi.SetPINDirection(const PIN: integer;
  const Direction: TGPIODirection);
var
  lPIN: string;
  lFile: integer;
begin
  ExportPIN(PIN);
  Sleep(FSetupDelay); // otherwise we cannot set the direction
  lPIN := IntToStr(PIN);
  lFile := fpopen('/sys/class/gpio/gpio' + lPIN + '/direction', O_WrOnly);
  try
    case Direction of
      dirINPUT:
        begin
          if fpwrite(lFile, PChar('in'), 2) = -1 then
            raise Exception.Create('Cannot setup ' + lPIN + ' as input');
        end;
      dirOUTPUT:
        begin
          if fpwrite(lFile, PChar('out'), 3) = -1 then
            raise Exception.Create('Cannot setup ' + lPIN + ' as output');
        end
    else
      raise Exception.Create('Invalid Direction');
    end;
  finally
    fpclose(lFile);
  end;

  SetLength(FExportedPINs, Length(FExportedPINs) + 1);
  FExportedPINs[Length(FExportedPINs) - 1] := PIN;
end;
```

Here's how to set the pin's value:

```
procedure TRaspberry Pi.SetPIN(const PIN: integer; const Status: boolean);
var
  fileDesc: integer;
  lValueToWrite: PChar;
  lRetCode: integer;
begin
  fileDesc := fpopen('/sys/class/gpio/gpio' + IntToStr(PIN) + '/value',
    O_WrOnly);
  try
    if Status then
      lValueToWrite := PIN_ON
```

```
      else
        lValueToWrite := PIN_OFF;
      lRetCode := fpwrite(fileDesc, lValueToWrite, 1);
      if lRetCode = -1 then
        raise Exception.Create('Cannot write to the GPIO sysfs descriptor');
    finally
      fpclose(fileDesc);
    end;
  end;
```

How to do it...

For the part related to the Raspberry Pi code, you can find everything in the `Raspberry Pi` directory of this recipe. In the next section, we will go into the details of how it works.

Now, you must properly configure the electronic components in order to perform all the necessary steps to build the circuit, as shown in the following diagram:

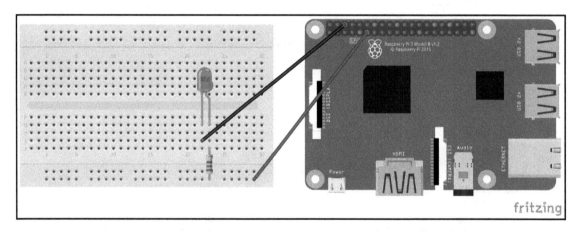

Figure 10.8: Raspberry Pi Blink LED circuit

Let's move on to the Delphi part and see what the necessary steps are to build a TCP/IP client that's able to send commands to switch the LED on the Raspberry Pi ON and OFF. Perform the following steps:

1. Create a new **Multi-Device Application**: **File** | **New** | **Multi-Device Application**.

2. Click on the **Save All** icon and use these names: `MainFMX.pas` for the form unit file and `RPIBlinkLED` for the project name.

3. On the form surface, put one **TToolbar**, two **TLabel**, one **TMemo**, and one **TSwitch**. Make all the necessary changes to make sure that your form looks like the following screenshot (or you can copy the previous recipe):

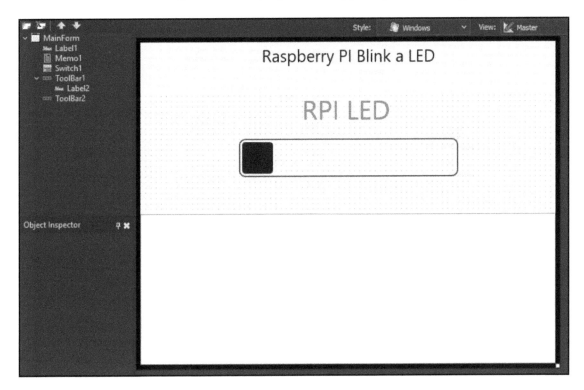

Figure 10.9: RPI Blink LED form at design time

4. Now, we have to write the form code. This is very simple, as it is only necessary to write a private method to send the command to the TCP/IP server—`SendCmd(PCMD: string)`. For this, I used the TCP/IP client made available by Indy: `TIdTCPClient`. Then, we have to implement the **OnSwitch** event of the **TSwitch** component to actually send the command:

```
unit MainFMX;

interface

uses
    System.SysUtils, System.Types, System.UITypes, System.Classes,
```

```
   System.Variants,
   FMX.Types, FMX.Controls, FMX.Forms, FMX.Graphics, FMX.Dialogs,
FMX.StdCtrls,
   FMX.Controls.Presentation, IdBaseComponent, IdComponent,
IdTCPConnection,
   IdTCPClient, FMX.ScrollBox, FMX.Memo;

type
  TMainForm = class(TForm)
    Switch1: TSwitch;
    ToolBar1: TToolBar;
    Label1: TLabel;
    Label2: TLabel;
    Memo1: TMemo;
    ToolBar2: TToolBar;
    procedure Switch1Switch(Sender: TObject);
  private
    function SendCmd(PCMD: string): string;
  end;

var
  MainForm: TMainForm;

implementation

uses
  System.StrUtils, IdException;

{$R *.fmx}

function TMainForm.SendCmd(PCMD: string): string;
var
  LTCPClient: TIdTCPClient;
begin
  Result := '';
  LTCPClient := TIdTCPClient.Create(nil);
  try
    LTCPClient.Host := '192.168.1.83'; // change with your RPI ip
address
    LTCPClient.Port := 8008;
    try
      LTCPClient.Connect;
      LTCPClient.SendCmd(PCMD);
    except
      on E: EIdConnClosedGracefully do
      begin
      end;
      on E: Exception do
```

```
      Memo1.Lines.Add(E.Message);
    end;
    if not LTCPClient.IOHandler.InputBufferIsEmpty then
      Result := LTCPClient.IOHandler.AllData;
  finally
    LTCPClient.Disconnect;
    LTCPClient.Free;
  end;
end;

procedure TMainForm.Switch1Switch(Sender: TObject);
var
  LCMD: String;
begin
  LCMD := 'L' + ifthen(Switch1.IsChecked, '1', '0');
  SendCmd(LCMD);
end;

end.
```

5. Since the type of project is a multi-device application, it is possible to choose your target platform from Android, iOS, Windows, and macOS.

How it works...

At this point, starting with Raspberry Pi, the TCP/IP server of this recipe (in the `Raspberry Pi` folder of this recipe), also starts the Delphi application that we've just written. If you have correctly followed all the steps, by clicking on the switch on the Delphi app, you will be able to turn the LED on the circuit associated with the Raspberry Pi ON and OFF. Great!

Let's understand how the various elements involved work. The server performs these operations:

1. It initializes the GPIO pins; that is, it makes the `UnExport` and sets the pin direction to `OUTPUT`. The first point is used to start from a clean situation, and the second is used to specify that the direction of the pin is out because we have to turn the LED on and off (using the `TRaspberry Pi` class seen previously). As you can see, the reference pin is `17`:

```
procedure InitializeGPIO(PRaspberry Pi: TRaspberry Pi);
begin
  PRaspberry Pi.UnExport([17]);
  PRaspberry Pi.SetPINDirection(17, TGPIODirection.dirOUTPUT);
end;
```

2. It starts listening at port 8008 and when it reads something, it passes it to the HandleClient method:

```
procedure RunServer;
var
  ClientSock: TSocket;
  lSocket: TTCPBlockSocket;
  LRPi: TRaspberry Pi;
begin
  LRPi := TRaspberry Pi.Create;
  try
    // initialize GPIO
    InitializeGPIO(LRPi);
    lSocket := TTCPBlockSocket.Create;
    try
      lSocket.CreateSocket;
      lSocket.setLinger(True, 10000);
      lSocket.Bind('0.0.0.0', '8008');
      lSocket.Listen;
      while True do
      begin
        if lSocket.CanRead(1000) then
        begin
          ClientSock := lSocket.Accept;
          if lSocket.LastError = 0 then
            HandleClient(ClientSock, LRPi);
        end;
      end;
    finally
      lSocket.Free;
    end;
  finally
    LRPi.Free;
  end;
end;
```

3. The HandleClient method understands which command is passed by the TCP/IP client. Once the value is understood, it is passed to the HandleGPIOCmd method:

```
procedure HandleClient(aSocket: TSocket; PRaspberry: TRaspberry
Pi);
var
  lCmd, LSS: string;
  sock: TTCPBlockSocket;
begin
  sock := TTCPBlockSocket.Create;
```

```
try
  Sock.socket := aSocket;
  sock.GetSins;
  lCmd := Trim(Sock.RecvTerminated(60000, #13));
  if Sock.LastError <> 0 then
  begin
    Sock.SendString('error:' + sock.LastErrorDesc);
    Exit;
  end;

  // do something on the rpi
  LSS := HandleGPIOCmd(PRaspberry, lCmd);
  Sock.SendString(LSS);
  if Sock.LastError <> 0 then
  begin
    Sock.SendString('error:' + sock.LastErrorDesc);
    Exit;
  end;
finally
  Sock.Free;
end;
end;
```

4. The `HandleGPIOCmd` method defines the logic to interpret the client command and makes changes to the filesystem through the `TRaspberry Pi` class:

```
function HandleGPIOCmd(PRaspberry Pi: TRaspberry Pi; PCmd: string):
string;
var
  LPin: string;
  LDir, LPinI: integer;
begin
  // command accept 2 chars: PIN and Direction ex. L1, B1, C0.
  LPin := PCmd[1];
  if LPin = 'L' then
    LPinI := 17
  else
    exit('');

  LDir := StrToInt(PCmd[2]);
  PRaspberry Pi.SetPIN(LPinI, LDir > 0);
  exit('ok');
end;
```

In this case, the logic adopted is the following. The command consists of two characters: pin and direction. Since we have to control an LED, I have decided that by sending `L1`, the LED will be ON and sending `L0` the LED will be OFF. `L` represents pin `17` and `0` or `1` represent OFF and ON.

The Delphi application is very simple to understand. When you click on the switch, the command to be sent is constructed, with `L1` or `L0`, depending on the status of the switch, and this value is passed to the `SendCmd` method that sets up a class for the TCP/IP communication and sends the command:

```
procedure TMainForm.Switch1Switch(Sender: TObject);
var
  LCMD: String;
begin
  LCMD := 'L' + ifthen(Switch1.IsChecked, '1', '0');
  SendCmd(LCMD);
end;
```

If exceptions are raised during TCP/IP communication, they are written to the `Memo1` component.

See also

In this recipe, we have seen how to access the GPIO pins by writing to the file system, but there are excellent alternatives, and one of these is the famous **Wiring Pi**: `http://wiringpi.com/`.

I will also leave you with some excellent links, ranging from documentation to the management of peripherals and GPIO:

- Official website: `http://www.Raspberry Pi.org/`
- Documentantion and schema: `http://elinux.org/R-Pi_Hub`
- Peripherals: `http://elinux.org/RPi_VerifiedPeripherals`
- Troubleshooting: `http://elinux.org/R-Pi_Troubleshooting`
- Official forum: `http://www.Raspberry Pi.org/phpBB3/index.php`
- Everything about GPIO: `https://pinout.xyz/`

How to drive multiple relays with Raspberry Pi and Delphi

The recipe series on Raspberry Pi continues with the same route seen for Arduino. So, let's see how you can open and close a circuit using a Relay Module and how to make it programmable through a Delphi application.

Getting ready

The purpose of this recipe is to turn two different light bulbs ON or OFF through the use of a 2 Relay Module and make it controllable through an application written in Delphi.

In this recipe, we are going to use the following:

- Raspberry Pi 3 Model B
- 2 Channel Relay Module (http://wiki.sunfounder.cc/index.php?title=2_Channel_5V_Relay_Module)
- Two light bulbs and a relative supply power (the important thing is that the light bulbs have their own power supply)
- Jumper wires

How to do it...

For the part related to the Raspberry Pi code, you can find everything in the Raspberry Pi directory of this recipe. In the next section, we will go into detail regarding how it works.

Now, you must properly configure the electronic components, so let's perform all the necessary steps to build the circuit, as shown in the following diagram:

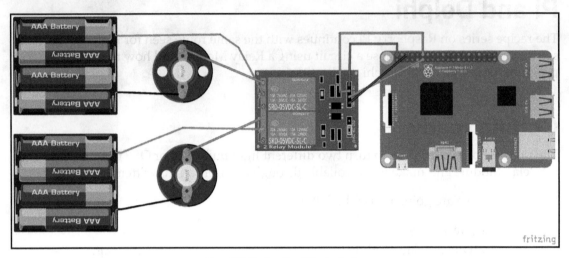

Figure 10.10: Raspberry Pi multiple relay circuits

Let's move on to the Delphi part and see what are the necessary steps to build a TCP/IP client that's able to send commands to switch the specific light bulb on and off on Raspberry Pi. Perform the following steps:

1. Create a new **Multi-Device Application**: **File** | **New** | **Multi-Device Application**.
2. Click on the **Save All** icon and use these names: `MainFMX.pas` for the form unit file and `RPIMultipleRelays` for the project name.

3. On the form surface, put one **TToolbar**, three **TLabel**, one **TMemo**, and two **TSwitch**. Make all the necessary changes to make sure that your form looks like the following screenshot (or you can copy the previous recipe):

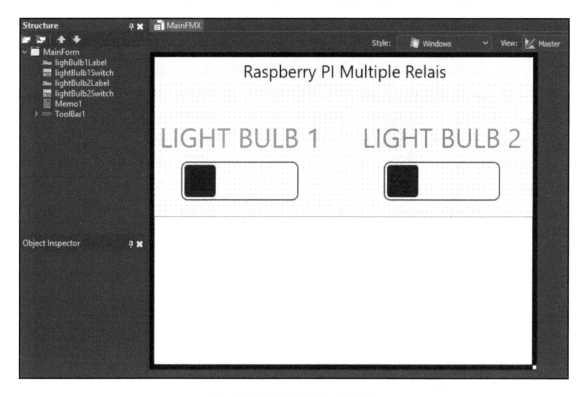

Figure 10.11: RPI multiple relays form at design time

4. Now, we have to write the form code. As for the previous recipe, we have to write the function to send the command when one of the two switches related to the light bulb is clicked. Each switch uses its own code plus its status for the command. For example, if the switch for light bulb 1 is clicked and the switch is in the ON state, the command will be R1:

```
unit MainFMX;

interface

uses
    System.SysUtils, System.Types, System.UITypes, System.Classes,
    System.Variants,
    FMX.Types, FMX.Controls, FMX.Forms, FMX.Graphics, FMX.Dialogs,
```

```
FMX.StdCtrls,
  FMX.Controls.Presentation, IdBaseComponent, IdComponent,
IdTCPConnection,
  IdTCPClient, FMX.ScrollBox, FMX.Memo;

type
  TMainForm = class(TForm)
    lightBulb1Switch: TSwitch;
    ToolBar1: TToolBar;
    lighBulb1Label: TLabel;
    Label2: TLabel;
    Memo1: TMemo;
    lightBulb2Switch: TSwitch;
    lightBulb2Label: TLabel;
    procedure lightBulb1SwitchSwitch(Sender: TObject);
    procedure lightBulb2SwitchSwitch(Sender: TObject);
  private
    function SendCmd(PCMD: string): string;
  end;

var
  MainForm: TMainForm;

implementation

uses
  System.StrUtils, IdException;

{$R *.fmx}

function TMainForm.SendCmd(PCMD: string): string;
var
  LTCPClient: TIdTCPClient;
begin
  Result := '';
  LTCPClient := TIdTCPClient.Create(nil);
  try
    LTCPClient.Host := '192.168.1.83'; // change with your RPI ip
address
    LTCPClient.Port := 8008;
    try
      LTCPClient.Connect;
      LTCPClient.SendCmd(PCMD);
    except
      on E: EIdConnClosedGracefully do
      begin
      end;
      on E: Exception do
```

```
        Memo1.Lines.Add(E.Message);
    end;
    if not LTCPClient.IOHandler.InputBufferIsEmpty then
      Result := LTCPClient.IOHandler.AllData;
  finally
    LTCPClient.Disconnect;
    LTCPClient.Free;
  end;
end;

procedure TMainForm.lightBulb2SwitchSwitch(Sender: TObject);
var
  LCMD: String;
begin
  LCMD := 'L' + ifthen(lightBulb2Switch.IsChecked, '0', '1');
  SendCmd(LCMD);
end;

procedure TMainForm.lightBulb1SwitchSwitch(Sender: TObject);
var
  LCMD: String;
begin
  LCMD := 'R' + ifthen(lightBulb1Switch.IsChecked, '0', '1');
  SendCmd(LCMD);
end;

end.
```

5. Since the type of project is a multi-device application, it is possible to choose Android, iOS, Windows, or macOS as your target platform.

How it works...

Starting with Raspberry Pi, the TCP/IP server of this recipe (in the `Raspberry Pi` folder of this recipe), starts the same as the Delphi application that we've just written. If you have followed all the necessary steps, by clicking on the switches on the Delphi app, you will be able to turn the relative light bulb ON and OFF on the circuit associated with the Raspberry Pi.

Let's understand how the various elements involved work. The server is very similar to the one in the previous recipe, so I will only focus on the points that are different:

1. It initializes the **GPIO** pins, that is, it makes the UnExport and sets the pin direction to OUTPUT. The first point is used to start from a clean situation, and the second is used to specify that the direction of the pin is out because we have to turn the relays on and off:

```
const
  RELAY_PIN_1: integer = 17;
  RELAY_PIN_2: integer = 18;

procedure InitializeGPIO(PRaspberry Pi: TRaspberry Pi);
begin
  PRaspberry Pi.UnExport([RELAY_PIN_1, RELAY_PIN_2]);
  PRaspberry Pi.SetPINDirection(RELAY_PIN_1,
TGPIODirection.dirOUTPUT);
  PRaspberry Pi.SetPINDirection(RELAY_PIN_2,
TGPIODirection.dirOUTPUT);
  PRaspberry Pi.SetPIN(RELAY_PIN_1, True);
  PRaspberry Pi.SetPIN(RELAY_PIN_2, True);
end;
```

2. It starts listening at port 8008 and when it reads something, it passes it to the HandleClient method.

3. In the HandleClient method, it understands which command is passed by the TCP/IP client. Once the value is understood, it is passed to the HandleGPIOCmd method.

4. The HandleGPIOCmd method defines the logic to interpret the client command, and this makes changes to the filesystem through the TRaspberry Pi class:

```
function HandleGPIOCmd(PRaspberry Pi: TRaspberry Pi; PCmd: string):
string;
var
  LPin: string;
  LDir, LPinI: integer;
begin
  // command accept 2 chars: PIN and Direction ex. R1, L1, R0.
  LPin := PCmd[1];
  // light bulb 1
  if LPin = 'R' then
    LPinI := RELAY_PIN_1
  // light bulb 2
  else if LPin = 'L' then
    LPinI := RELAY_PIN_2
  else
```

```
      exit('');

    LDir := StrToInt(PCmd[2]);
    PRaspberry Pi.SetPIN(LPinI, LDir > 0);
    exit('ok');
  end;
```

The following table is used to describe the commands and actions for this:

Command	Action
R1	Turns light bulb 1 ON
R0	Turns light bulb 1 OFF
L1	Turns light bulb 2 ON
L0	Turns light bulb 2 OFF

The Delphi application is very simple to understand. When you click on the switch, the command to be sent is constructed (R1-R0-G1-G0), depending on which switch and its status. This value is passed to the SendCmd method that sets up a class for the TCP/IP communication and sends the command:

```
procedure TMainForm.greenLEDSwitchSwitch(Sender: TObject);
var
  LCMD: String;
begin
  LCMD := 'G' + ifthen(greenLEDSwitch.IsChecked, '1', '0');
  SendCmd(LCMD);
end;

procedure TMainForm.redLEDSwitchSwitch(Sender: TObject);
var
  LCMD: String;
begin
  LCMD := 'R' + ifthen(redLEDSwitch.IsChecked, '1', '0');
  SendCmd(LCMD);
end;
```

If exceptions are raised during TCP/IP communication, they are written to the Memo1 component.

Reading data from Raspberry PI

In the previous recipes we saw how to communicate something to Raspberry PI, in this recipe we are going to close the recipes about Raspberry PI with the opposite operation: how to read data that Raspberry PI controls.

Getting ready

In this recipe we will use the same circuit built previously. The goal is to make small changes to the TCP / IP server, so that at the request of a specific light bulb status it responds with a code to identify whether it is ON or OFF.

We are going to use:

- Raspberry PI 3 Model B
- Two Channel Relay Modules (`http://wiki.sunfounder.cc/index.php?title=2_Channel_5V_Relay_Module`)
- Two light bulbs and relative supply power (the important thing is that the light bulbs have their own power supply)
- Jumper wires

For the management of the GPIO the `TRaspberryPI` class is used, for any need refer to the recipe *How to blink a LED using Raspberry PI* where there is a good explanation.

How to do it...

For the part related to the Raspberry PI code, you can find everything in the `raspberrypi` directory of this recipe. In the next section we will go into the details of how it works.

Now you must properly configure the electronic components. The circuit is identical to that of the previous recipe, so if you need, you can refer to the previous recipe, *How to drive multiple relays with Raspberry PI and Delphi*.

Let's move on to the Delphi part and see what are the steps necessary to build a TCP / IP client able to send the commands necessary to switch a specific light bulb on Raspberry PI on and off. Perform the following steps:

1. Create a new **Multi-Device Application**: **File** | **New** | **Multi-Device Application**.
2. Click on **Save All** icon and use these names: `MainFMX.pas` for the Form unit file and `RPIReadData` for project name.

3. Put on the Form surface one **TTimer**, one **TToolbar**, three **TLabel**, one **TMemo**, two **TSwitch**. Make all the necessary steps to make sure that your form looks like the following image (or you can copy the previous recipe and add a **TTimer**):

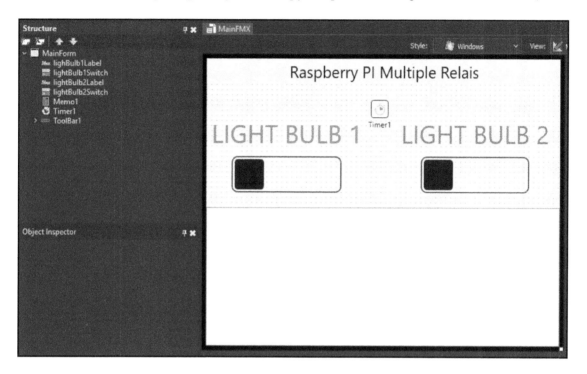

Fig 10.12: RPI Read data form at design time

4. Now we have to write the form code. A `TTimer` is used to make requests every second, to find out the status of the light bulbs. The function to send the command is the same as the previous recipe, but having changed the server, now we expect a response on the state:

```
unit MainFMX;

...

implementation

uses
  System.StrUtils, IdException;

{$R *.fmx}
```

```
function TMainForm.SendCmd(PCMD: string): string;
var
  LTCPClient: TIdTCPClient;
begin
  Result := '';
  LTCPClient := TIdTCPClient.Create(nil);
  try
    LTCPClient.Host := '192.168.1.83'; // change with your RPI ip
address
    LTCPClient.Port := 8008;
    try
      LTCPClient.Connect;
      LTCPClient.SendCmd(PCMD);
    except
      on E: EIdConnClosedGracefully do
      begin
      end;
      on E: Exception do
        Memo1.Lines.Add(E.Message);
    end;
    if not LTCPClient.IOHandler.InputBufferIsEmpty then
      Result := LTCPClient.IOHandler.AllData;
  finally
    LTCPClient.Disconnect;
    LTCPClient.Free;
  end;
end;


procedure TMainForm.Timer1Timer(Sender: TObject);
begin
  TTask.Run(
    procedure
    var
      LStatus: string;
    begin
      LStatus := SendCmd('R');
      UpdateLightBulb('R', LStatus);
      LStatus := SendCmd('L');
      UpdateLightBulb('L', LStatus);
    end);
end;


procedure TMainForm.UpdateLightBulb(const ACMD, AStatus: String);
begin
  TThread.Synchronize(nil,
    procedure
    begin
      if ACMD = 'R' then
```

```
            lightBulb1Switch.IsChecked := AStatus = '1'
        else if ACMD = 'L' then
            lightBulb2Switch.IsChecked := AStatus = '1'
    end);
  end;

  end.
```

5. Since the type of project is a multi-device application, it is possible to choose between Android, iOS, Windows and macOS as target platform.

How it works...

This recipe contains two FreePascal/Lazarus projects: the TCP / IP server (folder `raspberrypi\RECIPE06\tcpserver.lpi` of this recipe) and a console application (folder `raspberrypi\RECIPE06\console` of this recipe) to turn the light bulbs on and off cyclically.

Start the Raspberry PI and the console app, then also start the TCP/IP server and the Delphi application just written. If you have correctly followed all the steps, you will see in the circuit associated with the Raspberry Pi the lighting and the extinguishing of the light bulbs in a cyclical way and the Delphi application that through the switches graphically reports these changes.

Let's understand how the various elements involved work. We have already defined that the console application is used to turn the light bulbs ON and OFF (and GPIO initializations). The logic is very simple and the comments explain the logic well, so I leave you with the code:

```
var
  LRPI: TRaspberryPI;

begin

  LRPI := TRaspberryPI.Create;
  LRPI.UnExport([RELAY_PIN_1, RELAY_PIN_2]);
  LRPI.SetPINDirection(RELAY_PIN_1, TGPIODirection.dirOUTPUT);
  LRPI.SetPINDirection(RELAY_PIN_2, TGPIODirection.dirOUTPUT);
  LRPI.SetPIN(RELAY_PIN_1, True);
  LRPI.SetPIN(RELAY_PIN_2, True);

  while True do
  begin
```

```
// insert a delay
TThread.Sleep(2000);

// turn light bulb 1 ON
LRPI.SetPIN(RELAY_PIN_1, False);
TThread.Sleep(5000);

// turn light bulb 1 OFF
LRPI.SetPIN(RELAY_PIN_1, True);
TThread.Sleep(5000);

// turn light bulb 2 ON
LRPI.SetPIN(RELAY_PIN_2, False);
TThread.Sleep(5000);

// turn light bulb 2 ON
LRPI.SetPIN(RELAY_PIN_2, True);
TThread.Sleep(5000);

  end;
end.
```

The server is very similar to the previous recipe, so I will only focus on the points that are different:

1. It starts listening at port 8008 and when it reads something, passes it to the HandleClient method.

2. In the HandleClient method, find out which command is passed by the TCP/IP client. Once the value is understood, it is passed to the HandleGPIOCmd method.

3. The HandleGPIOCmd method defines the logic to interpret the client command and returns the status of a specific light bulb through the TRaspberryPI class:

```
function HandleGPIOCmd(PRaspberryPI: TRaspberryPI; PCmd: string):
string;
var
  LPin: string;
  LDir, LPinI: integer;
begin
  // command accept 1 char: 'R' light bulb 1 or 'L' light bulb 2
  LPin := PCmd[1];
  // light bulb 1
  if LPin = 'R' then
    LPinI := RELAY_PIN_1
  // light bulb 2
  else if LPin = 'L' then
    LPinI := RELAY_PIN_2
```

```
  else
    exit('');

// being a NO (Normally open) circuit when the value is LOW (false)
the light bulb is on
  Result := BoolToStr(PRaspberryPI.GetPIN(LPinI), '0', '1');
end;
```

4. The Delphi application, through a timer with one second intervals, makes requests to the TCP / IP server to find out the status of the bulbs, and based on the response, graphically updates the corresponding switch:

```
procedure TMainForm.Timer1Timer(Sender: TObject);
begin
  TTask.Run(
    procedure
    var
      LStatus: string;
    begin
      LStatus := SendCmd('R');
      UpdateLightBulb('R', LStatus);
      LStatus := SendCmd('L');
      UpdateLightBulb('L', LStatus);
    end);
end;

procedure TMainForm.UpdateLightBulb(const ACMD, AStatus: String);
begin
  TThread.Synchronize(nil,
    procedure
    begin
      if ACMD = 'R' then
        lightBulb1Switch.IsChecked := AStatus = '1'
      else if ACMD = 'L' then
        lightBulb2Switch.IsChecked := AStatus = '1'
    end);
end;
```

Other Books You May Enjoy

If you enjoyed this book, you may be interested in these other books by Packt:

Delphi High Performance
Primož Gabrijelčič

ISBN: 978-1-78862-545-6

- Find performance bottlenecks and easily mitigate them
- Discover different approaches to fix algorithms
- Understand parallel programming and work with various tools included with Delphi
- Master the RTL for code optimization
- Explore memory managers and their implementation
- Leverage external libraries to write better performing programs

Expert Delphi

Paweł Głowacki

ISBN: 978-1-78646-016-5

- Understand the basics of Delphi and the FireMonkey application platform as well as the specifics of Android and iOS platforms
- Complete complex apps quickly with access to platform features and APIs using a single, easy-to-maintain code base
- Work with local data sources, including embedded SQL databases, REST servers, and Backend-as-a-Service providers
- Take full advantage of mobile hardware capabilities by working with sensors and Internet of Things gadgets and devices
- Integrate with cloud services and data using REST APIs and scalable multi-tier frameworks for outstanding multi-user and social experience
- Architect and deploy powerful mobile back-end services and get super-productive by leveraging Delphi IDE agile functionality
- Get to know the best practices for writing a high-quality, reliable, and maintainable code base in the Delphi Object Pascal language

Leave a review - let other readers know what you think

Please share your thoughts on this book with others by leaving a review on the site that you bought it from. If you purchased the book from Amazon, please leave us an honest review on this book's Amazon page. This is vital so that other potential readers can see and use your unbiased opinion to make purchasing decisions, we can understand what our customers think about our products, and our authors can see your feedback on the title that they have worked with Packt to create. It will only take a few minutes of your time, but is valuable to other potential customers, our authors, and Packt. Thank you!

Index

S

Scalable Vector Graphics (SVG) 217
separation of concerns (SoC) 131
Serial.print()
  reference 608
Serial.write()
  reference 608
server
  PDF file, downloading from 560
service
  bound service 578
  started service 578
shared resources
  synchronizing, with TMonitor 240, 242, 243,
    244, 246
signals 397
SMS sending service
  building, with REST API 528
SpeechToText engine 566, 568
SQLite databases
  used, for handling to-do list 467, 468, 472
stack of embedded forms
  creating 36, 37, 38
started service 578
stream
  overview 56, 58, 59
  reference 57
strings
  checking, with regular expression (RegEx) 152,
    153, 154, 155, 156
styled TListBox
  creating 197, 198, 201
styled TListView
  used, for handling long list of data 480, 482,
    485, 487, 490
SuperObject
  reference 45
Synapse library 613
System.NetEncodings
  classes 172
  used, for coping with encoded internet world
    172, 173, 174, 176, 177
System.Zip
  space, saving 178, 179, 180

systemd
  about 409
  reference 410, 411

T

Table Data Gateway (TDG)
  reference 330
tasks
  using 271, 272, 273, 275, 276, 278
TCP/IP Linux server
  building 411, 412, 413, 414
  daemonizing 411, 412, 413, 414
TCustomStyleServices
  reference 36
TDBGrid
  customizing 18, 19, 20, 21, 22, 25
TDD
  reference 255
TEnumerable
  reference 117
TEnumerator
  reference 117
TEvent
  multiple threads, synchronizing 251, 252, 253,
    254
TFDBatchMove
  about 91, 92, 93, 95, 97, 99
  reference 100
TFDLocalSQL 100, 102, 105
TFDManager class
  reference 291
TFDTable
  about 82
  configuring 83
  using 83, 85, 88, 90
thread-safe queue
  using 247, 248, 249, 250
THTTPClient's methods
  HTTP verbs, mapping 169, 170
TIdTCPServer 393
TListView
  customizing 490, 491, 494, 498, 499
  used, for displaying local data 459, 460, 464,
    465
  used, for searching local data 459, 460, 464,